Titles in This Series

Volume

Titles in This Series

Volume

35 **Four-manifold theory,** Cameron Gordon and Robion Kirby, Editors

36 **Group actions on manifolds,** Reinhard Schultz, Editor

37 **Conference on algebraic topology in honor of Peter Hilton,** Renzo Piccinini and Denis Sjerve, Editors

38 **Topics in complex analysis,** Dorothy Browne Shaffer, Editor

39 **Errett Bishop: Reflections on him and his research,** Murray Rosenblatt, Editor

40 **Integral bases for affine Lie algebras and their universal enveloping algebras,** David Mitzman

41 **Particle systems, random media and large deviations,** Richard Durrett, Editor

42 **Classical real analysis,** Daniel Waterman, Editor

43 **Group actions on rings,** Susan Montgomery, Editor

44 **Combinatorial methods in topology and algebraic geometry,** John R. Harper and Richard Mandelbaum, Editors

45 **Finite groups–coming of age,** John McKay, Editor

46 **Structure of the standard modules for the affine Lie algebra $A_1^{(1)}$,** James Lepowsky and Mirko Primc

47 **Linear algebra and its role in systems theory,** Richard A. Brualdi, David H. Carlson, Biswa Nath Datta, Charles R. Johnson, and Robert J. Plemmons, Editors

48 **Analytic functions of one complex variable,** Chung-chun Yang and Chi-tai Chuang, Editors

49 **Complex differential geometry and nonlinear differential equations,** Yum-Tong Siu, Editor

50 **Random matrices and their applications,** Joel E. Cohen, Harry Kesten, and Charles M. Newman, Editors

51 **Nonlinear problems in geometry,** Dennis M. DeTurck, Editor

52 **Geometry of normed linear spaces,** R. G. Bartle, N. T. Peck, A. L. Peressini, and J. J. Uhl, Editors

53 **The Selberg trace formula and related topics,** Dennis A. Hejhal, Peter Sarnak, and Audrey Anne Terras, Editors

54 **Differential analysis and infinite dimensional spaces,** Kondagunta Sundaresan and Srinivasa Swaminathan, Editors

55 **Applications of algebraic K-theory to algebraic geometry and number theory,** Spencer J. Bloch, R. Keith Dennis, Eric M. Friedlander, and Michael R. Stein, Editors

56 **Multiparameter bifurcation theory,** Martin Golubitsky and John Guckenheimer, Editors

57 **Combinatorics and ordered sets,** Ivan Rival, Editor

58.I **The Lefschetz centennial conference. Part I: Proceedings on algebraic geometry,** D. Sundararaman, Editor

58.II **The Lefschetz centennial conference. Part II: Proceedings on algebraic topology,** S. Gitler, Editor

58.III **The Lefschetz centennial conference. Part III: Proceedings on differential equations,** A. Verjovsky, Editor

59 **Function estimates,** J. S. Marron, Editor

60 **Nonstrictly hyperbolic conservation laws,** Barbara Lee Keyfitz and Herbert C. Kranzer, Editors

61 **Residues and traces of differential forms via Hochschild homology,** Joseph Lipman

62 **Operator algebras and mathematical physics,** Palle E. T. Jorgensen and Paul S. Muhly, Editors

63 **Integral geometry,** Robert L. Bryant, Victor Guillemin, Sigurdur Helgason, and R. O. Wells, Jr., Editors

64 **The legacy of Sonya Kovalevskaya,** Linda Keen, Editor

65 **Logic and combinatorics,** Stephen G. Simpson, Editor

66 **Free group rings,** Narian Gupta

67 **Current trends in arithmetical algebraic geometry,** Kenneth A. Ribet, Editor

68 **Differential geometry: The interface between pure and applied mathematics,**

Titles in This Series

Volume

Mladen Luksic, Clyde Martin, and William Shadwick, Editors

69 **Methods and applications of mathematical logic,** Walter A. Carnielli and Luiz Paulo de Alcantara, Editors

70 **Index theory of elliptic operators, foliations, and operator algebras,** Jerome Kaminker, Kenneth C. Millett, and Claude Schochet, Editors

71 **Mathematics and general relativity,** James A. Isenberg, Editor

72 **Fixed point theory and its applications,** R. F. Brown, Editor

73 **Geometry of random motion,** Rick Durrett and Mark A. Pinsky, Editors

74 **Geometry of group representations,** William M. Goldman and Andy R. Magid, Editors

75 **The finite calculus associated with Bessel functions,** Frank M. Cholewinski

76 **The structure of finite algebras,** David C. Hobby and Ralph Mckenzie

77 **Number theory and its applications in China,** Wang Yuan, Yang Chung-chun, and Pan Chengbiao, Editors

78 **Braids,** Joan S. Birman and Anatoly Libgober, Editors

79 **Regular differential forms,** Ernst Kunz and Rolf Waldi

80 **Statistical inference from stochastic processes,** N. U. Prabhu, Editor

81 **Hamiltonian dynamical systems,** Kenneth R. Meyer and Donald G. Saari, Editors

82 **Classical groups and related topics,** Alexander J. Hahn, Donald G. James, and Zhe-xian Wan, Editors

83 **Algebraic K-theory and algebraic number theory,** Michael R. Stein and R. Keith Dennis, Editors

84 **Partition problems in topology,** Stevo Todorcevic

85 **Banach space theory,** Bor-Luh Lin, Editor

86 **Representation theory and number theory in connection with the local Langlands conjecture,** J. Ritter, Editor

87 **Abelian group theory,** Laszlo Fuchs, Rüdiger Göbel, and Phillip Schultz, Editors

88 **Invariant theory,** R. Fossum, W. Haboush, M. Hochster, and V. Lakshmibai, Editors

89 **Graphs and algorithms,** R. Bruce Richter, Editor

90 **Singularities,** Richard Randell, Editor

91 **Commutative harmonic analysis,** David Colella, Editor

92 **Categories in computer science and logic,** John W. Gray and Andre Scedrov, Editors

93 **Representation theory, group rings, and coding theory,** M. Isaacs, A. Lichtman, D. Passman, S. Sehgal, N. J. A. Sloane, and H. Zassenhaus, Editors

94 **Measure and measurable dynamics,** R. Daniel Mauldin, R. M. Shortt, and Cesar E. Silva, Editors

95 **Infinite algebraic extensions of finite fields,** Joel V. Brawley and George E. Schnibben

96 **Algebraic topology,** Mark Mahowald and Stewart Priddy, Editors

97 **Dynamics and control of multibody systems,** J. E. Marsden, P. S. Krishnaprasad, and J. C. Simo, Editors

98 **Every planar map is four colorable,** Kenneth Appel and Wolfgang Haken

99 **The connection between infinite dimensional and finite dimensional dynamical systems,** Basil Nicolaenko, Ciprian Foias, and Roger Temam, Editors

100 **Current progress in hyperbolic systems: Riemann problems and computations,** W. Brent Lindquist, Editor

101 **Recent developments in geometry,** S.-Y. Cheng, H. Choi, and Robert E. Greene, Editors

Recent Developments in Geometry

CONTEMPORARY MATHEMATICS

101

Recent Developments in Geometry

Proceedings of the AMS Special Session in Geometry
November 14–15, 1987

S.-Y. Cheng, H. Choi, and
Robert E. Greene, Editors

AMERICAN MATHEMATICAL SOCIETY • PROVIDENCE, RHODE ISLAND

The Proceedings of the Special Sessions on Recent Developments in Geometry held at the 838th meeting of the American Mathematical Society at the University of California in Los Angeles, California, Novembrer 14–15, 1987.

1980 *Mathematics Subject Classification* (1985 *Revision*). Primary 53C20, 53C42, 53C55, 53A10, 58E20, 58G25, 32B30, 32H20.

Library of Congress Cataloging-in-Publication Data

AMS Special Session in Geometry (1987: University of California, Los Angeles)
Recent developments in geometry: proceedings of the AMS Special Session in Geometry, November 14–15, 1987/S.-Y. Cheng, H. Choi, and R.E. Greene, editors.
p. cm. — (Contemporary mathematics, ISSN 0271-4132; v. 101)
Includes bibliographical references.
ISBN 0-8218-5107-1 (alk. paper)
1. Geometry—Congresses. I. Cheng, S.-Y. II. Choi, H. III. Greene, R. E. IV. American Mathematical Society. V. Title. VI. Series: Contemporary mathematics (American Mathematical Society); v. 101.
QA440.A47 1987
516–dc20 89-18039
CIP

Printed in the United States of America.
This volume was printed directly from author-prepared copy.
The paper used in this book is acid-free and falls within the guidelines established to ensure permanence and durability. ∞

10 9 8 7 6 5 4 3 2 1   94 93 92 91 90 89

Contents

Preface

This volume of papers is an outgrowth of a special session on geometry at the November 1987 meeting at UCLA of the American Mathematical Society (meeting No. 838). The organizers of the special session were S. Y. Cheng, H. Choi and R. E. Greene. This special session was unusually well attended. More than forty addresses were given, and the audience (including the lecturers themselves) numbered over sixty. During the meeting, Alan Weinstein, who was serving as an editor of *Contemporary Mathematics*, suggested to the organizers that, in view of the breadth of the coverage of geometry in session, an interesting volume might be prepared by soliciting manuscripts from the participants; and in particular he expressed interest in such a volume for publication in the *Contemporary Mathematics* series. In the event, the participants greeted this idea with enthusiasm; by common consent, it was decided that the papers should be surveys of relatively broad areas of geometry, rather than detailed presentations of new research results as such.

Given the startling diversity and extent of contemporary geometry, it is perhaps not to be expected that a volume such as this, nor indeed a weekend special session, should cover anything like the field in toto. On the other hand, it is certainly our hope that this volume will serve the purpose of providing geometers, whatever their own specialties, with some insight into recent developments in a variety of divergent specialities in geometry. Weinstein had originally suggested that the volume might appropriately begin with a general survey of the field as a whole and of the organizers' views of future prospects for geometry. In practice, contemplating the enormousness of geometry's recent progress we decided that such a survey to be complete and uniform would have to be so extended in scope as to be disproportionate to the size of the volume itself. And at the same time it seemed to us that such a survey might be more appropriate to the truly systematic coverage of the field that will occur in the Proceedings of the American Mathematical Society Summer Institute on Geometry, scheduled to occur in 1990.

For all our hesitancy to undertake a general survey, it did occur to us that this might be an appropriate moment to survey some recent developments in a most classical and fundamental general problem of Riemannian geometry in its purest form: what complete Riemannian manifolds exist that satisfy a given set of geometric conditions, i.e. conditions on curvature and other geometric invariants? The slogan is curvature controls topology. The problem is to make the slogan into specific mathematical results. This problem is so general as to encompass a great deal of Riemannian geometry from its inception to the present. But as it happens, the last year or two have seen some exciting new developments along these lines. The first article to follow summarizes some of these; while it is without pretence of completeness, we hope that it will draw the reader's attention to the continuing vitality of the purely geodesic-geometric aspect of the subject while at the same time noting some of the remarkable interactions between these classical topics and recent developments in partial differential equations, interactions which are transforming the face of geometry.

As always, the editors are indebted to their contributors, without whom there would have been nothing to edit. We are also indebted to Alan Weinstein, who as noted was the originator of the concept of this volume. Without his encouragement, the November 1987 geometry session would have been memorialized at most in the memories of the participants. All the papers that follow are in final form in the sense specified by *Mathematical Reviews* and will not appear elsewhere.

S. Y. Cheng
H. Choi
R. E. Greene

Contemporary Mathematics
Volume **101**, 1989

SOME RECENT DEVELOPMENTS IN RIEMANNIAN GEOMETRY

ROBERT E. GREENE*

From the beginning, the idea that curvature should control global geometry and topology was fundamental in Riemannian geometry. Indeed, in Riemann's celebrated memoir [64], this idea is already indicated, and some form of it occurred even in the earlier work of Gauss. But it was only in relatively modern times, starting in the 1940's and 50's, that essentially optimal characterization results became possible. In particular, the basic program of Riemannian geometry in its purest form received new life from the well-known 1/4-pinching characterization of the sphere obtained by M. Berger and W. Klingenberg, following the pioneering work of H. Rauch. Since then, this form of geometry has progressed relatively rapidly, though it is still a field to which Gauss's motto "few but ripe" (pauca sed matura) applies compared to those disciplines with a less specific focus. As it happened, at the time of the American Mathematical Society meeting that this volume commemorates and in the year or so since, there have been a number of significant advances in Riemannian geometry that fit together in a coherent whole; roughly speaking, these results all deal with finiteness phenomena that arise from bounding curvature below. Philosophically if not literally, they arise from attempts to extend the 1/4-pinched sphere

---

1980 *Mathematics Subject Classification* (1985 Revision). 53C20.
*The author is supported by a grant from the National Science Foundation.

theorem in the compact case and the basic results on noncompact manifolds of nonnegative curvature of Cheeger-Gromoll-Meyer.

This article makes no attempt to survey in its entirety Riemannian geometry's recent progress; in particular, the spectacular advances made in the realm of negative curvature are omitted (with regret). As to the explosive development of the subject of partial differential equations in Riemannian manifolds, that is so clearly beyond the possibility of a brief survey as to require no further comment on its omission. It is interesting to note in advance, however, that some of the results that we shall discuss were proved by partial differential equations methods, and to date cannot be demonstrated otherwise, in spite of the purely geometric nature of their statement.

The author is indebted to Karsten Grove and Peter Petersen for a number of illuminating conversations about finiteness theorems and related matters, to Mladen Bestvina for sharing his expertise on manifold topology, and to H. Wu for numerous helpful suggestions.

1. Where things went from quarter pinched.

A compact, simply connected Riemannian manifold with sectional curvature strictly between 1/4 and 1 is homeomorphic to a sphere. This famous result of Berger [7] and Klingenberg [52], extending Rauch [63], invites, and invited historically, not only admiration but, as always with mathematicians, extension and generalization. The strictness of pinching, i.e., more than one-quarter pinching, is of course optimal (in even dimensions) as it stands: $P_n\mathbf{C}$, $n \geq 2$, with its standard Fubini-Study metric has curvature between 1/4 and 1 inclusive. (To fix our terminology in accord with usual conventions we shall say henceforth that a Riemannian manifold with $\alpha \leq$ curvature ≤ 1 is α-pinched.) But natural questions nonetheless abound: What examples arise with exactly quarter pinching? What is required to insure diffeomorphism, not just homeomorphism, to the standard sphere? What are the possibilities if one assumes pinching of α to 1 where $0 < \alpha < 1/4$? Can sectional curvature be replaced by some weaker curvature condition involving, say, Ricci curvature?

The first two of these questions were given substantial answers soon after: In connection with the first question, Berger [7] proved that a compact simply connected even-dimensional manifold with a quarter-pinched metric which is *not* homeomorphic to a sphere must be isometric to one of the following with a multiple of its (symmetric space) standard metric: a complex projective space, a quaternionic projective space, or the Cayley plane.

The second question, on diffeomorphism, was treated successively in [34], independently by E. Calabi, in [14], [72], and [74] and, in another direction [65]. The ultimate result (in [74]) was that there is a number α such that α-pinching implies diffeomorphism to the standard sphere for compact, simply connected Riemannian manifolds of any dimension. (The initial estimate of [74], $\alpha = 0.87$ was improved in [41] to $\alpha = 0.76$. Asymptotically, as dimension goes to infinity, smaller values suffice, decreasing to 0.68 as shown in [49].) The question remains open whether any further pinching beyond just more than 1/4 is actually required for diffeomorphism to a standard sphere to be universally true. At present, there is no known example of a metric of (strictly) positive sectional curvature on an exotic sphere. In [36] a metric of nonnegative sectional curvature on an exotic sphere of dimension seven was given, however.

The question of what examples arise among α-pinched manifolds as we reduce α below 1/4 has given rise to a whole new line of thought. Namely, it has given rise to what are now known as finiteness theorems. That is, in the absence of an actual list of possibilities, one might try simply to prove that there are only a finite number of examples possible, without knowing what these possibilities are. This viewpoint turns out to be useful and interesting not just in terms of pinching problems but in general. This will be discussed in detail in the next section.

Regarding the fourth question, of alternative curvature conditions, a considerable body of information has arisen. In [57], Min-Oo and E. Ruh have investigated successfully a new approach to the idea of geometric resemblance to symmetric spaces of compact type. E. Ruh has also treated pointwise pinching ([66]). The latter result in detail is that: There is a $\delta(n), 1/4 < \delta(n) < 1$, such that any compact manifold of dimension n which is locally $\delta(n)$-pinched is diffeomorphic to a spherical space form. Here local $\delta(n)$-pinching means that for each point p there is a positive number A_p such that each sectional curvature K at p satisfies $\delta(n)A_p \leq K \leq A_p$.

This result is conceptually related to the classical theorem of I. Schur that if, at each point of a Riemannian manifold if dimension $n \geq 3$, all sectional curvatures are equal, then in fact the manifold has constant curvature. But the "almost equal", pointwise-pinching result already quoted turns out to be surprisingly subtle in comparison. A related result is the theorem of [56] that a compact simply connected manifold of everywhere positive curvature operator is necessarily homeomorphic to a sphere. The relationship arises from the algebraic fact [51] that for each $n = 2, 3, 4 \ldots$ there is a number $\lambda(n), \lambda(n) < 1$, such that a locally $\lambda(n)$-pinched n-manifold necessarily has positive curvature operator. Actually in [56] it is shown that a weaker condition than positive curvature operator implies that the (compact, simply connected) manifold is homeomorphic to a sphere. This weaker condition (positive curvature operator on totally isotropic 2-planes) is implied pointwise algebraically by pointwise δ-pinching, $\delta > 1/4$; thus [56] is a true generalization of the Berger-Klingenberg 1/4-pinched Theorem.

Along the same general line of considering weaker curvature conditions, it is natural to investigate Ricci curvature, or even scalar curvature. One expects two features:

(1) that Ricci (or scalar) curvature will exert less control in higher dimensions than in lower; and

(2) that assuming negative Ricci (or scalar) curvature will impose fewer conditions than positive. Indeed, it is known that metrics of negative scalar curvature (even constant negative) always exist on compact manifolds of dimension ≥ 3 ([19], using [75]), and complete metrics of negative scalar curvature on noncompact manifolds ([30]) and in fact constant negative ([11], using [6]). In the case of negative Ricci, universal existence is so far known only for compact $3n$-manifolds ($n = 1, 2, 3...$) and certain classes of noncompact manifolds [24], cf. also [23], [13], [14]. But it is natural to conjecture ([24]) along the lines of principle (1) that:

(Conjecture): Every manifold of dimension ≥ 3 admits a complete metric of negative Ricci curvature.

It has long been known that not all (compact) manifolds can admit metrics of positive Ricci curvature: e.g., such a manifold necessarily has a finite fundamental group by Myers' Theorem applied to its universal cover. Indeed, it is also well known that not all compact manifolds admit metrics of positive scalar curvature [55] (see also [39], [68]). But recent results have shown that the control exerted by Ricci curvature has been simultaneously underestimated in low dimensions and overestimated in high dimensions (in regard to finiteness results) in the past. These results will be discussed in §4 and §5.

The existence of sphere theorems in which pinched sectional curvature is replaced by other conditions should be mentioned. These are particularly interesting in light of the recent developments in finiteness theorem discussed in the next section, because they dispense with upper bounds on the curvature. Specifically, the following hold:

1. ([44]) If a compact Riemannian manifold M has sectional curvatures ≥ 1 and diameter $> \pi/2$, then M is homeomorphic to a sphere.

2. ([73]) For each positive integer n, and positive real number Λ, there is a positive number $\varepsilon = \varepsilon(n, \Lambda)$ such that every compact Riemannian manifold of dimension n that has Ricci curvature $\geq (n-1)$, sectional curvature $\geq -\Lambda^2$, and volume $\geq \mathrm{vol}(S^n) - \varepsilon$ is homeomorphic to the n-sphere S^n. Here $\mathrm{vol}(S^n) =$ volume of S^n in its standard metric. (See [59], [79] for generalization)

2. From pinching to finiteness.

One of the crucial ingredients in the proof of the quarter-pinched sphere theorem is the estimation from below of the injectivity radius. For odd-dimensional manifolds, this estimation requires a subtle Morse theory argument in which the full force of the curvature hypotheses is used. But, for even dimensions, a much simpler argument suffices. In particular, a slight modification of the standard argument for proving Synge's Theorem yields the following (cf., e.g. [16] or [53]): If M is a compact even-dimensional simply connected manifold of positive sectional curvature, then the injectivity radius is realized by conjugate points; in particular, $i(M) \geq \pi/\sqrt{K}$, where $K =$ the maximum of the sectional curvatures of M. (Here it of course suffices by Synge's Theorem itself to suppose M orientable, from which simple connectivity follows in this case.)

It was a pivotal observation of A. Weinstein [78] that, armed with this injectivity radius estimate, one could estimate the number of homotopy types of simply connected compact manifolds M of fixed even dimension and with sectional curvatures between α and 1, for each fixed positive α. In particular, this number of homotopy types is finite. The proof of this result became the prototype for subsequent finiteness theorems:

First, note that the hypotheses yield that $i(M) \geq \pi$ and $\operatorname{diam}(M) \leq \pi/\sqrt{\alpha}$. From $i(M) \geq \pi$ and curvature ≤ 1, one sees by standard second variation arguments that open balls of radius $\leq \pi/2$ are strictly convex (i.e., if $p, q \in$ such a ball, then there is a unique arclength parameter minimal geodesic from p to q and it lies entirely in the same ball). Next, one sees that there is a set S of no more than N points, N depending only on α and $\dim M$, such that every point of M is within $< \pi/2$ of some point of S: This can be done by a simple packing argument using volume comparisons. Namely, choose a maximal family of disjoint open balls of disjoint open balls of radius $\pi/4$. Then the family of open balls with the same centers and radius $\pi/2$ covers M. On the other hand, the number of radius $\pi/4$-balls is clearly $\leq$ (volume of M)/(minimum possible volume of such a ball). The volume of M is bounded above in terms of n and α. The volume of a $\pi/4$-ball in M is bounded below on account of the curvature being ≤ 1 and the injectivity radius $\geq \pi > \pi/2$. This argument on the number of $\pi/2$-balls needed to cover can be done alternatively in terms of finding coverings of the (euclidean) interior of the cut locus in the tangent space at some point, and then using the distance-nonincreasing property of the exponential map. This approach uses only the lower curvature bound. The packing argument on M can also be modified to use only the lower curvature bound, or even a lower Ricci curvature bound: see §4. From this, one obtains a cover of M by no more than $N(n, \alpha)$ strictly convex balls. From this, the finite number of homotopy types follows easily.

The argument just given (from [78]) is subject to generalization simply by examining the proof. The exact ingredients needed are:

(1) a lower bound on the injectivity radius;

(2) an upper bound on curvature so that point (1) yields a lower bound C on the size of convex balls;

(3) an upper bound on diameter and lower bound on curvature so that a $\frac{1}{2}C$-dense set can be found in M with an estimatable number of points.

To get the estimated number of points for (3), one can use the same packing argument as before. (cf. [15]). Alternatively, in the argument using covers of the interior of the cut locus, replaces the distance-nonincreasing property of the exponential map by the fact that on a ball of fixed radius in the tangent space and with curvature bounded below, the exponential map has bounded magnification (on the interior of the cut locus). In particular, it follows easily that: The set of all compact Riemannian manifolds of fixed dimension n with $|\text{sectional curvature}| \leq \Lambda$, diameter $\leq d_0$ and injectively radius $\geq i_0 > 0$, contains only finitely many homotopy types.

This line of thought was pursued by J. Cheeger in [15]. In particular, he noted that, in the presence of the other bounds, the lower bound on injectivity was implied by, and indeed equivalent to, the more geometrically natural sounding condition of a lower bound on volume; also, he showed that the argument could be strengthened to yield the conclusion that there were finitely many homeomorphism types. (This is a nontrivial improvement since in higher dimensions there can be infinitely many homeomorphically inequivalent manifolds which are homotopy equivalent). In detail, he established the following: Let $\mathcal{C}(n, \Lambda, d_0, v_0)$ be the set of compact Riemannian manifolds of dimension n satisfying $|\text{sectional curvature}| \leq \Lambda$, diameter $\leq d_0$ and volume $\geq v_0 > 0$. Then there is a finite set of manifolds $M_1, \dots M_N$, $N = N(n, \Lambda, d_0, v_0)$ in

$\mathcal{C}(n, \Lambda, d_0, v_0)$ such that each manifold in $\mathcal{C}(n, \Lambda, d_0, v_0)$ is homeomorphic to one of the M_i, $i = 1, \ldots, N$.

It follows from standard topological results that "homeomorphic" can be replaced by "diffeomorphic" in case $n \neq 4$: If $n \neq 4$, then a fixed topological manifold admits only a finite number of differentiable structures. But for $n = 4$, this statement is false; [21] provides explicit examples of 4-manifolds with infinitely many inequivalent differentiable structures. Nonetheless, Cheeger's result as quoted remains true in all dimensions for diffeomorphism, not just homeomorphism. This was proved by a geometric method (no exotic topology involved) in [60]; the method there uses the center-of-mass construction for Riemannian manifolds introduced by E. Cartan historically and developed in detail by K. Grove and H. Karcher in [40].

Once it is known that such a class $\mathcal{C}(n, \Lambda, d_0, v_0)$ contains only finitely many manifolds up to diffeomormphism, it becomes natural to consider the set of all metrics on a fixed manifold that belong to a fixed $\mathcal{C}(n, \Lambda, d_0, v_0)$. M. Gromov [37] suggested (with a method of proof) that this set should be precompact in some suitable topology. This was established in detail by the author and H. Wu [33] and independently by S. Peters [61]. The specific result proved is the following:

Lipschitz Convergence Theorem: Suppose that M is a compact manfold and that $\{g_i : i \equiv 1, 2, 3, \ldots\}$ is a sequence of C^∞ Riemannian metrics on M. Suppose also that there are positive numbers Λ, d_0 and v_0 such that, for each $i = 1, 2, 3, \ldots$, the Riemannian manifold (M, g_i) has |sectional curvature| $\leq \Lambda$, diameter $\leq d_0$ and volume $\geq v_0$. Then there is a subsequence $\{g_{i_j} : j = 1, 2, 3 \ldots\}$ of the sequence $\{g_i\}$ for which there is a sequence of C^∞ diffeomorphisms $\phi_j : M \to M$ with the following properties: (1) For each $\alpha \in (0, 1)$, the sequence of metrics $\{\phi_j^* g_{i_j} : j = 1, 2, 3, \ldots\}$ converges in $C^{1,\alpha}$ norm; and (2) the limit $\lim_{j \to \infty} \phi_j^* g_{i_j}$ is a positive definite symmetric tensor of class $C^{1,\alpha}$ for all $\alpha \in (0, 1)$.

Here $\phi_j^* g_{i_j}$ means as usual the pullback of the metric g_{i_j} by the diffeomorphism.

One thinks of this rather convoluted statement informally as saying that a sequence of C^∞ metrics satisfying the indicated geometric bounds (i.e., all in a fixed $\mathcal{C}(n, \Lambda, d_0, v_0)$ class) has a subsequence that converges, after correction by suitable diffeomorphisms, to a limit metric of class $C^{1,\alpha}$ for all $\alpha \in (0, 1)$, and with convergence occurring in all $C^{1,\alpha}$ norms at once. It is apparent that the correction by diffeomorphism is needed; otherwise, a single metric pulled back by diffeomorphisms with ever larger first derivatives at some point would provide a counterexample to convergence. Ever smaller first derivatives at some point would similarly provide a counterexample to positive definiteness of the limit.

Obvious examples, such as smooth approximations of a finite cylinder capped at both ends by a hemisphere show easily that the convergence could not be in general better than $C^{1,1}$. In [61], it was shown that the convergence rate as stated – $C^{1,\alpha}$, all $\alpha \in (0, 1)$ – is in fact optimal, at least as far as harmonic coordinates are concerned. It is elementary to see that C^2 convergence cannot generally occur, even with best possible coordinate choices. But the $C^{1,1}$ question for best possible coordinates coordinates is apparently unresolved.

A natural approach to prove this Lipschitz Convergence Theorem, and indeed the method used in [33] and [61], is to choose "good" coverings (by convex balls) of M as in the finiteness theorems and to use coordinates on each ball in which the metrics satisfy the best possible bounds on their derivatives (and for which coordinates the overlap maps satisfy the best bounds). This suggests the use of harmonic coordinates, since these give maximal regularity of the metric, as shown in [18] and developed further in [50a], see also [50]. (See also [67] and [58] for related developments in the Soviet literature). The appearance thus on the scene of coordinates obtained from the theory of elliptic equations explains, in general terms at least, how the $C^{1,\alpha}$, $\alpha \in (0,1)$ convergence rate could arise. This type of convergence in Hölder norms is unusual in purely geometric situations but is of course standard in elliptic theory.

The convergence idea is of natural utility in examining what happens at the limit behavior of geometric results with open-interval bounds as hypotheses. For example, it occurs (in a slightly weaker form) in the proof in [8] of a characterization of manifolds which admit almost 1/4-pinched metrics; the precise results are: If (M, g_i) is a sequence of (C^∞) metrics on a compact, even-dimensional, simply connected manifold and if g_i is δ_i-pinched and $\lim \delta_i = 1/4$, then the manifold M admits a 1/4-pinched metric. In particular M is either homeomorphic to a sphere or diffeomorphic to a compact symmetric space of rank 1 (complex projective space, quaternionic projective space, or the Cayley plane). From this, it follows via the Cheeger-Peters finiteness theorem that: There is, for each $n \in \mathbf{Z}^+$, an $\varepsilon(n) > 0$ such that every compact, simply connected Riemannian manifold of dimension $2n$ that is $1/4$-$\varepsilon(n)$ pinched is either homeomorphic to a sphere or diffeomorphic to a symmetric space of rank 1.

It is natural to conjecture, as was suggested to the author by K. Grove, that such a phenomenon might hold more generally. That is, there might exist, for each even dimension $2n$, a sequence of positive numbers δ_i, $\lim \delta_i = 0$, $\delta_{i+1} < \delta_i$, such that: (1) $\delta_1 = 1/4$; (2) if M is a compact simply connected

δ-pinched manifold, $\delta_{i+1} < \delta \leq \delta_i$ then M is either diffeomorphic to a manifold with a δ'-pinched metric, $\delta' > \delta_i$, or diffeomorphic to one of a (finite) set of manifolds which admit δ_i-pinched metrics but no metric with pinching greater than δ_i and (3) each of the finite set of manifolds which admit δ_i-pinched metrics but no greater pinching admits a unique δ_i-pinched metric, up to diffeomorphism. Informally, one might thus divide even-dimensional simply connected manifolds of positive curvature into an increasing set of possibilities according to requiring less and less pinching, and each time new possibilities arise, they arise rigidly. Note that this situation does not follow trivially from the Finiteness and Convergence Theorems. Our present state of knowledge allows the possibility that an (even-dimensional simply connected) manifold M could admit a sequence of metrics g_i, each λ_i-pinched, with $\lambda_i < \lambda_{i+1}$, $\lim \lambda_i = \lambda_0, \lambda_0 < 1/4$, but that M might admit no C^∞ metric of λ_0 pinching. The Convergence Theorem guarantees a metric of regularity $C^{1,\alpha}$, all $\alpha \in [0,1)$, on M: the upper bound on diameter and lower bound on injectivity radius (and hence volume) are automatic in this case. This $C^{1,\alpha}$ metric will be λ_0-pinched, in the generalized sense, say, of the Toponogov Comparison Theorem holding as if it were λ_0-pinched. But it is not at all clear that this metric will be C^∞ or even that a C^∞ λ_0-pinched metric can be found. This is of course the reason that the almost 1/4-pinched result of [8] is not an immediate corollary of the Convergence Theorem. The odd-dimensional cases are complicated by the absence of a bound below on injectivity radius, e.g., the "Berger spheres" (see, e.g. [53] and [47], cf. also [54]).

§3. Finiteness Theorems without Injectivity Radius.

The proofs of the results on finiteness of topology and on convergence of metrics for the manifolds of class $\mathcal{C}(n, \Lambda, d_0, v_0)$ in the previous section depend directly on the possibility of bounding the injectivity radius below with a positive bound depending only on n, Λ, d_0, v_0: The first step in all the proofs is to find covers by metric balls of a size which is *a priori* small enough to make them contractible and even to have contractible multiple intersections with each other. At first sight, this restriction seems necessary: If the building blocks themselves are complex, or intersect in complex ways, how can one hope to estimate the structure that is built from them? But some remarkable recent developments have shown that, in some circumstances, estimates of topological complexity can be obtained in circumstances where a lower bound on injectivity radius is not available.

One result of this type is the theorem of M. Gromov [38] that there is an a priori bound, depending only on dimension, on the Betti numbers of (compact) manifolds of nonnegative curvature. In this case, a lower bound on injectivity radius is obviously not available, e.g., lens spaces, or, in the simply connected case just spheres with sharp points in the obvious sense. Moreover, even with the assumption of simple connectivity, the conclusion cannot be obtained by estimation of numbers of homotopy types since it was shown in [20] that there are infinitely many homotopy types among the compact simply connected manifolds of dimension seven (arising by modifying the examples in [3]) that admit metrics of positive sectional curvature, even of fixed pinching, see also [47]. Gromov's proof of the result uses the generalized (non)critical point theory of nonsmooth functions (cf. [44], and also, e.g., [29] [27] and [28]) combined with new arguments of extraordinary subtlety concerning the detailed behavior of the distance function.

From the general viewpoint so far discussed, it is particularly striking in Gromov's Betti number estimate that no use is made of any upper bound on curvature. In this context it becomes very natural to ask whether the upper bound on sectional curvature in the definition of $\mathcal{C}(n, \Lambda, d_0, v_0)$ is needed for the finiteness conclusions. It definitely is needed for the Lipschitz Convergence Theorem, as the sharp-pointed spheres show. But no examples were known showing the need for an upper bound on curvature as far as the finiteness conclusions were concerned.

In [42], K. Grove and P. Petersen showed, in a surprising development, that in fact the upper bound was not required. Specifically they proved:

> The set of all compact Riemannian manifolds with dimension n, sectional curvature $\geq -\Lambda$, diameter $\leq d_0$ and volume $\geq v_0$ contains only finitely many homotopy types.

In [43], Grove, Petersen and Y. Wu improved the result to assert finitely many diffeormorphism types. It can be seen by straightforward examples that all the bounds used as hypotheses are essential even for the homotopy type conclusion. Thus this latter diffeomorphism statement is indeed optimal, the end of the road for this type of finiteness theorem. Part of the viewpoint, however, holds in a more general context [62].

The essential ingredient in [42] is the realization that the homotopy equivalence of two manifolds can be proved without using "nice" covers by "good" (i.e., convex) balls. It was already known from an observation of Gromov [38] that a (compact) manifold with sectional curvature bounded below by $-\Lambda$ and diameter bounded above by d_0 could be covered by a number N of balls of radius $\varepsilon > 0$, with $N = N(n, \varepsilon, \Lambda, d_0)$ depending only on n, r, Λ, d_0. Indeed, this holds with only a lower bound on Ricci curvature, as will be discussed in the next section. Thus with the Grove-Petersen hypotheses the number of ε-balls needed to cover each manifold could be estimated in terms of ε. What definitely could not be estimated was how small ε needed to be chosen to make the balls convex. As a substitute for that, Grove and Petersen used instead the center-of-mass approach, along the lines of [60].

Specifically, they noted that if there were an $\varepsilon_0 > 0$ such that each set of points all with ε_0 of each other had a well-defined (weighted) center of mass, then homotopy equivalences could be obtained. (Here ε_0 has to be uniform over all the manifolds.) Roughly, the map from one manifold to the other could be obtained as follows: Suppose M_1 and M_2 have covers by balls $U_1, \ldots, U_N$ and $V_1, \ldots, V_N$ with the same intersection pattern. Let $\rho_1, \ldots, \rho_N$ be a partition-of-unity subordinate to $\{U_i\}$. Then a map from M_1 to M_2 can be obtained by sending $p \in M$, to the center-of-mass of $q_1, \ldots, q_N$ with weights $\rho_1(p) \ldots \rho_N(p)$ where $q_1, \ldots, q_N$ are the centers of the balls $V_1, \ldots, V_N$. For small U's and V's, most of the weights will be zero, and all the non-zero weighted points will be close together. Thus one only has to deal in effect with the uniform estimate on closeness needed to get a well-defined and continuously varying center-of-mass.

To treat this latter point, it was noted in [42] that the question reduces to introducing a flow on an (estimatably from below) small neighborhood of the diagonal, with the flow ending in the diagonal. A suitable flow is then obtained by a geodesic construction that uses a refined version of an argument in [15] for ruling out short smoothly closed geodesics in the class $\mathcal{C}(n, \Lambda, d_0, v_0)$ in that case: the relevant argument in [15] is noted not to depend on the upper curvature bound.

The refinement of the homotopy type result of [42] to the diffeomorphism result of [43] involves use of deep theorems of topology, rather than extensive additional geometric arguments. It would be of interest to find a direct, geometric proof of the finiteness of the number of diffeomorphism classes.

§4. Ricci Curvature: Compact Manifolds

In [38], M. Gromov suggested that estimates on Betti numbers should probably be establishable in the presence of nonnegative Ricci curvature alone, without assumption on sectional curvature; i.e., that there should exist a number $N(n)$, depending on the dimension n alone, such that each compact manifold of dimension n with a metric of nonnegative Ricci curvature would have each of its Betti numbers $\leq N(n)$. Such an estimate is well known to hold for the first Betti number: By the classical Bochner technique, the harmonic 1-forms on a compact manifold of nonnegative Ricci curvature are necessarily parallel; thus the space of harmonic 1-forms can have dimension at most n, and so the first Betti number is at most n. However, the corresponding Bochner technique for k-forms, $k \geq 2$, involves stronger conditions on the curvature tensor than nonnegative Ricci or even nonnegative sectional curvature, cf. [12]. (It is interesting to note that this method can be used to show that a manifold with α-pinched (positive) sectional curvature, α sufficiently close enough to 1, has all Betti numbers below the top dimension zero, and even that this holds if the sectional curvature is sufficiently pointwise pinched. This approach does not seem to have been heavily utilized historically. For one thing, $\alpha > 1/4$ is needed and this already implies that the manifold is a sphere; however, the Bochner technique was known long before the 1/4-pinched theorem, so at some point the Bochner technique was relevant to the issue, and before [7] and [52] it was certainly relevant to the pointwise pinching hypothesis, cf. [22].

As it happens, the failure of either the generalized Morse-theoretic technique or the Bochner technique to produce a priori ith Betti numbers estimates, $i \geq 2$, using Ricci curvature alone is inevitable. In [70] Sha and Yang showed by examples that no such a priori bounds are possible. Specifically, they gave a method using surgery by which certain manifolds of fixed dimension ≥ 7 with arbitrarily large (total) Betti number can be given a metric of positive Ricci curvature. This possiblity is surprising in terms of the general philosophy expressed in [38] and elsewhere concerning the importance of Ricci curvature.

It is worth noting here that one of the basic covering estimates needed for finiteness theorems along the lines of [78], [15] and [60] does in fact hold for nonnegative Ricci curvature or, indeed, for Ricci curvature just bounded below, and diameter bounded above. Specifically, Gromov in [38] pointed out that R. Bishop's result ([10]) on volume comparison combined with an elementary argument about ε-nets can be used to establish the following: For each $\varepsilon > 0$, $d_0 > 0$, $n \in \mathbb{Z}^+$, and Λ, there is a number $N(\varepsilon, d_0, \Lambda, n)$ such that each compact n-dimensional manifold M with Ricci curvature $\geq -\Lambda$ and diameter $\leq d_0$ has a covering by $N(\varepsilon, d_0, \Lambda, n)$ open balls of radius ε. Of course in case of a positive lower bound, i.e. Ricci curvature $\geq \Lambda > 0$, an automatic upper bound on diameter arises from Myers Theorem; and even if Ricci curvature is only supposed ≥ 0, the diameter bound can be arranged by rescaling. The difficulty in passing to finiteness results is thus concentrated on the impossibility of estimating the topological complexity of ε-balls for any a priori ε and the topological complexity of intersections of such balls. In particular, the Sha-Yang examples present a fundamental obstacle to pursuit of Ricci curvature finiteness theorems along these lines. The possibility apparently still exists, however, that some type of finiteness results, e.g., for Betti numbers, might hold for (compact) manifolds with positive Ricci curvature bounded above and bounded (above) diameter, or similarily for Ricci curvature bounded above and below by positive constants, which of course implies diameter bounded above.

In dimension 3, the relationship between Ricci curvature and sectional curvature is more intimate than in higher dimensions. In particular, it is elementary to see that constant Ricci curvature $2R$ implies constant sectional curvature R in three dimensions; this follows by pointwise calculation: If e_1, e_2, e_3 are an orthonormal basis at a point and $K(e_i, e_j)+K(e_i, e_k) = 2a$, $i = 1, 2, 3, j, k \neq i$ then $K(e_i, e_j) = a$ by linear algebra. (By contrast, there are compact 4-manifolds with Ricci curvature identically 0 but which have no metric of sectional curvature zero: these global examples arise from the solution of the Calabi conjecture by S. T. Yau [81], [82].) However, no such algebraic consideration applies with just the hypothesis of Ricci curvature of constant sign in dimension 3; that is, a curvature tensor at a point can have negative (positive) Ricci curvature without all the sectional curvatures at that point being negative (positive), as one can see easily by direct calculation. This possibility also holds globally: As shown in [23] and already noted every compact 3-manifold has a metric of negative Ricci curvature, but of course not all compact 3-manifolds admit a metric of even nonpositive sectional curvature, e.g., S^3.

These remarks make it quite surprising that in case of positive Ricci curvature, some conclusions about sectional curvature and topology are possible. For complete noncompact 3-manifolds, R. Schoen and S. T. Yau proved that positive Ricci curvature implies diffeomorphism to $\mathbf{R}^3$. Concerning the compact case, R. Hamilton proved in [45] that:

A compact 3-manifold with positive Ricci curvature necessarily admits another metric of constant positive sectional curvature.

The proof of this theorem involves deforming the given metric via a process analogous to heat flow; this process yields in the limit a metric of constant positive Ricci curvature and hence constant positive sectional curvature. In [46], the method is extended to encompass nonnegative curvature for 3-manifolds; it is shown that a compact manifold with nonnegative Ricci curvature is diffeomorphic either S^3 or to a quotient of $S^3, S^2 \times \mathbf{R}^1$, or $\mathbf{R}^3$ by a fixed-point free group of isometrics of the standard metric in each case.

The deformation process itself can be shown to converge ([48], see also [58a]) in all dimensions, and indeed to converge to a (C^∞) metric of constant positive sectional curvature, provided that the original metric is close enough in a suitable sense to a metric of constant curvature. The precise sense needed is that at each point the norm of the Weyl conformal curvature tensor and of the traceless Ricci tensor should be not larger than a (small) constant times the scalar curvature. Since this is a pointwise estimate, [48] gives an alternative approach to the pointwise-pinching sphere theorems of [66] and [56], although the methods of [48] and [58a] require at present a stronger hypothesis than the just greater-than -1/4 pointwise pinching that suffices for [56]. (It is an algebraic fact that sufficient pointwise pinching implies the conditions needed for the deformation process in [48], cf. [51].)

In dimension 4, Hamilton showed that the deformation process of [45] and [48] was in fact valid in the presence of just nonnegative curvature operator; moreover, he proved that in the presence of everywhere positive curvature operator, the limit metric had constant positive sectional curvature so that the manifold must be diffeomorphic to S^4 or RP^4. And he was able to classify the possibilites for compact 4-manifolds in case of nonnegative curvature operator; the manifold in that case must be diffeomorphic either to $S^4, \mathbf{R}P^4, \mathbf{C}P^2$ or to a quotient by a fixed-point free group action of $S^3 \times \mathbf{R}, S^2 \times S^2, S^2 \times \mathbf{R}^2$ or $\mathbf{R}^4$, with the group action being by isometrics relative to the respective standard metrics. In the case of positive curvature operator, it was already known ([57]) that the manifold, if simply connected, must be homeomorphic to a 4-sphere. But it is at present unknown whether the 4-sphere admits exotic differentiable structures, so that, in our present state of knowledge, the result of [46] on diffeomorphism gives additional information.

Hamilton's Theorem implies in particular that a compact simply connected 3-manifold which admits a metric of positive Ricci curvature is diffeomorphic to a sphere. Of course if the 3-dimensional Poincare conjecture is true, then the metric hypothesis is irrelevant. In the non-simply connected case, Hamilton's Theorem interacts with another unsolved topological problem. Namely, suppose M is a nonsimply connected compact manifold with positive Ricci curvature. By Hamilton's Theorem, M has a metric of constant positive sectional curvature; in particular, M is a quotient of S^3 by a group action which is isometric relative to the standard metric. Such a group action is of course linear. Thus no "pathological" nonlinear possibilities occur in the positive Ricci curvature case in dimension 3. Whether any of them occur at all is an unsolved problem of topology. The results of [46] and [48] have similar relevance to the corresponding question for quotients of higher dimensional spheres. However, in dimensions > 4, the type of hypothesis needed in [48] was already known to imply that the manifold was a spherical space form: [66], as discussed in section 1. A corresponding result directly concerning pinched curvature was given in [41], where it is shown that 0.98-pinching again implies that the manifold is a spherical space form, but this is global not pointwise-local pinching, and is in that sense weaker than the results of [48] or [66].

Results on everywhere positive Ricci curvature are in effect results on Ricci curvature nonnegative, but positive at one point, because it was shown in [5] that: If M is a compact manifold with nonnegative Ricci curvature and if the Ricci curvature tensor is (strictly) positive at one point, then M admits a metric of everywhere positive Ricci curvature.

The deformation method of [45] and [46] can be applied in the noncompact case as well [71], under some mild hypotheses on the curvature, e.g., bounded, nonnegative curvature operator in dimension 4. This yields in [71] a classification of complete noncompact 4-manifolds of bounded nonnegative curvature operator. The content of this classification was in fact already known indeed under weaker hypotheses, in [76], see also [77] using refinements of the original Cheeger–Gromoll-Meyer arguments ([35] and [17]). But the method in [71] remains of interest; it is extended further in [71a] to (noncompact) manifolds of higher dimension.

§5. Ricci Curvature: Noncompact Manifolds

Philosophically, there seems to be some relationship between the two types of topological finiteness theorems that arise naturally in geometry, namely, between: (1) a priori bounds on the topological complexity of compact manifolds satisfying certain geometric conditions and (2) the finiteness of topological type of (each) complete noncompact manifold satisfying some particular set of geometric bounds. Choose on M a C^∞ exhaustion function $\varphi : M \to \mathbf{R}$ with, say, the positive integers being noncritical values. Then, if M has infinite topological complexity in some sense, the sublevels $\varphi^{-1}((-\infty, j]), j = 1, 2, 3 \ldots$ are compact manifolds-with-boundary if unbounded complexity. Of course, it may not be possible to "cap off" these manifolds-with-boundary to be compact without boundary, while still satisfying the geometric conditions. But if it is possible, then one obtains a sequence providing a counterexample to a priori bounds on topological complexity for compact manifolds of type (1).

Viewing the matter the other way around, if counterexamples to type (1) results are obtained by connected sums of simpler manifolds, then one might expect to find a counterexample to type (2) results by stringing together an infinite connected sum of compact examples.

This indefinite though suggestive picture is illustrated by substantial mathematics in [70]. First, as already discussed, it is shown there that there is a sequence of fixed-dimensional compact manifolds having positive Ricci curvature but unbounded Betti numbers. This sequence is in fact constructed by successive modifications by surgery. Then examples of complete noncompact manifolds of positive Ricci curvature but not all Betti numbers finite are produced in effect by an infinite iteration of the construction. In practice, this infinite iteration is more delicate than this summary has suggested. The construction of the metric on the modified manifold involves a not completely localized rescaling process, and some care is necessary to see that a noncompact limit can indeed be obtained by an infinite iteration of the process. The specific result obtained is for dimension ≥ 7, but this is presumably only a technical restriction and it is expected that similar conclusions hold in all dimensions ≥ 4 (conjectured in [70]).

A careful analysis of the construction of the noncompact examples in [70] shows that, on account of the required rescaling, there is a certain sense in which the metric size of the manifold is growing as a function of distance. To make the sense of this growth precise, one needs some definitions, introduced in [2]. First, for each open (not necessarily connected) set U in a Riemannian manifold M and each connected subset C of U, set $\operatorname{diam}(C, U) =$ the diameter of C in the Riemannian manifold U, i.e., $\operatorname{diam}(C, U) = \sup(d_U(p, q), p, q \in C$ where $d_U(p, q) =$ the infimum of the lengths of piecewise C^∞ curves from p to q

in U. Now, if M is a (connected) noncompact complete Riemannian manifold and $p_0 \in M$, set $C(p_0, r) =$ the union of the unbounded connected components of $M - B(p_0, r)^-$, where $-$ denotes closure. Next, choose a number $\zeta \in (\frac{1}{2}, 1)$ and set $\operatorname{diam}(p_0, r) = \sup_{\Sigma} \operatorname{diam}(\Sigma, C(p_0, \zeta r))$ where the supremum is taken over all components Σ of $\partial C(p_0, r)$. Finally, let $f : \mathbf{R}^+ \to \mathbf{R}^+$ be a monotone nondecreasing function. A Riemannian manifold M with $p_0 \in M$ is said to have *diameter growth of order* $o(f)$ if and only if $f(r)^{-1} \operatorname{diam}(p_0, r)$ converges to 0 as $r \to +\infty$; it has growth $O(f)$ if $f(r)^{-1} \operatorname{diam}(p_0, r)$ is bounded as $r \to +\infty$. With these definitions from [2], the results of [2] are :

(1) A complete noncompact Riemannian manifold of nonnegative Ricci curvature and diameter growth $o(r^{1/n})$ and sectional curvature bounded below is homotopy equivalent to the interior of a compact manifold with boundary.

(2) Let M^n be a complete noncompact Riemannian manifold, $p_0 \in M$. Suppose that (a) there is a nonincreasing function $\lambda : [0, \infty) \to [0, \infty)$ such that $\int_0^\infty r \cdot \lambda(r) dr$ is finite and such that $Ric_q \geq -(n-1)\lambda(\operatorname{dis}(p_0, q))$ for all $q \in M$.

(b) The sectional curvatures of M are bounded below.

(c) M^n has a diameter growth $o(r^{1/n})$ with respect to p_0.

Then M^n is homotopy equivalent to the interior of a compact manifold with boundary.

The $o(r^{1/n})$ restriction on diameter growth should be compared to the Sha-Yang examples: these (seven-dimensional) examples have diameter growth $O(r^{2/3})$, and bounded sectional curvature as noted in [70] in connection with [2] (see also the remarks in [2]).

The condition involving the finiteness of the integral $\int_0^\infty r\lambda(r)dr$ has been known for some time to arise naturally in considering the geometry of noncompact manifolds. Philosophically, its underlying significance begins with the fact that the curvature scales inverse-quadratically with respect to distance when the metric is multiplied by a constant. Thus in some general sense a metric which has curvature decaying faster than quadratically with distance might be thought of as being flat at infinity. Alternately, if the curvature is bounded below by a (negative) lower bound that decays faster than quadratically, it might well be thought of as nonnegatively curved at infinity. These philosophical ideas have been given concrete substance in [31], where $\int_0^\infty r\lambda(r)dr$, $\lambda(r)$ a sectional curvature bound at distance r, is shown to control quasi-isometry of the exponential map. Also in [32], the $\int r\lambda(r)$ integral gives the boundary between rates of curvature that can occur and those that imply flatness if the curvature has one sign only (or is zero); the general view here is that: If curvature ≥ 0 everywhere (with pole) or ≤ 0 everywhere (and simply connected) and if |curvature| at distance $r \leq \lambda(r)$ with $\int r\lambda(r)$ finite, then the manifold must be flat. (See [32] for details). Also, in [1], it was shown that asymptotic nonnegativity of curvature at infinity in the sense indicated implies that the manifold has the homotopy type of a compact manifold with boundary. Ideas of this type also occur in Gromov's work on Tits metrics of noncompact manifolds.

To put these results in perspective, it should be noted that the methods of [25] can be modified to show that, for any monotone function of slower than quadratic decay and any noncompact manifold M, there is a complete Riemannian metric on M the absolute value of the curvature of which decays faster than the given function, as a function of distance ([26]).

REFERENCES

[1] U. Abresch, *Lower curvature bounds, Toponogov's theorem, and bounded topology*, Ann. Scient. Ec. Norm. Sup. **4** serie, t. 18 (1985) 651–670.

[2] U. Abresch and D. Gromoll, *On complete manifolds with nonnegative Ricci curvature*, preprint.

[3] S. Aloff and N. Wallach, *An infinie family of distinct 7-manifolds admitting positively curved Riemannian structures*, Bull. Amer. Math. Soc. **81** (1975), 93–97.

[4] M. Anderson, P. Kronheimer, and C. LeBrun, *Complete Ricci-flat Kähler manifolds of infinite topological type*, (preprint, to appear).

[5] T. Aubin, *Metriques riemanniennes et courbre*, J. Diff. Geom. **4** (1970), 383–424.

[6] P. Aviles and R. McOwen, *Conformal deformation to constant negative scalar curvature on noncompact Riemannian manifolds*, J. Diff. Geom. **27** (1988), 225–239.

[7] M. Berger, *Les varietes riemanniennes 1/4-pincees*, Ann. Scuola. Norm. Sup. Pisa Sci. Fis Mat. **(3)14** (1960), 161–170.

[8] M. Berger, *Sur les varietes riemanniennes pincees juste au-dessous de 1/4*, Ann. Inst. Fourier (Grenoble) **33** (1983), 135–150.

[9] M. Berger, *Sur quelques varietes riemanniennes suffisament pincees*, Bull. Soc. Math. France **88** (1960), 57–71.

[10] R. L. Bishop and R. J. Crittenden, *Geometry of manifolds*, Academic Press, 1964.

[11] J. Bland and M. Kalka, *Complete metrics of negative scalar curvature on noncompact manifolds*, Contemporary Math. **51**, (1986), 31 35.

[12] S. Bochner and K. Yano, *Curvature and Betti Numbers*, Princeton University Press, 1953.

[13] R. Brooks, *A construction of metrics of negative Ricci curvature*, Jour. Diff. Geom. **29** (1989), 85–94.

[14] R. Brooks, *Designer metrics on Riemannian manifolds*, Contemporary Math. (this volume).

[15] J. Cheeger, *Finiteness theorems for Riemannian manifolds*, Amer. J. Math. **92** (1970), 61–74.

[16] J. Cheeger, and D. Ebin, *Comparison Theorems in Riemannian Geometry*, North-Holland, 1975.

[17] J. Cheeger, and D. Gromoll, *On the structure of complete manifolds of nonnegative curvature*, Ann. Math. **96** (1972), 413–443.

[18] D. DeTurck and J. Kazdan, *Some regularity theorems in Riemannian geometry*, Ann. Sci. Ecol. Norm. Sup. Paris 4e, serie **14** (1981), 249–260.

[19] H. Eliasson, *On variation of metrics*, Math. Scand. **29** (1971), 317–327.

[20] J. Eschenburg, *New examples of manifolds with strictly positive curvature*, Inv. Math. **66** (1982) no. 3, 469–480.

[21] R. Friedman and J. W. Morgan, *On the diffeomorphism types of certain algebraic surfaces.* I, J. Diff. Geom. **27** (1988) 297-369.

[22] S. Gallot and D. Meyer, *Operateur de scourbure et Laplacien des formes differentielles d'une variete riemannienne*, J. Math. Pure Appl. **54** (1975), 259–284.

[23] L. Z. Gao, *The construction of negatively Ricci curved manifolds*, Math. Ann. **271** (1985), 185–208.

[24] L. Gao and S. T. Yau, *The existence of negatively Ricci curved metrics of 3-manifolds*, Inv. Math. **85** (1986), 637–652.

[25] R. E. Greene, *Complete metrics of bounded curvature on noncompact manifolds*, Arch. Math. **31** (1978), 89–95.

[26] R. E. Greene, *Metrics of rapid curvature decay*, Amer. Math. Soc. Abstracts, **53** (1987), 420.

[27] R. E. Greene and K. Shiohama, *Convex functions on complete noncompact manifolds: differentiable structure*, Ann. Sci. Ecole Norm. Sup (4) **14** (1981), 357–367.

[28] R. E. Greene and K. Shiohama, *Convex functions on complete noncompact manifolds: topological structure*, Inv. Math. **63**, (1981), 129–157.

[29] R. E. Greene and H. Wu, *Integrals of subharmonic functions on manifolds of nonnegative curvature*, Inv. Math. **27** (1974), 265–298.

[30] R. E. Greene and H. Wu, *Whitney's imbedding theorem by solutions of elliptic equations and geometric consequences*, Proc. Sym. Pure Math., Amer. Math. Soc. **27** (1975), 287–295.

[31] R. E. Greene and H. Wu, *Function Theory on Manifolds Which Possess a Pole*, Springer Lecture Notes in Math. no. **699**, (1977).

[32] R. E. Greene and H. Wu, *Gap theorems for noncompact Riemannian manifolds*, Duke Math. J. **49** (1982)3, 731–756.

[33] R. E. Greene and H. Wu, *Lipschitz convergence of Riemannian manifolds*, Pacific J. of Math. **131** (1988), no. 1, 119–142.

[34] D. Gromoll, *Differentiation Strukturen und Metriken positiver Krümmung auf Sphären*, Math. Ann. **164** (1966), 353–371.

[35] D. Gromoll and W. Meyer, *On complete open manifolds of positive curvature*, Ann. of Math. **90** (1969), 75–90.

[36] D. Gromoll and W. Meyer, *An exotic sphere with non-negative sectional curvature*, Ann. Math. **100(1)** (1974), 401–406.

[37] M. Gromov, *Structure metriques pour les varietes riemanniennes*, redige par. J. Lafontaine et P. Pansu, Textes math. n°1 Cedic-Nathan, Paris, 1981.

[38] M. Gromov, *Curvature, diameter, and Betti numbers*, Commentarii Math. Helv. **56** (1981), 179–195.

[39] M. Gromov and B. Lawson, *The classification of simply connected manifolds of positive scalar curvature*, Ann. Math. **111** (1980), 423–435.

[40] K. Grove and H. Karcher, *How to conjugate C' close group actions*, Math. Z. **132** (1973), 11–20.

[41] K. Grove, H. Karcher, and E. Ruh, *Jacobi fields and Finsler Metrics on compact Lie groups with an application to differentiable pinching problems*, Math. Ann. **211** (1974), 7–21.

[42] K. Grove and P. Petersen, *Bounding homotopy types by geometry*, Ann. Math. **128** (1988), 195–206.

[43] K. Grove, P. Petersen, J. Wu, *Geometric finiteness theorems via controlled topology*, preprint, 1989.

[44] K. Grove and K. Shiohama, *A generalized sphere theorem*, Ann. Math. **106** (1977), 201–211.

[45] R. S. Hamilton, *Three-Manifolds with positive Ricci curvature*, J. Diff. Geom. **17** (1982), 255–306.

[46] R. S. Hamilton, *Four-Manifolds with positive curvature operator*, J. Diff. Geom. **24** (1986), 153–179.

[47] H. M. Huang, *Some remarks on the pinching problem*, Bull. Inst. Math. Acad. Sinica **9** (1981), 321–340.

[48] G. Huisken, *Ricci deformation of the metric on a Riemannian manifold*, J. Diff. Geom. **21** (1985), 47–62.

[49] H. C. Im Hof and E. A. Ruh, *An equivariant pinching theorem*, Comment. Math. Helvetici **50** (1975) 389-401.

[50] J. Jost, *Harmonic mappings between Riemannian manifolds*, Proc. Centre for Math. Analysis, Australian National Univ. **4** (1983).

[50a] J. Jost and H. Karcher, *Geometrische Methoden zur Gewinnung von a-priori-Schranker für harmonische Abbildungen Manuscripta*, Math. **40** (1982), 27–77.

[51] H. Karcher, *Pinching Implies Strong Pinching*, Comm. Math. Helv. **46** (1971)1, 124–126.

[52] W. Klingenberg, *Über Riemannsche Mannigfaltigkeiten mit positiven Krümmung*, Comm. Math. Helv. **35** (1961), 47–54.

[53] W. Klingenberg, *Riemannian Geometry*, de Gruyter Studies in Math., no. 1, Berlin-New York, 1982.

[54] W. Klingenberg and T. Sakai, *Remarks on the injectivity radius estimate for almost 1/4-pinched manifolds*, Springer Lec. Notes in Math. **1201**, 1986.

[55] A. Lichnerowicz, *Spineurs harmoniques*, C. R. Ac. Sci. Ser. A–B **257** (1963), 7–9.

[56] M. J. Micallef and J. D. Moore, *Minimal two-spheres and the topology of manifolds with positive curvature on totally isotropic two-planes*, Ann. Math. **127** (1988) 199-227. 8pt

[57] Min-Oo and E. Ruh, *Comparison theorems for compact symmetric spaces*, Ann. Sci. Ecole Norm. Sup. (4) **12** (1979)3, 335–353.

[58] I. G. Nikolaev, *Parallel translation and smoothness of the metric of space of bounded curvature*, Dokl. Akad. Nauk SSSR, **250** (1980), 1056–1058.

[58a] S. Nishikawa, *Deformation of Riemannian metrics and manifolds of bounded curvature, ratios*, Proc. Sym. Pure Math. **44** (1986) 343-352.

[59] Y. Otsu, K. Shiohama, and T. Yamaguchi, *A new version of the differentiable sphere theorem*, (preprint), to appear.

[60] S. Peters, *Cheeger's finiteness theorem for diffeomorphism classes of*

Riemannian manifolds, J. Reine Angew. Math. **349** (1984), 77–82.

[61] S. Peters, *Convergence of Riemannian manifolds*, Compositio Math. **62** (1987), 3–16.

[62] P. Petersen, *A finiteness theorem for metric spaces*, (to appear).

[63] H. E. Rauch, A contribution to differential geometry in the large, Ann. Math. **54** (1951), 38-55.

[64] B. Riemann, *Über die Hypothesen, welche der Geometrie zu Grunde liegen in Collected Works of Bernhard Riemann*, Dover Publications (1953), pp. 272–287.

[65] E. Ruh, *Curvature and differentiable structures on spheres*, Comm. Math. Helv. **46** (1971), 127–136.

[66] E. Ruh, *Riemannian manifolds with bounded curvature ratios*, Jour. Diff. Geom. (1982).

[67] I. K. Sabitov and S. Z. Sefel, *Connections between the order of smoothness of a surface and that of its metric*, Sibirsk. Math. Zh. **17** (1976), 687–694.

[68] R. Schoen and S. T. Yau, *On the structure of manifolds with positive scalar curvature*, Manuscripta Math. **28** (1979), 159–183.

[69] R. Schoen and S. T. Yau, *Complete three dimensional manifolds with positive Ricci curvature and scalar curvature*, Seminar on Differential Geometry, Princeton University Press, Princeton, N.J. (1982), 209–228.

[70] J.-P. Sha and D. G. Yang, *Examples of manifolds of positive Ricci curvature*, J. Diff. Geom. **29** (1989), 95–103.

[71] W. X. Shi, *Complete noncompact three-manifolds with nonnegative Ricci curvature*, J. Diff. Geom. **29** (1989), 353–360.

[71a] W. X. Shi, *Deforming the metric on complete Riemannian manifolds*, J. Diff. Geom. **30** (1989), 223-301.

[72] Y. Shikata, *On the differentiable pinching problem*, Osaka Math. J. **4** (1967), 279–287.

[73] K. Shiohama, *A sphere theorem for manifolds of positive Ricci curvature*, Trans. Amer. Math. Soc. **275** (1983), 811–819.

[74] K. Shiohama and M. Sugimoto, *On the differentiable pinching problem, with improvement by H. Karcher*, Math. Ann **195** (1971), 1–16.

[75] N. Trudinger, *Remarks concerning the conformal deformation of Riemannian structures on compact manifolds*, Ann. Scuola Norm. Sup. Pisa **22** (1968), 265–274.

[76] G. Walschap, *A splitting theorem for 4-dimensional manifolds of nonnegative curvature*, Proc. Amer. Math. Soc. **104** (1988) no. 1, 265–268.

[77] G. Walschap, *Open manifolds of nonnegative curvature*, Contemporary Math., (this volume).

[78] A. Weinstein, *On the homotopy type of positively pinched manifolds*, Arch. Math. (Basel) **18** (1967), 523–524.

[79] T. Yamaguchi, *A differentiable sphere theorem for volume-pinched manifolds*, Advanced Studies in Pure Math. **3** (1984), 183–192.

[80] T. Yamaguchi, *Manifolds of almost nonnegative Ricci curvature*, J. Diff. Geom. **28** (1988), 157–167.

[81] S. T. Yau, *Calabi's conjecture and some new results in algebraic geometry*, Proc. Nat. Acad. Sci. USA **71** (1974), 1798–1799.

[82] S. T. Yau, *On The Ricci curvature of a compact Kähler manifold and the complex Monge-Ampere equation I*, Comm. Pure Appl. Math **31** (1978), 339–411.

Robert E. Greene
Department of Mathematics
University of California at Los Angeles
Los Angeles, CA 90024

Contemporary Mathematics
Volume **101**, 1989

Designer Metrics on Riemannian Manifolds

Robert Brooks*

ABSTRACT. We show how to construct various special types of metrics on Riemannian manifolds, starting from hyperbolic orbifold metrics. Such metrics are fairly common, and exist on all compact 3-manifolds. Constructing the desired metrics - negatively Ricci curved, ergodic and pinched negatively curved - then reduces to satisfying some simple ordinary differential inequalities in one variable.

A standard question in Riemannian geometry asks whether a given manifold (or class of manifolds) caries a Riemannian metric of a certain type. The type under consideration generally falls into one of two categories - either it is a type which is stable under small perturbations (for instance, negative curvature) or it is quite rigid (for instance, constant negative curvature).

In recent years, there has been increasing interest in the first category. This interest comes from two quite separate sources. First of all, problems in dynamical systems have raised the question of which manifolds admit metrics which are ergodic. Strictly speaking, this is a global property of the metric, but in practice it is deduced from local or semi-local properties of the metric in question. Thus the theorem of Anosov tells us that metrics of negative sectional curvature are ergodic, but until relatively recently it was an unsolved problem to construct an ergodic smooth metric on S^2. Of course, one believes that a suitably generic metric on any manifold will be ergodic.

The solution to this problem on S^2, which we will discuss below, involved starting with a metric on S^2-points, which carries a metric of negative curvature to which one may start to apply Anosov's theorem, and then patching this metric over the points in such a way that ergodicity is preserved. The heart of the problem was to establish a

* AMS (MOS) Classification 53C25 - Partially supported by NSF grant DMS-8801158

criterion for ergodicity in such a way that one may see that the patching can be carried out.

A second direction is the question of whether manifolds carry metrics for which various types of curvature (sectional, Ricci, or scalar) carry sign conditions. From the work of Gromov-Lawson [6] the obstructions to positive scalar curvature are fairly well understood, and obstructions to positive Ricci curvature are classical. On the other hand, there are no obvious obstructions to negative Ricci curvature (in dimensions bigger than two), and every manifold (again in dimensions bigger than two) admits a metric of negative scalar curvature. In [5], Gao and Yau raised the question of whether every compact manifold of dimension greater than two admits a metric of negative Ricci curvature, and they showed that this was the case for 3-manifolds.

In a parallel development, Gromov and Thurston [7] addressed the question of whether metrics of suitably pinched negative sectional curvature could be made to have constant negative curvature. In dimension 3, the question of whether a manifold which has negative sectional curvature can be made to have constant sectional curvature is still open, but it falls sufficiently well within the Thurston theory [16] that one may claim a fair amount of understanding of this question. On the other hand, in dimensions four and above the list of manifolds of constant negative curvature is quite small (by a theorem of Wang), and so the issue is: how easy is it to construct metrics of pinched negative sectional curvature? Gromov and Thurston then showed how to do this, providing examples of manifolds of arbitrarily pinched negative sectional curvature which do not carry metrics of constant negative curvature.

All of these developments were, or could be seen as, variations on a similar idea. That idea, which is the main theme of this paper, can be explained as follows: while constant curvature metrics are, in general, somewhat rare, constant curvature orbifold metrics are significantly more common. Indeed, as we will see below, every compact 3-manifold has an orbifold hyperbolic metric, and the author knows of no obstruction for any manifold to carry such a metric (although surely there must be some). On the other hand, the hyperbolic metric behaves in a standard fashion in a standard neighborhood of the orbifold singularity, so that in principle one can reduce the problem of constructing the desired metric to a problem of one variable (distance to the singularity), at which point hopefully it becomes readily computable.

We will carry out this program for the examples cited above.

We would like to thank Larry Zalcman and Bar Ilan University for their generous hospitality when most of this paper was written.

§1: Some Hyperbolic Geometry

Let us first recall some basic results from the geometry of the hyperbolic plane, together with their geometric interpretation.

Theorem ([19]): Let $\begin{pmatrix} 1 & 1 \\ 0 & 1 \end{pmatrix}$ and $\begin{pmatrix} a & a \\ c & d \end{pmatrix}$ generate a discrete subgroup of $PSL(2,\mathbb{R})$. Then either $c = 0$ or $/c/ \geq 1$.

Proof: We compute

$$\begin{pmatrix} a & b \\ c & d \end{pmatrix}\begin{pmatrix} 1 & 1 \\ 0 & 1 \end{pmatrix}\begin{pmatrix} d & -b \\ -c & a \end{pmatrix} = \begin{pmatrix} 1-ac & a^2 \\ -c^2 & 1+ac \end{pmatrix}$$

Replacing $\begin{pmatrix} a & b \\ c & d \end{pmatrix}$ by the right hand matrix and iterating, we see that if $c \neq 0$ and , $/c/ < 1$, we obtain matrices converging to the identity matrix, contradicting the discreteness of the group generated by $\begin{pmatrix} 1 & 1 \\ 0 & 1 \end{pmatrix}$ and $\begin{pmatrix} a & b \\ c & d \end{pmatrix}$.

To give this theorem geometric content, recall the action of $PSL(2,R)$ on the upper half plane $\mathbb{H}^2 = \{(x,y) \varepsilon \mathbb{R}^2, y > 0\}$. This action is an isometry of the hyperbolic metric . Note that $\begin{pmatrix} 1 & 1 \\ 0 & 1 \end{pmatrix}$ fixes the point at infinity, and preserves all the level surfaces $y = const$. If $A = \begin{pmatrix} a & b \\ c & d \end{pmatrix}\begin{pmatrix} 1 & 1 \\ 0 & 1 \end{pmatrix}\begin{pmatrix} d & -b \\ -c & a \end{pmatrix}$ is any other parabolic motion, then $\begin{pmatrix} 1 & 1 \\ 0 & 1 \end{pmatrix}$ sends the point at infinity to $\frac{a}{c}$, and the line $y = (const)$ to a circle tangent to the x-axis at , whose Euclidean diameter is $\frac{1}{c2\ (const)}$.

If we now set $(const) = 1$, it follows from $/c/ \geq 1$ that the regions $y > 1$ and the interior of the above circle are disjoint. As a result, if Γ is any discrete group containing $\begin{pmatrix} 1 & 1 \\ 0 & 1 \end{pmatrix}$ then the region $\{y > 1\}/\begin{pmatrix} 1 & 1 \\ 0 & 1 \end{pmatrix}$ descends to $\mathbb{H}^2/\Gamma$ in a 1 - to -1 manner, and furthermore, the corresponding regions for any two cusps on $\mathbb{H}^2/\Gamma$ are disjoint.

The region $\{y > 1\}$ is often called the standard horocycle.

It was shown by Troels Jorgensen in [11] that this fact generalizes in a nice explicit way for matrices $\begin{pmatrix} a & b \\ c & d \end{pmatrix}$ in $PSL(2,\mathbb{C})$ whose traces are small. Indeed, one has:

Theorem (Jorgensen [11]): Suppose that X and Y generate a discrete, non-elementary subgroup of $PSL(2,\mathbb{C})$. Then

$$\left|tr^2(X)-4\right|+\left|tr\left(XYX^{-1}Y^{-1}\right)-2\right|\geq 1$$

Note that $tr^2(X)-4$ has a natural interpretation in terms of the length of the geodesic left invariant by X, and $\left|tr\left(XYX^{-1}Y^{-1}\right)-2\right|$ has a similar interpretation in terms of the distance between the geodesic left invariant by X and that left invariant by YXY^{-1} (see [20] for details). We conclude as above that if g is a closed geodesic in $\mathbb{H}^3/\Gamma$ which is sufficiently short, then there is a tubular neighborhood about γ whose width $r(\gamma)$ may be chosen independent of Γ, and tubular neighborhoods of disjoint short geodesics are disjoint.

The beauty of Jorgensen's inequality is its explicit nature, and indeed one may work out explicit estimates for $r(\gamma)$. This is done in [3], [13] .

It is more or less clear that there is an analogous theory for tubular neighborhoods for short geodesics in hyperbolic spaces of all dimensions. A rigorous but non-effective argument to this effect could be based in the Kazhdan-Margulis Theorem. We remark, however, that it is an interesting problem to give effective estimates in higher dimensions, see [17] for some ideas.

It is elementary, however, and important for what follows, that in higher dimensions we can be explicit in the following special case: Let us say that an isometry X of $\mathbb{H}^n$ is elliptic of order k if the fixed point set of X is of codimension 2, and X is a rotation about angle $\frac{2\pi}{k}$ about the fixed point set. Then:

Theorem(see [7]): Let Γ be a discrete group of isometries of $\mathbb{H}^n$, and X elliptic of order k. Then the fixed point set of X has a tubular neighborhood in $\mathbb{H}^n/\Gamma$ of radius at least r, where

$$\cosh(r)=\frac{1}{2\sin\left(\frac{\pi}{k}\right)}$$

Proof: In $\mathbb{H}^n$, let g be the shortest geodesic joining $A=$ the fixed point set of X with a translate of A by Γ, and let 2λ be its length. Then acting on γ by X, and joining the two endpoints of γ not on A, we obtain an isosceles triangle whose common sides are 2λ, included angle is $\frac{2\pi}{k}$, and whose opposite side has length 2ℓ, greater than 2λ. By the law of sines in hyperbolic geometry, $\sin\left(\frac{\Pi}{n}\right)=\frac{\sinh(\ell)}{\sinh(2\lambda)}=\frac{\sinh(\ell)}{2\sinh(\lambda)\cosh(\lambda)}$ so $\cosh(\lambda)=\frac{1}{2\sin\left(\frac{\Pi}{n}\right)}\left(\frac{\sinh(\ell)}{\sinh(\lambda)}\right)\geq\frac{1}{2\sin\left(\frac{\Pi}{n}\right)}$.

It is an instructive exercise to understand why this argument is a direct translation of the iteration argument we applied to the matrix $\begin{pmatrix} a & b \\ c & d \end{pmatrix}$.

§2: Some 3-Dimensional Topology

In this section we will show there exists k_1 and k_2 such that:

Theorem: Every 3-manifold M has a link L such that M has a hyperbolic structure which is branched of order between k_1 and k_2 about L.

This theorem has had a rather interesting development, which we will sketch below.

It was proved by Montesinos in [14] that every 3-manifold M arises as a branched covering of S^3. It was subsequently observed by Thurston that there is a link L in S^3 such that every 3-manifold arises as a branched covering over S^3 whose branch locus in S^3 is L. Such a link he referred to as a universal link, and it became a natural question to produce universal links with particularly nice properties.

It was therefore highly significant when it was shown by Hilden, Lozano, and Montesinos [8] that the figure - 8 knot, pictured below, was such a universal knot. This is true for a variety of reasons. First of all, the figure - 8 knot is one of the simplest knots imaginable, after the Girl Scout Knot. Secondly, the complement of the figure - 8 knot has a well-known hyperbolic structure, which has been studied in great detail by Riley [15], Jorgensen [10], and others. In effect, one may write down the fundamental group of the complement as

$$\Pi_1 = \{X, Y : (YXY^{-1})\, X\, (YX^{-1}Y^{-1}) = XYX^{-1}\}$$

- see [2] for a discussion - and look explicitly in $PSL(2, C)$ for a discrete representation of Π_1. One is led to some complicated but basically straightforward algebraic equations, and the only hard part is to check that there are solutions which are discrete groups. This is done by a delicate use of the Poincaré Polyhedron Theorem in [10].

Arguing similarly, one may show, as in [10], that there is a hyperbolic structure on S^3 which is branched of order k about the figure - 8 knot for all sufficiently large k - say $k > 3$.

It now follows that every 3-manifold has a hyperbolic structure which is branched about some link, which furthermore is explicitly known once one knows the Hilden-

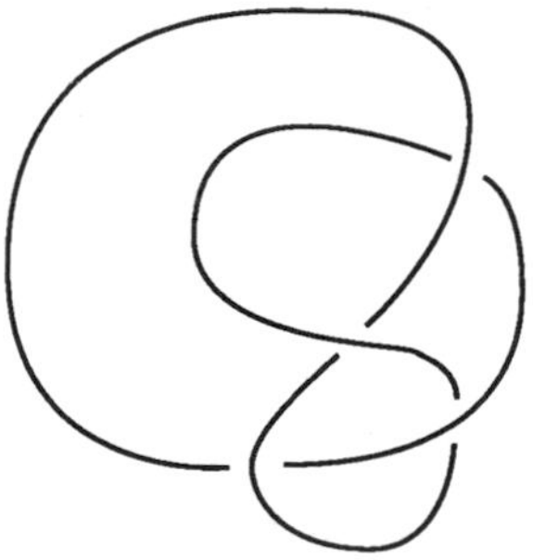

Fig. 1: The Figure-8 Knot

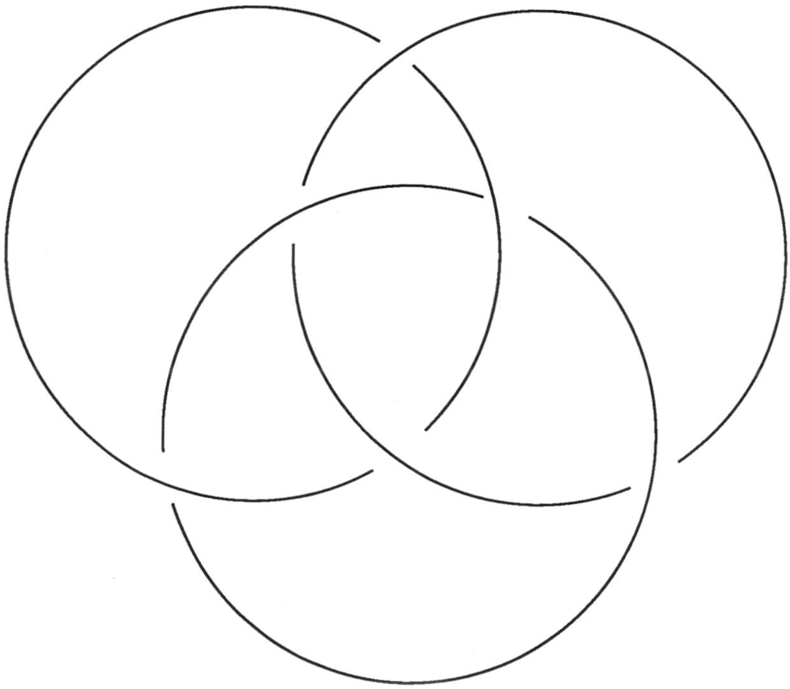

Fig. 2: The Borromean Rings

Montesinos branched covering (this may be a tall order in practice, of course). Indeed, one simply pulls back the branched hyperbolic metric on S^3 branched along the figure - 8 knot, choosing k to be divisible by all the branching orders of all the links.

This still falls somewhat short of the theorem, as we would like uniform bounds on k. To that end, the following striking refinement of [8] was found in [9]:

Theorem [9]: Every 3-manifold arises as a branched covering of S^3, whose branching locus is the Borromean rings, and whose branching order is 1,2 or 4 on each component of f^{-1} (Borromean rings).

One may now use either the fact that the Borromean rings complement has a well-understood hyperbolic structure, or the universal property of the figure - 8 knot, to bound k, and hence complete the proof of the theorem.

§3: Negative Ricci Curvature

In this section, we will prove:

Theorem ([1]): There exist positive constants a and b with the property that any compact hyperbolic 3-manifold has a metric all of whose Ricci curvatures are between $-a$ and $-b$. Our exposition here follows closely the discussion of [1]. In fact, our main observation here is how well the technique of [1] fits into a larger context.

Let us begin with a geodesic γ in the upper half-space $\mathbb{H}^3 = \{x, y, Z : Z > 0\}$, which we might as well choose to be the z-axis. Then we may choose Fermi coordiantes $\{t, r, \theta\}$ for $\mathbb{H}^3$ about γ, where t is geodesic distance along γ, r is distance from γ, and θ is an angular coordinate about γ. With respect to these coordinates, the hyperbolic metric on $\mathbb{H}^3$ is

$$ds^2 = \cosh^2(r)\,dt^2 + dr^2 + \sinh^2(r)\,d\theta^2.$$

Now let M be a hyperbolic orbifold, and γ a component of the set where the orbifold metric is singular of order k. We may then choose Fermi coordinates about γ, and one sees easily that this metric is given by

$$ds^2 = \cosh^2(r)\,dt^2 + dr^2 + \left(\frac{\sinh(r)}{k}\right)^2 d\theta^2.$$

Let us first consider a new metric

$$ds^2 = g^2(r)\,dt^2 + dr^2 + f^2(r)d\theta^2$$

and calculate the Ricci curvature in this metric. One sees after some calculation that the Ricci curvatures are:

$$-\left(\frac{g''}{g}+\frac{f''}{f}\right)$$
$$-\left(\frac{g''}{g}+\frac{g'}{g}\cdot\frac{f'}{f}\right)$$
$$-\left(\frac{f''}{f}+\frac{g}{g}\cdot\frac{f'}{f}\right)$$

Our theorem will follow once we prove:

Lemma: For k sufficiently large, there are functions f and g satisfying:

(i) $g(0)>, g'(0)=0 \qquad f'(0)=0 \quad f(0)=1$

(ii) (a) $\frac{g''}{g}+\frac{f''}{f}>0$

(b) $\frac{g''}{g}+\frac{g'}{g}\cdot\frac{f'}{f}>0$

(c) $\frac{f''}{f}+\frac{g'}{g}\cdot\frac{f'}{f}>0$

(iii) For $r \geq r_0$

$$g(r)=\cosh(r) \qquad f(r)=\frac{\sinh(r)}{k}$$

Notice the condition that $f'(0)=1$ is necessary and sufficient for the metric to be smooth, and is the main obstacle to the proof of the theorem.

There are two main steps in constructing f and g as in the lemma. The first step is a general patching lemma, which tells us how to to construct a solution of differential inequalities from piecewise solutions. A good case to bear in mind is the following special case: if g_1 and g_2 are increasing convex functions, g_1 (a) $= g_2$(a), then there us a convex function g agreeing with g_1 for $r<a-\epsilon$ and with g_2 for $r>a+\epsilon$, provided $g_1' < g_2'$ [5]. The general case we will need is not much more difficult ([1]):

Lemma: Suppose f, g, and g_2 are increasing function such that (g_1, f) and $(g2, f)$ satisfy $g_1(a)=g_2(a)$ and $g_1'(a)<g_2'(a)$ and conditions (ii) $(a-c)$. Then for all ϵ there exists g such that (g,f) satisfies (ii) $(a-c)$, and $g(r)\equiv g_1(r)$ for $r<a-\epsilon, g(r)\equiv g_2(r)$ for $r>a+\epsilon$.

We now must choose our desired piecewise solutions. Note that our tubular neighborhood estimate gives a lower bound of about $\frac{1}{2\Pi}$ for $\frac{\sinh(r)}{k}$ at $r=r_0$, so we must choose f

with the property that it has at most this value for this value of r_0. One may check easily that

$$f = \frac{1}{\lambda_1} \arctan(\sinh(\lambda_1 r))$$
$$g = \alpha \cosh(\lambda_2 r)$$

will be a solution of (i) - (ii) provided $\lambda_2 > \lambda_1$ and $\alpha > 0$. Since arctan is bounded by $\Pi/2$, by choosing λ_1 sufficiently large we may make f arbitrarily small arbitrarily far out.

We now must patch f with $\frac{\sinh(r)}{k}$ and g with $\cosh(r)$. The first patching is trivial, since, because $f'' < 0$, where f meets $\frac{\sinh(r)}{k}$ we must have $f'' < \frac{\cosh(r)}{k}$.

The second patching is more problematic, since g'' is positive. We proceed in two steps: letting $g_1 = \alpha \cosh(\lambda_2 r), g_3 = \cosh(r)$, let g_2 be close to the tangent line approximation to g_3:

$$g_2(r) = g_3(r_0) + \beta \cdot (r - r_0)$$

where $\beta < g_3'(r_0)$.

Note that $g_2(r)$ crosses the x-axis at $r_0 - r = \frac{g_3(r)}{\beta}$ which is arbitrarily close to $r = r_0 - \tanh(r_0)$. Since $g_2'' = 0$ and $g_1'' > 0$, by choosing α small enough, we can make g_1 cross g_2 with $g_1' < g_2'$. It is now a simple matter to complete the patchings to prove the lemma, and hence the theorem.

§4: Ergodic Metrics

In this section, we will show how to construct ergodic metrics on manifolds of various types. Our main criterion for ergodicity is due to Wojtkowski [18], Burns and Gerber [4], and Katok [12], generalizing the theorem of Anosov that metrics of negative sectional curvature are ergodic.

For simplicity, let us restrict to the 2-dimensional case. In this case, the geodesic flow on a surface S is best understood as a flow on the unit tangent bundle $T^1(S)$, which is 3-dimensional. To understand this flow, recall that a point $\mu \varepsilon T^1(S)$ describes a unit speed geodesic γ_μ, and a 4-dimensional family of vector-fields Y_t along γ_μ, called Jacobi vector fields, which satisfy the Jacobi equation

$$Y_t'' + K(Y_t, \gamma_{\mu,t}) Y = 0$$

and which represent infinitesimal deformations of γ_μ through families of geodesics. These are determined by the initial conditions $Y_t(0), Y_t'(0)$, and it is easy to see that the infinitesimal deformation γ_μ of will remain of unit speed if and only if $Y_t'(0)$ is perpendicular to μ.

We are then left with a 3-dimensional space which we can identify with $T^1_\mu(S)$, along which the geodesic flow ξ is given by the Jacobi equation.

The main point now is the following criterion for ergodicity ([4], [12], [18]):

Theorem: Let M be a compact 3-manifold with a C_2 flow φ^t with the following properties:

(a) φ^t is volume preserving

(b) There is a plane-field P transverse to the flow preserved by φ^t . Suppose further that there is a continuous family of cones $K \subseteq P$ and an open subset $U \subset M$

(i) $\bigcup_{t \in \mathbb{R}} \varphi^t(U)$ has full measure in M

(ii) $\varphi^t(K)_\mu$ is strictly contained in $K_{\varphi^t(\mu)}$ whenever $\mu, \varphi^t(\mu)$ lie in U.

Then φ^t is ergodic.

In the two-dimensional case, we may set $X_t = \gamma_t\left(\frac{\partial}{\partial t}\right)$, and N_t the orthogonal vector to X_t satisfying the right-hand rule. Identifying $\eta \,\varepsilon\, T^1_\mu(S)$ with corresponding initial conditions to the Jacobi equation, $\eta = (\eta_0, \eta_0')$ we have seen that $\eta \,\varepsilon\, T^1_\mu(S)$ provided that $\eta_0' \perp \mu$. We may now set $P = \{\eta \varepsilon T^1_\mu(S) : \eta_0 \perp \mu\}$. We may thus identify η_0 and η_0' as multiples of N, and we will define

$$K = \{\eta \varepsilon P : \eta_0 \cdot \eta_0' \geq 0\}$$

For the sake of argument, let us assume that S has everywhere negative curvature. Taking $U = T^1(S) = M$, we will show that such U and P satisfy (i) and (ii). But (i) is obvious, and to show (ii) it suffices to show that, for any geodesic γ, solutions $Y_t = y(t) \cdot N_t$ of the Jacobi equation along g satisfy: if $y(0) \cdot y'(0) \geq 0$, then $y(t) \cdot y'(t) \geq 0$ for $t > 0$, with equality of and only if $y \equiv 0$.

Proof: It suffices to show $y\,y'$ is increasing. But

$$\left(y\,y''\right) = y\,y'' + (y')^2 = y(-Ky) + (y')^2$$

by the Jacobi equation.

But if $K < 0$, this is > 0 unless $y = y' = 0$. But by the uniqueness of the Jacobi equation, if at any point $y = y' = 0$, then $y \equiv 0$.

We now let S be an arbitrary compact surface, with an orbifold hyperbolic metric-that is, there are finitely many points where the hyperbolic metric looks like the quotient of $\mathbb{H}^2$ by a rotation through angle $\frac{2\Pi}{k}$ for some k, and everywhere else S is locally isometric to $\mathbb{H}^2$. Any S has such a metric after picking sufficiently many points.

We now modify the metric in a neighborhood of each of these points to look like $ds^2 = dr^2 + f^2(r)\, d\theta^2$

We may choose f so that ds^2 has decreasing curvature $K(r)$ as a function of r, f agrees with $\frac{\sinh(r)}{k}$ for r large, and for some r_0, the circle of radius r_0 is a geodesic, see [4]. Note that this r_0 is where $K(r_0) = 0$.

Given this, we wish to check the invariant cone condition. We need only worry about geodesics which enter the region of positive curvature, and by symmetry they will have a closest point to 0, which we label $g(0)$, and be symmetric about the line connecting 0 and $g(0)$. Let T_0 denote the value of r at which $\gamma(T_0)$ hits the geodesic circle.

We now observe that $\gamma(T_0) \cdot \gamma'(T_0) \geq 0$ if and only if $\frac{y'}{y}(T_0) \geq 0$, and to that end we set $\mu = \frac{y'}{y}$. It then satisfies the Ricatti equation

$$\mu' = \frac{y''y - (y')^2}{y^2} = \frac{-K(t)y^2 - (y')2}{y^2} = -K(t) - \mu^2$$

or $\mu' + \mu^2 + K(t) = 0$.

As this is a first-order differential equation, it is evident that two solutions never cross.

Now let Y_S and Y_C be the Jacobi vector fields with y_s and y_c satisfying $y_S(0) = 0, y_S'(0) = 1,\quad y_c(0) = 1, y_c'(0) = 0$. It is evident that Y_S is the Jacobi field given by rotation about 0, and from this we see that $ys(t) \neq 0$ for $t \neq 0$. y_s (resp. y_c) is evidently an odd (resp. even) function.

We now claim:

<u>**Claim:**</u> $y_S'(\pm T_0) = 0$.

<u>**Proof:**</u> Since the circle of radius r_0 is a geodesic we must have $\nabla_{\frac{\partial}{\partial\theta}}\left(\frac{\partial}{\partial\theta}\right) = \nabla_{\frac{\partial}{\partial r}}\left(\frac{\partial}{\partial\theta}\right) = 0$, so in particular $\nabla_\nu\left(\frac{\partial}{\partial\theta}\right) = 0$ for all ν based along the circle of radius r_0. Thus $y_s'(T_0) = \nabla_{\frac{\partial}{\partial t}}\left(\frac{\partial}{\partial\theta}\right) = 0$.

Now let $\mu_s = \frac{y_s'}{y_s}$ and $\mu_c = \frac{y_c'}{y_c}$. Then $\mu_s(\neq T_0) = 0$, and μ_s blows up at one point.

We claim:

<u>**Claim:**</u> For some $\tau \varepsilon (0, T_0], \mu_c(\tau) = -\infty$

<u>**Proof:**</u> We show that for τ positive, we have

$$\mu_c(\tau) < \mu_s(\tau - T_0)$$

until τ passes a point at which μ_s blows up.

But $\mu_c(0) = \mu_s(-T_0) = 0$ and $\mu_c'(0) = -K(0) < 0 = \mu_s'(-T_0)$, and if $\mu_c(t) < \mu_s(t - T_0)$ for $t < \tau$, then

$$\begin{aligned} \mu_c(\tau) &= -\int_0^\tau \left(K(t) + \mu_c^2(t)\right) dt \\ &< -\int_0^\tau \left(K(\tau - T_0) + \mu_s^2(\tau - T_0)\right) dt \\ &= \mu_s(\tau - T_0) \end{aligned}$$

where we have used

$$\begin{aligned} \int_0^\tau K(t - T_0)\, dt &= \int_0^\tau K(T_0 - t)\, dt \\ &= \int_{T_0 - t}^{T_0} K(t)\, dt < \int_0^\tau K(t)\, dt \end{aligned}$$

since K is a decreasing function of t in $(0, T_0)$. Since $\mu_s(\tau - \tau_0)$ blows up at $t = T_0$, our claim is proved.

We now claim that if h is any solution of the Riccati equation with $\mu(-T_0)$ positive, then $u(T_0)$ is also positive. To see this, let us first assume that the τ_1, above is $< T_0$. In this case, we see that $\mu(t) > \mu_s(t)$ for $t \varepsilon [-T_0, 0]$ and $\mu(t) < \mu_c(t)$ for $t > -\tau_1$, until $u(t) \to -\infty$, which then must happen between 0 and τ_1. After $\mu(t)$ blows up, it stays $> \mu_s(t) \to 0$.

If $\tau_1 = T_0$, we need only add that $\mu(t)$ must blow up before T_0, since if it blows up at T_0, then writing $\tau\mu = \frac{y'}{y}$, we see $y(T_0) = 0$, and so $\mu = \mu_c$.

It now follows that the cone condition is satisfied, and the construction is complete.

§5: Pinched Matrices of Negative Sectional Curvature.

In this section, we follow [7] in constructing manifolds of dimensions 4 which have matrices of negative sectional curvature pinched between -1 and $-(1+\epsilon)$. The proof that many of these manifolds do not carry matrices of constant negative curvature is carried out by a clever argument in [7], and we do not have anything to add to the beautiful exposition there.

Let us recall the basic construction: let M be a hyperbolic manifold, and V a totally geodesic submanifold of M, which has a tubular neighborhood of radius r, assumed large.

If the homology class $[V] \varepsilon H_{n-2}(M)$ vanishes, then for each n we may form the m-fold branched cyclic cover M^n, branched along V. In a r-neighborhood of V in M^n, the lifted metric looks like $ds^2 = \cosh^2(r)\, dx^2 + dr^2 + (n \sinh(r))^2 \, d\theta^2$ where dx^2 denotes the hyperbolic metric on V.

We observe that such M and V can be constructed by standard arithmetic constructions, see [7].

Suppose we change the metric to $ds^2 = \cosh^2(r)\, dx^2 + dr^2 + f^2 d\theta^2$. Then the sectional curvatures all lie between $-1, -\frac{f''}{f}$ and $-\frac{f'}{f} \cdot \frac{\sinh(r)}{\cosh(r)}$. It is evident that we are done once we show the following:

Lemma: Given r and n, such that $\rho \sinh\left(\frac{\rho}{6}\right) > 6\,[n-1]$, there is a function f such that

(i) f' and f'' are positive.

(ii) $f(0) = 0, f'(0) = 1$ and $f(r) \equiv n \sinh(r)$ for $r > \rho$.

(iii) $\left|\frac{f''}{f} - 1\right| < 3\delta$ and $\left|\frac{f'}{f} \frac{\sinh(r)}{\cosh(r)} - 1\right| < 2\delta$ where δ satisfies

$$(\delta\rho) \sinh\left(\frac{\delta\rho}{2}\right) = 2\,[n-1]$$

Proof: Let us write $f(r) = h(r) \sinh(r)$. Then

$$f'(r) = h(r) \cosh(r) + h'(r) \sinh(r)$$
$$f''(r) = h(r) \sinh(r) + 2h'(r) \cosh(r) + h''(r) \sinh(r)$$

and we have

$$\frac{f'}{f}\frac{\sinh(r)}{\cosh(r)} = 1 + \frac{h'}{h}\frac{\sinh(r)}{\cosh(r)}$$

$$\frac{f''}{f} = 1 + 2\frac{h'}{h}\frac{\cosh(r)}{\sinh(r)} + \frac{h''}{h}$$

so we need to choose h so that $h(0) = 1, h(r) \equiv n$ for $r > \rho$, and $\frac{h''}{h}$ and $\frac{h'}{h}$ small.

We first observe that it is fairly easy to choose h so that M^n has negative curvature (ignoring pinching conditions). If we choose h so that $h' \geq 0$, we need only verify that

$$\frac{h''}{h} + 2\frac{h'}{h}\frac{\cosh(r)}{\sinh(r)} + 1 > 0.$$

Using that $\frac{\cosh}{\sinh} > 1$, it suffices to solve

$$h'' + 2h' + h \geq 0, \quad h(0) = 1, h(\rho) = n$$

but if we set

$$\begin{aligned} h(r) &= ae^{-r} + bre^{-r} && \text{for } r < r_1 \\ &= n && \text{for } r > r_1. \end{aligned}$$

with $h'(r_1) = 0$, we find $a = 1$ and $b = \frac{1}{1-r_1}$ so that $h(r_1) = \frac{e^{-r_1}}{(1-r_1)}$, which goes from 1 to $+\infty$ as r goes from 0 to 1, so we may find the desired r, as long as $\rho > 1$.

To control the pinching, we will construct h in the following way: let $r_1 = \frac{\rho}{2}$. We will describe h by first prescribing h'' , and then integrating to obtain h. After some trial and error, we set

$$\begin{aligned} h'' &= \delta^2 \cosh(\delta r) && \text{for } 0 < r < r_1 \\ &= \delta^2 \cosh(\delta(r - r_1)) && \text{for } r_1 < r < 2r_1 = \rho \\ &= 0 && \text{for } r > 2r_1 \end{aligned}$$

Of course, when the dust clears, we will approximate this h" by a smooth one.

Then we may construct the following table:

	$0 < r < r_1$	$r_1 < r < r_2$	$2r_1 < r$
h''	$\delta^2 \cosh(\delta r)$	$-\delta^2 \cosh(\delta(r - r_1))$	0
h'	$\delta^2 \cosh(\delta r)$	$\delta \sinh(\delta_1) - \delta \sinh(\delta(r - r_1))$	0
h	$\cosh(\delta r)$	$\cosh(\delta r_1) + (\delta(r - r_1)) \sinh(\delta r_1) - \cosh(\delta(r - r_1)) + 1$	$\delta r_1 \sinh(\delta r_1) + 1$

We may now complete the table as follows: for $0 < r < r_1$, we have

$$\frac{h''}{h} = \delta^2 \frac{h'}{h}\frac{\sinh(r)}{\cosh(r)} = \frac{\delta \sinh(\delta r)\sinh(\delta r)}{\cosh(\delta r)\cosh(r)} < \delta$$

(since $\sinh(x) < cosh(x)$)

$$\frac{h'}{h}\frac{\cosh(r)}{\sinh(r)} = \frac{\delta \tanh(\delta r)}{\tanh(r)} < \delta \text{ for} \delta < 1$$

since tanh is an increasing function. For $r_1 < r < r_2$, we argue as follows: h' is decreasing on this interval and h is increasing, so $\frac{h'}{h}$ takes its largest value at r_1, where it is $< \delta$. The same is true of $\frac{h'}{h} \cdot \frac{\cosh(r)}{\sinh(r)}$, and $\frac{\sinh}{\cosh}$ is < 1. Hence $\frac{h'}{h}\frac{\sinh(r)}{\cosh(r)} < \delta$ and $\frac{h'}{h}\frac{\cosh(r)}{\sinh(r)} < \delta$. Furthermore, $\left|\frac{h''}{h}\right| < \left|\frac{h''(r)}{h(r_1)}\right| = \frac{\delta^2\cosh(\delta(r-r_1))}{\cosh(\delta r_1)} < \delta^2$, since $\cosh(\delta(r-r_1)) < \cosh(\delta r_1)$ on $r_1 < r < r_2$. This completes the proof of the lemma, and hence the theorem.

REFERENCES

1. R. Brooks, "A Construction of Metrics of Negative Ricci Cuevature," to appear J. Diff. Geom.

2. R. Brooks and W. Goldman, "Volumes in Seifert Space," Duke Math. J. 51 (1984), pp. 529- 545.

3. R. Brooks and P. Matelski, "Collars in Kleinian Groups," Duke Math. J. 49 (1982), pp. 163 - 182.

4. K. Burns and M. Gerber, "Real Analytic Bernoulli Geodesic Flows on S^2," preprint.

5. L.Z. Gao and S. T. Yau, "The Existence of Negatively Ricci Curved Metrics of 3-Manifolds.

6. M. Gromov and H.B. Lawson, "The Classification of Simply - Connected Manifolds of Positive Scalar Curvature," Ann. Math. 11 (1980), pp. 423 - 434.

7. M. Gromov and W. Thurston, "Pinching Constants for Hyperbolic Manifolds", Invent. Math. 89 (9187), pp. 1 - 12.

8. H.M. Hilden, M.T. Lozano, and J.M. Montesinos, "On Knots that are Universal," Topology 24 (1985), pp. 499 - 504.

9. H.M. Hilden, M.T. Lozano, J.M. Montesinos, and W.C. Whitten, "On Universal Group and Three-Manifolds", Invent. Math. 87 (1984), pp. 411 - 456

10. T. Jorgensen, "Compact Manifolds of Constant Negative Curvature Fibering over a Circle", Ann. Math. 106 (1977), pp. 61 - 72.

11. T. Jorgensen, "On Discrete Groups of Mobius Transformations", Amer. J. Math. 98, pp. 739 - 749.

12. A. Katok, "Invariant Cone Families and Stochastic Properties of Smooth Dynamical Systems", preprint.

13. R. Meyerhoff, "A Lower Bound for the Volume of Hyperbolic 3-Manifolds", Can. J. Math. XXXIX (1987), pp. 1038 - 1056.

14. J.M. Montesinos

15. R. Riley, "A Quadratic Parabolic Group", Math. Proc. Camb. Phil. Soc. 77 (1975), pp. 281 - 288.

16. W. Thurston, Geometry and Topology of 3-Manifolds, Princeton Lecture Notes.

17. M. Wada, Ph.D. thesis, Cobumbia University 1987.

18. M. Wojtkowski, "Invariant Families of Cones and Lyapunov Exponents", Erg. Th. and Dyn-Syst. 5 (1985), pp. 145 - 161.

19. H. Farkas and I. Kra, Riemann Surfaces, Springer Verlag 1980.

20. R. Brooks and P. Matelski, " The Dynamies of 2-Generator Subgroups of $PSL(2,\mathbb{C})$", Ann Math. Stud. 97 (1980), pp. 65 - 71

Department of Mathematics
University of Southern California
Los Angeles, CA. 90089-1113

Contemporary Mathematics
Volume **101**, 1989

OPEN MANIFOLDS OF NONNEGATIVE CURVATURE

GERARD WALSCHAP

One of the traditional goals in Riemannian geometry is to understand what happens to a manifold when certain restrictions are imposed on its curvature. Here, we shall attempt a brief and by no means exhaustive survey of the main results concerning the structure of complete noncompact manifolds with sectional curvature $K \geq 0$.

From a chronological perspective, it was Cohn-Vossen in the thirties who first gave a classification in dimension 2, cf. [C-V]. In the late sixties, Gromoll and Meyer [GM] studied the strictly positive curvature case in arbitrary dimensions. Immediately thereafter, Cheeger and Gromoll [CG], in what today still constitutes the most significant paper on the subject, extended these results to the case $K \geq 0$. They gave, among other things, a complete topological description of the manifold (which in a sense reduces the noncompact case to the compact one), and a classification up to isometry in dimension 3. Finally, the last few years have witnessed a renewed interest in this area, as testified by the work of [ESS], [GW], [M 1-2], [Sh], [St], [SW], [W1-2] and [Y] among others, which sheds new light on the metric structure of these manifolds, particularly in dimension 4 and in some higher-dimensional special cases.

1. THE SOUL CONSTRUCTION. A fundamental concept in nonnegative curvature is that of convexity: a subset C of a Riemannian manifold M is said to be *convex* if any two points of C can be joined by some minimal geodesic lying entirely in C. If, moreover, this geodesic is unique in M, C is said to be *strongly convex*. Finally, C is *totally convex* (t.c.) if any geodesic of M with endpoints in C is entirely contained in C. Notice that M itself is

1980 Mathematics Subject Classification (1985 Revision). 53C20.

Supported in part by NSF grant DMS 88-01999
This paper is in final form and no version of it will be submitted for publication elsewhere.

always trivially totally convex, but in general need not have proper t.c. subsets, e.g. S^2. On the other hand, it is a standard fact that metric balls of sufficiently small radius in arbitrary Riemannian manifolds are strongly convex, cf.[GKM]. As the radius r of these balls increases however, curvature plays an important role: in the simply connected space forms for example, metric balls tend to remain convex when $K \leq 0$, whereas for $K > 0$, their complement becomes convex as r gets larger. This motivates the following construction:

Suppose M is open (i.e., complete, noncompact). Let $c : [0, \infty) \to M$ be a ray (meaning: the distance $d(c(0),\ c(t)) = t$ for all $t \geq 0$), $c(0) = p$. Define the open half-space

$$B_c = \bigcup_{t>0} B_t(c(t))$$

with respect to c, where $B_t(c(t))$ denotes the open metric ball of radius t centered at $c(t)$. By using Toponogov's triangle theorem, one can show that if $K \geq 0$, then the complement $M \backslash B_c$ is totally convex. It is then easily verified that for any $t > 0$, $C_t := \cap_c (M \backslash B_{c_t})$, where c runs over all rays emanating from p, and c_t is the restricted ray $c_t(s) := c(s + t)$, is a compact, t.c. set with boundary. To obtain the so called soul of M, we will need minimal t.c. sets. The key to constructing these is provided by the following:

THEOREM 1.1. *Let M have nonnegative curvature, and let C be a closed t.c. subset of M with boundary ∂C. Then the distance function $\psi : C \to \mathbf{R}$ to the boundary, $\psi(p) := d(p, \partial C)$, is weakly concave, i.e., for any geodesic $c : [a, b] \to M$ contained in C, the function $\psi \circ c : [a, b] \to \mathbf{R}$ is concave. Moreover, if $\psi \circ c \equiv d$ is constant, then c together with any minimal geodesic $\tau : [0, d] \to C$ from $c(a)$ to ∂C determines a flat totally geodesic rectangle $F : [a, b] \times [0, d] \to M$, $F(t, s) = \exp_{c(t)} sE(t)$.*

Here E is the parallel vector field along c with $E(a) = \mathring{\tau}(0)$.

The proof of 1.1 is based on the Rauch comparison theorem. It implies that if one contracts a t.c. set C, i.e., if one considers $C^\alpha := \{p \in C / d(p, \partial C) \geq \alpha\}$, then C^α is again t.c. Indeed, if $c : [a, b] \to C$ is a geodesic with endpoints in C^α, then by concavity of $\psi \circ c$, $(\psi \circ c)(t) \geq \min\{(\psi \circ c)(a),\ (\psi \circ c)(b)\} \geq \alpha$, for $a \leq t \leq b$.

To complete the soul construction, consider a t.c. set C as above, and let $C^{\max} = \bigcap_{C^a \neq \emptyset} C^a$. Notice that if $K > 0$ on C, then $C^{\max}$ is a point by the second part of 1.1. In any case, $\dim C^{\max} < \dim C$. If $C^{\max}$ has nonempty boundary, one obtains, by iterating the above procedure, a flag of t.c. sets, each of which consists of all points at maximal distance from the boundary of the preceding one, and is of lower dimension. Since the last one has empty boundary, we obtain:

THEOREM 1.2. *Let M be open with $K \geq 0$. Then M contains a compact,* *totally convex submanifold S without boundary, dim $S <$ dim M.*

Notice that since S is totally convex, it is also totally geodesic, and consequently has nonnegative sectional curvature. Moreover, as remarked earlier, if M has positive curvature, then S is a point. A manifold S constructed as above is called a *soul* of M.

Example 1.3. 1) Let M be the Riemannian product $S^1 \times \mathbf{R}$. Then there are exactly 2 rays emanating from any point p of M (in fact, the rays join in a line), and thus the first t.c. set in the basic construction, the circle through p orthogonal to the line, is a soul of M. More generally, let S be any compact manifold of nonnegative curvature, and consider the Riemannian product $M = S \times \mathbf{R}^k$, where $\mathbf{R}^k$ denotes flat Euclidean space. Then $S \times u$ is a soul of M for any $u \in \mathbf{R}^k$.

2) Let M be the paraboloid $z = x^2 + y^2$ in $\mathbf{R}^3$. For $p \neq 0$ in M, there is exactly one ray emanating from p, namely the restriction of the geodesic from the vertex that passes through p. Notice that any geodesic from the origin is a ray. The basic construction at $p \neq 0$ requires two steps, and only one at the origin. In any case, the soul is unique, and consists of the vertex.

3) Recall that a submersion $\pi : E \to M$ is said to be Riemannian if for any q in E, π_{*q} maps the orthogonal complement E_q^h of ker π_{*q} isometrically onto $M_{\pi(q)}$. By a formula of O'Neill [O'N], Riemannian submersions are curvature nondecreasing, i.e., for any 2-plane $P \subset E_q^h$, $K_{\pi_* P} \geq K_P$. Riemannian submersions, therefore, are a fundamental tool for constructing nontrivial examples in nonnegative curvature. For instance, suppose E has nonnegative curvature, and let I be a group of isometries acting freely on E so that $M := E/I$ is a manifold, and $\pi : E \to M$ a submersion. Then there exists a unique metric of nonnegative curvature on M such that π becomes a Riemannian submersion. Notice that this metric is well defined, for if $g \in I$, then $\pi \circ g = g$, and g is an isometry. A typical example is the two-dimensional vector bundle $M^4 = (S^3 \times \mathbf{R}^2)/S^1$ over S^2 associated to the Hopf fibration $\pi : S^3 \to S^2$, cf. also [C]. The vector bundle projection π_M is defined so that the following diagram commutes:

$$\begin{array}{ccc} S^3 \times \mathbf{R}^2 & \xrightarrow{\pi_1} & S^3 \\ {\scriptstyle\rho}\downarrow & & \downarrow{\scriptstyle\pi} \\ M & \xrightarrow[\pi_M]{} & S^2 \end{array}$$

Here, π_1 is projection onto the first factor, and $\rho(q, u)$ is the equivalence class of (q, u) under the equivalence relation $(q, u) \sim (qz, z^{-1}u)$, $q \in S^3 \subset \mathbf{C}^2$, $u \in \mathbf{R}^2 = \mathbf{C}$, $|z| = 1$. Since the action of S^1 on the Riemannian product $S^3 \times \mathbf{R}^2$ is by isometries, M inherits a metric of nonnegative curvature such that ρ

becomes a Riemannian submersion. It turns out that the soul of M is unique, and equals the 2-sphere of constant curvature 4. In fact, π_M is a Riemannian submersion onto the soul.

2. THE TOPOLOGICAL STRUCTURE OF M. All the examples in 1.3 are topologically vector bundles over their souls. This remains true in general:

THEOREM 2.1. *If S is a soul of M, then M is diffeomorphic to the normal bundle $\nu(S)$ of S in M.*

A relatively easy proof of 2.1 can be obtained by using the concept of generalized critical points of the distance function, introduced by [GS] : if $q \in M \backslash S$, then q belongs to the boundary of some t.c. set C containing S. By total convexity of C, there exists a vector $u \in M_q$ which makes an angle $> \pi/2$ with every minimal geodesic from q to S. Using a partition of unity argument, one obtains a smooth unit vector field on $M \backslash S$ with the same property, which coincides with the gradient of the distance function from S on a small neighborhood of S. The flow lines of this vector field then yield a diffeomorphism $\nu(S) \to M$.

Together with the fact that S is a point if $K > 0$, 2.1 immediately implies:

THEOREM 2.2. *If the curvature of M is strictly positive, then M is diffeomorphic to Euclidean space.*

Thus, up to diffeomorphism, open manifolds of nonnegative curvature are vector bundles over compact manifolds of nonnegative curvature. It is not known which vector bundles admit metrics of nonnegative curvature, even when the zero section is a sphere. For this, one would probably need more information on the metric interaction between M and S. A delicate argument based on properties of convex sets does however provide the following strong result, due to [Sh]:

THEOREM 2.3. *There exists a deformation retraction $M \to S$ which is distance nonincreasing.*

It should be noted that this retraction is in general not smooth. The concept of critical point is also used in the following theorem by Gromov [G]:

THEOREM 2.4. *For each integer n, there exists a constant $C(n)$ such that any n-dimensional complete manifold of nonnegative curvature has total Betti number bounded above by $C(n)$.*

For more information on the topology of M, in particular its fundamental group, see [CG].

3. THE METRIC STRUCTURE OF M. The first result concerning the metric structure of M is a difficult theorem due to Cheeger and Gromoll [CG]:

THEOREM 3.1. *If u is tangent to S and v is orthogonal to S, then the sectional curvature of the plane spanned by u and v is zero.*

Notice that since S is totally geodesic, the curvature tensor of the normal bundle $\nu(S)$ of S is just the restriction of the curvature tensor of M. One is therefore naturally led to consider the holonomy group of $\nu(S)$.

THEOREM 3.2. *If S has flat normal bundle, then there exists a (locally) isometrically trivial fibration $M \to S$.*

Theorem 3.2, due to [St], has been generalized by [Y]. The proof is based on 2.3: assuming without loss of generality that S is simply connected, 2.3 together with the rigidity part of the Rauch comparison theorem shows that exponentiating any global parallel section of $\nu(S)$ at constant distance yields a totally geodesic submanifold isometric to S. One thus obtains a Riemmanian submersion with vanishing O'Neill tensor on a neighborhood of the soul, given by metric projection π onto S. Each of these copies also satisfies 3.1, so that the fibers of π are totally geodesic, and therefore a neighborhood of S splits as a Riemannian product $S\times$ disk. This splitting is then extended to all of M.

3.2 can be used to classify M up to isometry in dimension ≤ 3. Indeed, the only possibilities for S are dim $S = 0, 1$, or codim $S = 1$, and in all these cases $\nu(S)$ must be flat. For a classification, see [CG], where it is obtained by different arguments.

In dimension 4, the only case where $\nu(S)$ need not be flat is when dim $S =$ codim $S = 2$. But then, since the fibers of $\nu(S)$ are 2-dimensional, the holonomy group of $\nu(S)$ must act transitively, and the second part of 1.1 implies that every direction orthogonal to S is a ray direction. It is not hard to show that in general, if every direction is a ray direction, then the exponential map $\exp_\nu : \nu(S) \to M$ is a diffeomorphism, and the projection $\pi : M \to S$ given by $\pi : \pi_\nu \circ \exp_\nu^{-1}$ (here $\pi_\nu : \nu(S) \to S$ is the vector bundle projection) is a Riemannian submersion, cf. [W1]. As an example, consider $(S^3 \times \mathbf{R}^2)/S^1$, discussed in 1.3. Combining all these results, one concludes ([W2]):

THEOREM 3.3. *Let M^n be an open manifold of nonnegative curvature with soul S. If $n \leq 4$ or if codim $S \leq 2$, then metric projection $\pi : M \to S$ is well defined, and is a Riemannian submersion. π splits (locally) isometrically if and only if S has flat normal bundle. Moreover, if $K > 0$ at some $p \in M$, there S is a point, and M is diffeomorphic to $\mathbf{R}^n$.*

Notice that even if S is simply connected, $\nu(S)$ may be a topologically trivial bundle over S with nonflat holonomy: it is not hard to construct a metric of nonnegative curvature on $S^2 \times \mathbf{R}^2$ which is not a product metric, cf. [W1]. Nevertheless, by 3.3, $S^2 \times \mathbf{R}^2$ does not admit any metric of nonnegative curvature with $K > 0$ at some point. The question of whether or not an arbitrary manifold M^n with $K \geq 0$, and $K > 0$ at some point, is diffeomorphic to $\mathbf{R}^n$, was first asked by Cheeger and Gromoll, and after twenty years, has still not been answered. In fact, one may conjecture that such a manifold satisfies a condition stronger than the Cheeger-Gromoll conjecture: Let M be an arbitrary complete Riemannian manifold, with totally geodesic submanifold S. M is said to be $S-flat$ if for any geodesic c in S and parallel vector field E along c with $E(0) \in \nu(S)$, the rectangle $F : \mathbf{R} \times \mathbf{R} \to M$ given by

$$F(s,t) = \exp_{c(s)} tE(s)$$

is flat and totally geodesic. If M has nonnegative curvature and S is a soul of M, then 3.1 may be interpreted as saying that M is "infinitesimally S-flat" (cf. also [M1], where a generalization of 3.1 is asserted). Moreover, it follows from the proof of 3.3 that M is S-flat whenever $\dim M \leq 4$ or $\operatorname{codim} S \leq 2$. In fact, it is not hard to show that if U is a sufficiently small neighborhood of S in M, then M is S-flat on U if and only if metric projection $\pi : U \to S$ is a Riemannian submersion. Constructing an example which is not S-flat seems to be difficult with the methods presently available, in view of the following:

THEOREM 3.4 [(SW]). *If M is S-flat for any soul S of M (e.g. if $M = S \times \mathbf{R}^k$), and if G is a group of isometrics acting freely on M, then M/G is also $\tilde{S}$-flat for some soul $\tilde{S}$ of M/G.*

The simplest examples of manifolds of nonnegative curvature are of course the Riemannian products $S \times P_k$, where P_k is $\mathbf{R}^k$ together with some nonnegatively curved metric. For instance, if the soul S of M is flat, then, by a theorem of Toponogov (cf. [CG]), the universal cover of M splits metrically. In this case K_p goes to 0 as p goes to ∞, or more precisely, $\lim_{r\to\infty} \kappa(r) = 0$, where $\kappa(r) := \sup\{K_p / d(p,S) \geq r\}$. Conversely, one has [ESS]:

THEOREM 3.5. *If codim $S \leq 3$, and if $\kappa(r) \to 0$, then S is flat.*

According to [M2], this theorem holds without any assumption on the codimension of S, but the arguments are somewhat unclear.

The key point in the proof of 3.5 is that, after reducing the problem to the simply connected case, there are only two possibilities for the holonomy of $\nu(S)$ if S has codimension ≤ 3 : either there is a global parallel section of $\nu(S)$, or the holonomy acts transitively even if one considers only curves of bounded length. In the first case, exponentiating the parallel section at constant distances yields totally geodesic submanifolds isometric to S arbitrarily far from S, and since

$\kappa(r) \to 0$, S must be flat. In the second case, it follows that the diameters of the sphere bundles over S are bounded by a fixed constant. Estimates on the second fundamental form show that their curvature goes to 0 as $r \to \infty$, cf. also [GW]. Since these sphere bundles are all diffeomorphic to the unit sphere bundle $\nu^1(S)$, $\nu^1(S)$ is an almost flat manifold (cf. [G2]), and homotopy considerations imply $\nu^1(S)$ is a circle, so that S must be a point.

Finally, if κ decays quickly enough, then under suitable topological conditions, M must actually be isometric to Euclidean space:

THEOREM 3.6. *Suppose $n := \dim M$ is odd, and the Euler characteristic of M is nonzero. Let $k := \dim S$, $a := 2 - 2k/(n-1)$. Then lim sup $\kappa(r)r^a > 0$ or M is isometric to $\mathbf{R}^n$.*

For a proof, see [ESS]. Two important special cases are worth mentioning:

(i) If $\kappa(r)r^\beta \to 0$ for some exponent $\beta \geq 2 - 2/(n-1)$, then S must be a point, and M is diffeomorphic to $\mathbf{R}^n$.

(ii) If $\kappa(r)r^2 \to 0$, then M is isometric to $\mathbf{R}^n$.

REFERENCES

[C] J. Cheeger, *Some examples of manifolds of nonnegative curvature*, J. Differential Geometry **8** (1972) 623-628.

[CG] J. Cheeger and D. Gromoll, *On the structure of complete manifolds of nonnegative curvature*, Ann. of Math. **96** (1972) 413-443.

[C-V] S. Cohn - Vossen, *Kürzeste Wege und Totalkrümmung auf Flächen*, Composito Math. **2** (1935) 69-133.

[ESS] J. Eschenburg, V. Schroeder and M. Strake, *Curvature at infinity of open nonnegatively curved manifolds*, to appear in J. Differential Geometry.

[GW] R. Greene and H. Wu, *Gap theorems for noncompact Riemannian manifolds*, Duke Math. J. **49** (1982) 731-756.

[GKM] D. Gromoll, W. Klingenberg, and W. Meyer, *Riemannsche Geometrie im Grossen*, Lecture Notes in Mathematics **55**, Springer-Verlag, 1975.

[GM] D. Gromoll and W. Meyer, *On complete open manifolds of positive curvature*, Ann. of Math. **90** (1969) 75-90.

[G1] M. Gromov, *Curvature, diameter and Betti numbers*, Comment. Math. Helvetici **56** (1981) 179-195.

[G2] __________, *Almost flat manifolds*, J. Differential Geometry **13** (1978) 231-242.

[GS] K. Grove and K. Shiohama, *A generalized sphere theorem*, Ann. of Math. **106** (1977) 201-211.

[M1] V. B. Marenich, *The structure of the curvature tensor of an open manifold of nonnegative curvature*, Soviet Math. Dokl. **28** (1983) 753-757.

[M2] __________, *The topological gap phenomenon for open manifolds of non-*

negative curvature, Soviet Math. Sokl. **32** (1985) 440-443.

[O'N] B. O'Neill, *The fundamental equations of a submersion*, Michigan Math. J. **13** (1966) 459-469.

[Sh] V. A. Sharafutdinov, *The Pogorelov-Klingenberg theorem for manifolds homeomorphic to* $\mathbf{R}^n$, Sib. Math. Zh. **18** (1977) 915-925.

[St] M. Strake, *A splitting theorem for open nonnegatively curved manifolds*, to appear in Manuscripta Math.

[SW] M. Strake and G. Walschap, *Σ-flat manifolds and Riemannian submersions*, preprint.

[W1] G. Walschap, *Nonnegatively curved manifolds with souls of codimension 2*, J. Differential Geometry **27** (1988) 525-537.

[W2] ________, *A splitting theorem for 4-dimensional manifolds of nonnegative curvature*, to appear in Proc. Amer. Math. Soc.

[Y] J. W. Yim, *Space of souls in a complete open manifold of nonnegative curvature*, preprint.

UNIVERSITY OF CALIFORNIA, LOS ANGELES

Contemporary Mathematics
Volume **101**, 1989

AN EXTRINSIC AVERAGE VARIATIONAL METHOD

S. Walter Wei

ABSTRACT. We propose an extrinsic, average variational method as an approach to confront and resolve problems in global, nonlinear analysis and geometry. This applications in topology, Liouville-type theorems and the regularity of minimizers are discussed, in particular super-strongly unstable manifolds and indices are found. Its interaction with an intrinsic, average variational method is also included.

§1. THE FORMULATION OF AN EXTRINSIC AVERAGE VARIATIONAL METHOD.

We have observed that many interesting topics and important advances in geometry, analysis, topology or physics are of a variational nature (for example, the work of S. Donaldson [D]). Critical points, local minima, or minimizers (i.e. global minima) for suitable elliptic functionals often serve as "catalysts" in ongoing processes of nonlinear analysis and global geometry. We, therefore, propose *an extrinsic, average variational method* ([SWW1,2,3],[HW1]) as an approach to confront and resolve problems in global, nonlinear analysis and geometry ([SWW4],[WY]). Progress has been made in the theory of minimal surfaces ([HW2][O1]) and Yang-Mills fields ([SWW5][KOT]) by utilizing this method (For instance, the classification of stable minimal surfaces (in fact, rectifiable currents) in rank one symmetric spaces was first established by R. Howard and the author [HW2][SWW1], which generalizes the work on spheres and complex projective spaces [LS].)

In the context of a functional $\mathcal{E}$ on mappings between Riemannian manifolds, e.g. given a smooth map $u : M^n \to N^k$, this extrinsic, average variational method means the following:

We isometrically immerse N^k into a Euclidean space R^q in which an orthonormal frame $V_1, \cdots, V_q$ gives rise to globally defined vector fields $V_1^T, \cdots, V_q^T$ on N via its tangential projections.

1980 Mathematics Subject Classification (1985 Revision) primary 58E15, 58E20, 58E30,58E35.
Research supported in part by N.S.F. Grant (No. DMS-8802745) and the University of Oklahoma Research Award

Then through the deformation $u_t^{\varphi V_j^T}$ of u along φV_j^T for each $1 \leqq j \leqq q$, (where φ is a smooth function on M with compact support) we take the average of the second variational formulas for the corresponding functional $\mathcal{E}$ over $1 \leqq j \leqq q$.

In brief, the proposed method, as distinct from other variational methods (c.f. [MS], [LS],[H]) based on an idea of Synge [SY], consists initially in choosing a set of generally *non-gradient* vector fields along the map through an isometric *immersion*, deforming the map in these indicated directions, and computing

$$(1) \qquad \sum_{j=1}^{q} \frac{d^2}{dt^2}\mathcal{E}(u_t^{\varphi V_j^T})|_{t=0}$$

If M is compact, we set $\varphi \equiv 1$ and employ a variational generalized principle of Synge ([SWW2]) – (generally speaking) for any analytic measurement $\mathcal{F}(f)$ (e.g. area or energy) of a geometric object f (e.g. surface or map) in an ambient space F, the existence of "average, $\mathcal{F}$-decreasing variation" of f, for each f in F implies the non-existence of $(\mathcal{F})$-stable f in F.

If M is complete, noncompact, we then fuse the ideas of Bochner [BS], De Giorgi [DE], Federer [F], Fleming [FW], Moser [MOJ], Simons [SJ] and Synge [SY], and employ the techniques or results of Cheng, Greene, Hardt, Howard, Karp, Lin, Schoen, Simon, Uhlenbeck, White, Wu and Yau (cf [CH],[CY],[GW],[HL],[H],[K],[SSY],[SU 1,2],[W1]).

Analogous to a Bernstein-type theorem and the regularity of codimension 1 minimizing currents, we establish Liouville-type theorems and the regularity of minimizers for $\mathcal{E}$.

§2. APPLICATIONS TO STABLE HARMONIC MAPS ON COMPACT MANIFOLDS. ([HW1])

We first discuss $\mathcal{E}$ in the setting of the energy functional E given by

$$E(u) = \tfrac{1}{2} \textstyle\int_M |du(x)|^2 dv$$

where $du(x)$ denotes the differential of u and dv is the volume element of M.

DEFINITION. A smooth map $u : M \to N$ is said to be harmonic if u is a critical point of the energy functional E on the set $C^\infty(M, N)$ of smooth maps.

DEFINITION. A smooth map $u : M \to N$ is said to be stable harmonic (resp. energy-minimizing) if u is a local (resp. global) minimum of the energy functional E on $C^\infty(M, N)$.

REMARK. A harmonic map $u : M \to N$ between Riemannian manifolds M and N can be defined to be a weak solution to the Euler-Lagrange equations for E on $L_1^2(M, N)$, the set of maps $u : M \to R^q$ whose component functions have first derivatives in L^2 and $u(x) \varepsilon N$ a.e. $x \varepsilon M$ (where N is isometrically embedded in R^q). Similarly, we can discuss maps that minimize energy, or are stable among $L_1^2(M,N)$.

Consider a C^2 map $F : M \times [0,1] \times [0,1] \to N$, and set $u_{s,t}(x) = F(x,s,t)$ for $x \in M$, such that $F(x,0,0) = u(x)$. Denote by ∇^F the pull-back connection on the pull-back bundle $F^{-1}TN$, and set $V = dF(\frac{\partial}{\partial s})$ and $W = dF(\frac{\partial}{\partial t})$. We will exploit the following fundamental formulas (c.f. [EL]) in which $\{e_1, \cdots, e_n\}$ is a local orthonormal frame field on M.

The first variational formula for the energy of $u : M \to N$

(2) $$\frac{\partial}{\partial s}E(u_{s,t}) = \sum_{i=1}^{n} \int_M < \nabla^F_{e_i}V,\ dF(e_i) > dv$$

The second variational formula of two parameters for the energy of u (u is not necessarily harmonic)

(3) $$\frac{\partial^2}{\partial t \partial s}F(u_{s,t}) = \sum_{i=1}^{n} \int_M < \nabla^F_{e_i}V,\ \nabla^F_{e_i}W > + < \nabla^F_{e_i}\nabla^F_{\frac{\partial}{\partial t}}V + R^N(W, dF(e_i))V,\ dF(e_i) > dv$$

When M is compact, by the proposed method, we find super-strongly unstable manifolds identified as follows:

DEFINITION. A complete k-dimensional Riemannian manifold N with Riemannian metric $<,>_N$ is said to be *super-strongly unstable (SSU)* if there exists an isometric immersion of N in Euclidean space R^q such that the operator Q_y^N, from the tangent space, $T_y(N)$ to N at y, into itself, given by

(4) $$Q_y^N = \sum_{\alpha=k+1}^{q} 2(A^{e_\alpha})^2 - (trA^{e_\alpha})A^{e_\alpha}$$

is *negative definite* for every $y \varepsilon N$, where $e_{k+1}, \cdots, e_q$ are local orthonormal vector fields (in R^q), perpendicular to N and A^{e_α} is the Weingarten map (defined by $A^{e_\alpha}(X) = -(\overline{\nabla}_X e_\alpha)^T$ in which $X \varepsilon T_y(N)$ and $\overline{\nabla}$ is the Euclidean connection on R^q).

That is, N is SSU if there exists a positive function ϕ on N such that

(5) $$< Q_y^N(X), X >_N \leqq -\phi(y)\langle X, X\rangle_N \text{ at every } y\varepsilon N \text{ and } X\varepsilon T_y(N).$$

REMARK. It follows from Meyers' theorem [MS] that an SSU manifold is compact if and only if there exists a positive constant c such that

(6) $$< Q_y^N(X), X >_N \leqq -c\langle X, X\rangle_N$$

for every $y \varepsilon N$ and $X \varepsilon T_y(N)$.

EXAMPLES OF SSU MANIFOLDS INCLUDE

1. A complete hypersurface N in R^{k+1} is SSU if and only if all principal curvatures satisfy

$$0 < \lambda_1 \leqq \cdots \leqq \lambda_k < \lambda_1 + \cdots + \lambda_{k-1}$$

In particular, an elliptic paraboloid $f(x) = |x|^2$ in R^{k+1}, a standard k-sphere, and a k-ellipsoid $E_a^k = \{(x_1, \cdots, x_{k+1}) \varepsilon R^{k+1} : ax_1^2 + x_2^2 + \cdots + x_{k+1}^2 = 1\}$, where $0 < a < k-1$ are SSU for $2 < k$.

2. (Classification Theorem [HW][O1]) A compact, connected symmetric space N is SSU if and only if N is one of the following: (A) sphere S^n, $n > 2$; (B) Grassman manifold $Sp(p+q)/Sp(p) \times SP(q)$, $1 \leqq p \leqq q, 3 \leqq p+q$ (including the quaternionic projective space $Qp(n)$, where $p = n$, and $q = 1$; (C) Cayley plane $F_4/\mathrm{Spin}(9)$; $(D) SU(2n)/Sp(n)$, $n \geqq 3$; (E) E_6/F_4; (F) compact simply-connected simple Lie groups of type A_n, $n \geqq 2$, B_2 and C_n, $n \geqq 3$ or (G) arbitrary finite product of any manifolds from (A), (B),(C),(D),(E),(F).

3. A compact minimal k-submanifold N of the unit sphere S^{q-1} with $\mathrm{Ricci}^N > k/2$.

4. Compact submanifolds N of the unit sphere S^{q-1} with the second fundamental form B satisfying
$$|B|^2 < \begin{cases} 4(k-2)/\sqrt{k} + 5 & \text{for} \quad 3 \leqq k \leqq 8 \\ 2(k-2)/\sqrt{k} + 1 & \text{for} \quad 9 \leqq k. \end{cases}$$

More examples are given in [HW1, SWW 3,4].

CRITERIA FOR SSU MANIFOLDS

Every SSU manifold N satisfies the following:

(i) Curvature condition: $\mathrm{Ric}^N > 0$ (by the Gauss curvature equation).

(ii) Dimension condition: $\dim N > 2$ (by the maximum principle and an averaging technique [WY], (the result is sharp, c.f. example 1 above)).

(iii) Topology condition: $\pi_1(N) = \pi_2(N) = 0$ (for compact N), due to a theorem of Sacks and Uhlenbeck [SU] and the following

MAIN THEOREM. (Howard and Wei [HW1]) Every compact SSU manifold N is strongly unstable (SU). Recall a compact Riemannian manifold N is defined to be SU if N is not the domain nor the target of any non-constant, smooth stable harmonic map (into or from an arbitrarily compact Riemannian manifolds).

PROOF. To show an SSU manifold N is not the target of any nonconstant, smooth stable harmonic map u, we employ the proposed method. (1), via (3) and (4), becomes

$$\sum_{j=1}^{q} \frac{d^2}{dt^2} E(u_t^{V_j^T})|_{t=0} = \sum_{i=1}^{n} \int_M < Q^N(\tilde{e}_i), \tilde{e}_i > dv < 0 \tag{7}$$

where $\{e_1, \cdots, e_n\}$ is a local orthonormal frame field on M and $\tilde{}$ *denotes the push forward vector* $du(\cdot)$.

If u were stable, then for each $1 \leqq j \leqq q$, $\frac{d^2}{dt^2} E(u^{V_j^T})_t|_{t=0}$ would be nonnegative, and so would be the sum on j, a contradiction. Similarly, one can show that an SSU manifold M is not the domain of any nonconstant, stable harmonic map u by applying the variational method to the domain M and obtain

$$\sum_{j=1}^{q} \frac{d^2}{dt^2} E(u_t^{du(V_j^T)})|_{t=0} = \sum_{i=1}^{n} \int_M < Q(\widetilde{e_i}), \widetilde{e}_i >_N dv < 0 \tag{8}$$

By an iteration method and the proposed approach, Howard and Wei show that *the homotopy class of any map into or from a compact SSU manifold N contains an element of arbitrarily small energy*; or equivalently, the identity map of a compact SSU manifold N is homotopic to a map of arbitrarily small energy. On the other hand, White [W] shows that the infimum of the energy is zero among maps homotopic to the identity on a compact manifold N if and only if $\pi_1(N) = \pi_2(N) = 0$. Putting them together, we draw the following diagram to summarize the results so far:

DIAGRAM

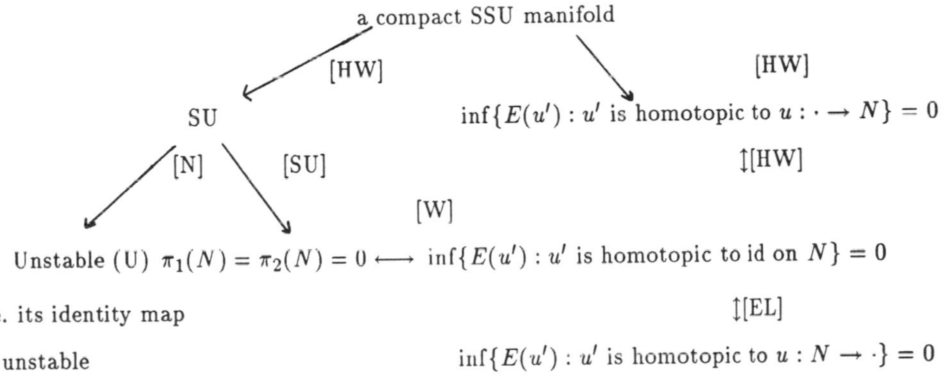

As compact (isotropy) irreducible homogeneous manifold N has "standard immersions" into Euclidean space, the above main theorem via Schur's lemma [HJ] and Takahashi's Theorem [TT] implies that $Q_y^N = (k\lambda_i - 2\tau)I$ for the i-th standard immersion where λ_k is the k-th eigenvalue of the Beltrami-Laplace operator on N, τ is the scalar curvature and N has the canonical metric. It follows that a compact (isotropy) irreducible homogeneous k-manifold N is SSU if and only if N is SU; if and only if N is Unstable (U); if and only if $k\lambda_1 < 2\tau$; if and only if $\lambda_1 < 1$. This, aided by the work of Smith [S] and Nagano [N], leads to classification theorem listed in example 2 above and generalizes some interesting work of Xin [X], Leung [L], Morrey [M], Eells and Sampson [ES], Min-oo [MO], and Eells and Lemaire [EL]. (c.f. [HW1] for details)

§3. APPLICATIONS TO STABLE HARMONIC MAPS ON COMPLETE, NONCOMPACT MANIFOLDS. ([SWW3,4])

When M is complete, via the proposed approach again, we find super-strongly unstable indices associated to SSU manifolds. For a complete, noncompact domain, the analogue of the *nonexistence* of stable harmonic maps on a compact manifold is a *Liouville-type theorem*. We will exhibit precisely how these SSU indices indicate the extent to which Liouville's theorems hold for harmonic maps (in this section), and also to the regularity of energy-minimizing maps into SSU manifolds (in the next section).

DEFINITION. The number w on an SSU k-manifold N given by $w = \inf_{y \varepsilon N} \phi(y)/k\mathcal{S}(y)$ is said to be an *SSU index*, where $\phi(y)$ is as in (5) and $\mathcal{S}$ is a positive function satisfying

(9) $\quad$ sectional curvature $\text{Riem}^N(y) \leqq \mathcal{S}(y)$ for each y in N.

REMARK. It follows from the Gauss curvature equation that

(10) $$-k\mathcal{S}(y)\langle X, X\rangle_N < \langle Q_y^N(X), X\rangle_N < -\phi(y)\langle X, X\rangle_N$$

and hence

($10\frac{1}{2}$) $$0 \leqq w < 1.$$

EXAMPLE OF AN SSU INDEX On a k-ellipsoid E_a^k, $0 < a < k-1$ and $3 \leq k$

$$\text{SSU index } w \geq \begin{cases} \frac{a(k-2)}{k} & \text{if } 0 < a \leq 1 \\ \frac{k-2}{ak} & \text{if } 1 \leq a \leq k-2 \\ \frac{k-1-a}{k} & \text{if } k-2 \leq a < k-1 \end{cases}$$

Through the variational method, we acquire

A General Average Formula:

(11) $$\sum_{j=1}^{q} \frac{d^2}{dt^2} E(u_t^{\varphi V_j^T})|_{t=0} = \int_M \sum_{i=1}^{n} < Q_y^N(\tilde{e}_i), \tilde{e}_i >_N \varphi^2 dV$$

and via (5) build on the following fundamental basis for all further development:

An Extrinsic Stability Inequality ([SWW3]). Let u be a stable harmonic map from a complete manifold M into an SSU manifold N. Then for any $\varphi \varepsilon C_0^\infty(M)$

(12) $$\textstyle\int_M (\phi(u)/k)\varphi^2 |du|^2 dV \leqq \int_M |\nabla\varphi|^2 dV$$

where φ is as in (5). Moreover, if N is compact or $u(M)$ is relatively compact in N then for any $\varphi \varepsilon C_0^\infty(M)$

(13) $$\textstyle\int_M \varphi^2 |du|^2 dV \leqq k/c \int_M |\nabla\varphi|^2 dV$$

On the other hand, if M is compact (11) yields (7) (by setting $\varphi \equiv 1$ as stated in §1).

Via an expanded technique of Bochner (c.f. [MOJ][SSY][SU2]), the stability inequality gives rise to

An L^p estimate for the energy density of a stable harmonic map u from a complete n-manifold into an SSU k-manifold N. For any $p \varepsilon [4, 2 + 2wn/(n-1)]$ and for each $\varphi \varepsilon C_0^\infty(M)$ we have

$$\int_M \varphi^p |du|^p dV \leqq c_1 \int_M K\varphi^p dV + c_2 \int_M |\nabla\varphi|^p dV$$

where K is a non-negative function on M, satisfying

(14) $\text{Ricci}^M \geqq -K$, w is an SSU index on M and c_1 and c_2 are constants depending only on φ (as in (5)), n, k, p, K and $\mathcal{S}$ (as in (9)).

As an application of these stability inequality and L^p estimates, we obtain, via [KW] and [SWW3]

Liouville theorem for stable harmonic maps. Every smooth, stable harmonic map on a complete n-manifold M (covered by a collection of geodesic balls B_r, of radius r, centered at a fixed point) into an SSU manifold N is constant provided

(i) M is parabolic, i.e. M admits no constant, negative subharmonic function or

(ii) $\text{Ricci}^M \geq 0$ and volume $(B_r) = 0(r^p)$ as $r \to \infty$, for some $p \varepsilon [4, 2 + 2wn/(n-1)]$. In particular, every smooth stable harmonic map on a complete n-manifold into an SSU-manifold w is constant provided

(i′) M has quadratic volume growth i.e. volume $(B_r) = 0(r^2)$ as $r \to \infty$ or (i″) M has moderate volume growth i.e. there is a point x_0 in a complete, noncompact manifold M and a positive nondecreasing function $f(r)$ such that $\lim_{r\to\infty} \sup \frac{1}{r^2 f(r)}$ volume $(B_r(x_0)) < \infty$. while some $0 < a$. (e.g. $f(r) = \log r$ c.f. [K]).

REMARK. Assumption (i′) implies assumption (ii″). Furthermore, (ii″) which is sharp yields assumption (i) (cf the work of Karp [K], Cheng and Yau [CY], Greene and Wu [GW]).

This generalizes the work of Schoen & Uhlenbeck [SU2] in which they treat the case $M^n = R^n$ and $N^k = S^k$ (where $w = (k-2)/k$).

§4 AN INTRINSIC AVERAGE VARIATIONAL METHOD.

If the target N of a stable harmonic map f is a compact, simply-connected, δ-pinched manifold (i.e. there exists a constant $r > 0$ such that $\text{Riem}^N \varepsilon (\delta r, r]$), then we normalize the δ-pinched metric of M by multiplication with $(1+\delta)/2$ as in [GKR] and construct a Euclidean vector bundle $\tilde{E}$ on M by $\tilde{E} = T(N) \oplus \varepsilon(N)$ where $T(N)$ is the tangent bundle and $\varepsilon(N)$ is a trivial line bundle on N. We then choose $\{V_1^T, \cdots, V_{k+1}^T\}$ to be the TN-component of an orthonormal basis $\{V_1, \cdots, V_{k+1}\}$ for the vector space of all parallel cross-sections of $\tilde{E}$ with respect to a flat metric connection. As before, we deform f along φV_j^T and take the second variational formula.

Then $\qquad 0 \leqq \sum_{j=1}^{k+1} \frac{d^2}{dt^2} E(u_t^{\varphi V_j^T})|_{t=0}$ implies

(15) $\qquad \int_M \varphi^2 |df|^2 dV \leqq C \int_M |\nabla\varphi|^2 dV$

for every $\varphi \varepsilon C_0^\infty(M)$, $8 \leqq k$ and $\delta = 0.83$, in which C is a constant.

On the other hand, in [SWW3] we deformed f along $\varphi\nabla(g \circ \varrho_y)$ (where ϱ_y is the geodesic distance from y in N and $g(t) = -\cos t$ for $|t| \leqq \pi$; otherwise $g(t) = 1$) and took the *integral average* of the second variational formula over a continuous family of deformation (instead of a finite sum) (c.f. [H]) $0 \leqq \int_N \frac{d^2}{dt^2} E(f_t^{\varphi\nabla(g\circ\varrho_y)})|_{t=0} dW$ via Fubini's theorem and the Hessian comparison theorem of Greene and Wu [GW], yielded an Intrinsic Stability Inequality ([SWW3],3.1,p.412). This means (15) holds for every $\varphi\varepsilon C_0^\infty(M)$, $3 \leqq k \leqq 7$, and $\delta_k < \delta = 0.83$ where δ_k is as in ([SWW3,(11),p.411]).

Combining these results, and proceeding as in [KW] we have proved

An Intrinsic Liouville Theorem. Every smooth stable harmonic map on a parabolic manifold into a compact, simply-connected, 0.83-pinched k-manifold is constant for $3 < k$.

COROLLARY. There does not exist any smooth, stable, nonconstant harmonic map from any compact manifold into a compact, simply connected, δ-pinched, k-manifold where $\delta = 0.83$ for $7 < k$, otherwise $\delta = \delta_k < 0.83$ as in [SWW3,(11),p.411].

This corollary slightly improves the work of Howard [H] and Okayasu [OT], where they gave partial affirmative answers to a harmonic-version of Lawson and Simons' conjecture [LS]. The conjecture asserts that there are no stable varifolds for any compact, simply-connected Riemannian manifold M which is $\frac{1}{4}$-pinched. Through the proposed extrinsic average variational method in the minimal surface theory (c.f. [HW2],[SWW1,Sec.3, p.536]), Howard and the author solved Lawson and Simons' conjecture in the case M is an odd dimensional hypersurface in R^{n+1}. In fact, more general results are found:

THEOREM ([HW2]). Let M^n be a complete hypersurface in the Euclidean space R^{n+1} which is pointwise δ-pinched for

$$\delta \geq \frac{1}{4} - \frac{4k^3 - 11k^2 - 12k - 3}{4(2k-1)^2(k^2+1)} \quad \text{if } n = 2k+1.$$

and

$$\delta \geq \frac{1}{4} + \frac{3}{4(k^2+1)} \quad \text{if } n = 2k \geq 4.$$

Then M has no stable varifolds.

REMARK. There is a neighborhood in the C^2 topology of the standard metric on the Euclidean sphere S^n such that for any metric g in this neighborhood, (S^n, g) has no stable varifolds. In particular, (S^n, g) has no stable rectifiable currents over any finitely generated abelian group G. (cf. [HW2]) This augments a theorem of Howard [H], through an intrinsic approach.

§5. APPLICATIONS TO ENERGY-MINIMIZING MAPS. ([SWW4])

Just as Bernstein's theorem for stable minimal cones ([SJ]) is closely tied to the regularity for codimension 1 mass-minimizing rectifiable currents ([F]), so is Liouvilles' theorem for stable tangent maps related to the regularity of energy-minimizing maps (Recall $\overline{u} : R^n \to N$ is said to be a tangent map if $\overline{u}$ is a homogeneous extension of $u : S^{n-1} \to N$). This makes the analogue of Simons' theorem – the Liouville theorem for stable tangent maps into an SSU manifold N – particularly interesting. In fact, the (Liouville) theorem (and its proof) which provides an ingredient for the regularity theorem, marks a solo appearance of differential geometry without measure theory.

To state our theorems, we set, for each SSU index w

$$d(w) = \begin{cases} [[2 + 2w + 2\sqrt{w^2 + w}]] & \text{if } 1 \leqq (w + \sqrt{w^2 + w})/2 \\ [(2 - w)/(1 - w)] & \text{otherwise} \end{cases} \tag{16}$$

where

$$[[t]] = [t], \text{ the greatest integer of } t, \text{ if } t \text{ is not an integer; otherwise } [[t]] = t - 1. \tag{17}$$

We denote $L_1^2(M, N)$ as in §2. Furthermore, for an energy-minimizing map u on M, a point $x \varepsilon M$ is said to be a *regular point* if u is continuous in a neighborhood of x. The singular set S is the complement of the set of all regular points in the interior of M. $\dim(S)$ denotes the Hausdorff dimension of singular set S in the interior of M.

Using the above notation, we have

Liouville Theorem for stable tangent maps. Every stable tangent map $\overline{u}$ defined on R^n into an SSU k-manifold N is constant provided $n \leqq d(w)$.

PROOF. Again the variational method or the extrinsic stability inequality (12) implies

$$\int_{R^n} (\phi(\overline{u})/k)\varphi^2 |d\overline{u}|^2 dx \leqq \int_{R^n} |\nabla \varphi|^2 dx \tag{18}$$

for any $\varphi \varepsilon C_0^\infty(R^n)$.

We then perform separation by variables, i.e. choose $\varphi : R^n \backslash \{0\} \to R$ to be $\varphi(x) = f(r) g(v)$ where $r = |x|$ and $v = \frac{x}{|x|} \varepsilon S^{n-1}$.

This splits the integrals in (18) over R^n into spherical and radial directions

$$(19)\qquad \int_{S^{n-1}} |\nabla g|^2 + (\phi(u)/k)g^2|du|^2)dv \leqq (n-2)^2 \int_{S^{n-1}} g^2 dv$$

$$(\leqq (\int_0^\infty f'^2 r^{n-1} dr / \int_0^\infty f^2 r^{n-3} dr) \int_{S^{n-1}} g^2 dv)$$

for $0 \neq f \varepsilon C_0^\infty(R)$ and $0 \neq g \varepsilon C^\infty(S^{n-1})$.

On the other hand, the Bochner method yields the lower bound of integral over S^{n-1}.

$$(20)\qquad (n-2)\beta \int_{S^{n-1}} g^2 dv \leqq \int_{S^{n-1}} |\nabla g|^2 - (\beta S(n-2)/(n-1))g^2|du|^2 dv \text{ where } \beta \geqq 1.$$

Combining (19) and (20), we show that if $\phi(u)/k > \beta S(n-2)/(n-1)$ (i.e. $n < 1 + \frac{\beta}{\beta - w}$) and $n \leqq 4\beta + 2$, where $\beta \geqq 1$, then u is constant.

Note that the minimum of $1 + \frac{\beta}{\beta - w}$ and $4\beta + 2$ attains a maximum value $2 + 2w + 2\sqrt{w^2 + w}$ when $1 + \frac{\beta}{\beta - w} = 4\beta + 2$ or $\beta = (w + \sqrt{w^2 + w})/2$ where $\beta \geqq 1$. (c.f. the diagram in the end of this section.

Consequently $\overline{u}$ is constant if $n \leqq d(w)$ where $d(w)$ is as in (16). This proves the Liouville Theorem and yields immediately.

Regularity theorem for energy-minimizing maps. If $u \varepsilon L_1^2(M, N)$ is energy-minimizing with $u(x) \varepsilon N_0$ a.e. for a compact subset N_0 of an SSU manifold N, then

$$\dim(S) \leqq n - d(w) - 1 = \begin{cases} n-7 & 4/5 < w \\ n-6 & 3/4 \leq w \leqq 4/5 \\ n-5 \quad \text{if} & 2/3 \leq w < 3/4 \\ n-4 & 1/2 \leq w < 2/3 \\ n-3 & w < 1/2 \end{cases}$$

If $n = d(w) + 1$, u has at most isolated singularities and if $n \leqq d(w)$, u is smooth in the interior of M.

COROLLARY. Let E_a^k be a k-ellipsoid (cf. example 1 in §2) and $u \in L_1^2(M, E_a^k)$ be energy-minimizing. Then

$$
\dim(S) \le \begin{cases} n-7 \\ n-6 \\ n-5 \\ n-4 \\ n-3 \end{cases} \quad if \quad
\begin{array}{ccccc}
0 < a \le 1 & & 1 \le a \le k-2 & & k-2 \le a < k-1 \\
\text{and:} & & \text{and:} & & \text{and:} \\
\frac{10a}{5a-4} < k; & & \frac{10}{5-4a} < k; & & 5+5a < k \\
\frac{8a}{4a-3} \le k \le \frac{10a}{5a-4}; & & \frac{8}{4-3a} \le k \le \frac{10}{5-4a}; & & 4+4a \le k \le 5+5a \\
\frac{6a}{3a-2} \le k < \frac{8a}{4a-3}; & or & \frac{6}{3-2a} \le k < \frac{8}{4-3a}; & or & 3+3a \le k < 4+4a \\
\frac{4a}{2a-1} \le k < \frac{6a}{3a-2}; & & \frac{4}{2-a} \le k < \frac{6}{3-2a}; & & 2+2a \le k < 3+3a \\
k < \frac{4a}{2a-1}; & & k < \frac{4}{2-a}; & & k < 2+2a
\end{array}
$$

DEFINITION ([HS]p.25). A Riemannian manifold M is said to be simple if there exist positive constants λ, μ and a diffeomorphic chart from M onto R^n such that the metric $ds^2 = g_{ij}dx^i dx^j$ of M with respect to these global coordinates satisfies

$$\lambda|\xi|^2 \le g_{ij}(x)\xi_i\xi_j \le \mu|\xi|^2 \ , \qquad \text{for all } \ x, \xi \in R^n.$$

As an application, we have

Liouville theorem for energy-minimizing map. Every energy-minimizing map u defined on a simple Riemannian manifold into an SSU manifold is constant for $n \leqq d(w)$.

Our regularity theorem augments immediately an existence and regularity theorem of White [W1], in the case N is an SSU manifold.

THEOREM. In every 1-homotopy class of maps in $L_1^2(M, N)$, there is a map of least energy and $\dim(S) \leqq n - d(w) - 1$.

It should be pointed out that Giaquinta and Giusti [GG] and Schoen and Uhlenbeck [SU1] have independently shown that a minimizing harmonic map into a Riemannian manifold is smooth in the interior off a closed set of at least Hausdorff codimension three i.e. $\dim(S) \leqq n-3$. Furthermore, the reduction in $\dim(S)$ for minimizing maps into S^k and related Liouville's theorem are due to Schoen and Uhlembeck [SU2]. Many other important special cases are due to Nakajima [NH].

DIAGRAM. SSU index w serves as an indication of the regularity of energy minimizing harmonic maps into SSU manifolds.

(i) When SSU index w is close to 1

$d(w) = [[2 + 2w + 2\sqrt{w^2 + w}]]$

(where $[[\cdot]]$ is as in (17))

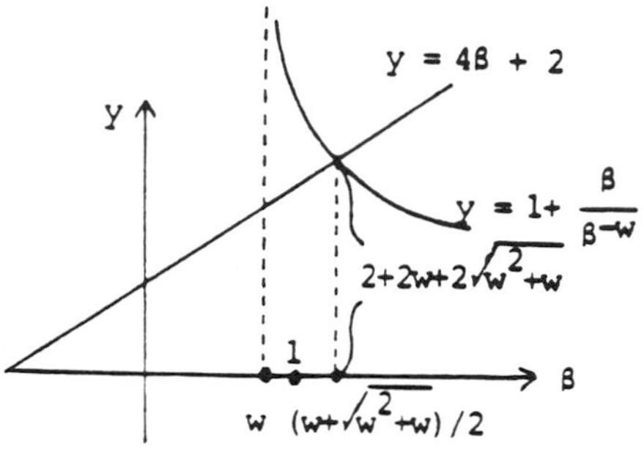

(ii) When SSU index w is much less than 1

$d(w) = [(2 - w)/(1 - w)]$

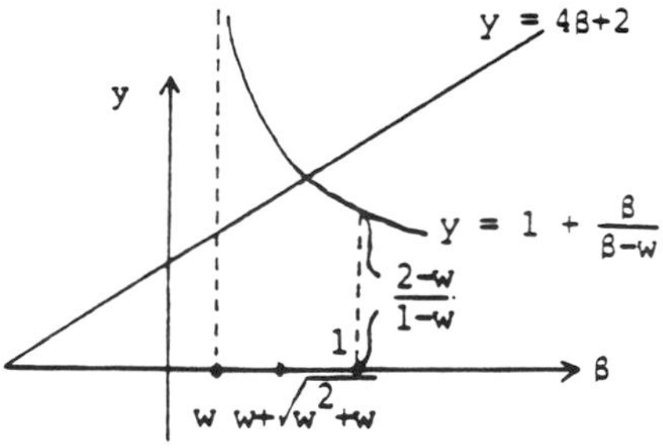

§6. APPLICATIONS TO p-STABLE AND p-MINIMIZING MAPS. ([WY])

In §3 and §5, the proposed methods lead to the discovery of SSU manifolds and their associated SSU indices (which lie in [0,1)). Furthermore, we observed how quantitatively, the closer this SSU index w is to 1, the easier it is to establish the Liouville theorem for a smooth, stable harmonic map (in terms of the volume growth condition in the domain), and moreover, the "smoother" the energy-minimizing map is (in terms of the Hausdorff dimension of the singular set in the domain). In both settings, the targets of these maps are SSU manifolds.

Similar phenomena occur when we apply the same variational method to the following generalized energy functional:

DEFINITION. The p-energy ($p > 0$) for a smooth map $u : M^n \to N^k$ between compact Riemannian manifolds of dimension n and k is given by

$$E_p(u) = \frac{1}{p} \int_M |du(x)|^p dv$$

where $du(x)$ denotes the differential of u, and dv is the volume element of M.

Analogously, we can define smooth maps to be p-harmonic if they are critical points, p-stable if they are local minima, and p-minimizing if they are global minima of p-energy functional respectively. We can also discuss maps that minimize p-energy or, p-harmonic, p-stable among $L_1^p(M, N)$. Furthermore, we have

DEFINITION. A compact Riemannian manifold N is p-strongly unstable ($p - SU$) if N is neither the domain nor the target of any nonconstant, smooth, stable p-harmonic maps.

In [WY] we computed the first and second variational formulas for p-energy, by differentiating the following perturbed p-energy functional E_p^n given by

$$E_p^n(u_t) = \frac{1}{p} \int_M (|du_t|^2 + \tfrac{1}{n})^{\frac{p}{2}} dv$$

where n is a positive integer. In this manner we can remove the obstacles caused by the zero set of du_t in differentiating the p-energy ($p > 0$) and apply the Lebesque Convergence Theorem, monotonicity, and uniform convergence theorem for differentiation to obtain the variational formulas:

$$\frac{d}{dt} E_p(u_t) = \frac{d}{dt} \lim_{n\to\infty} E_p^n(u_t) = \lim_{n\to\infty} \frac{d}{dt} E_p^n(u_t)$$

$$\text{and} \quad \frac{\partial^2}{\partial t \partial s} E_p(u_{s,t})|_{s,t=0} = \frac{\partial}{\partial t} \lim_{n\to\infty} \frac{\partial}{\partial s} E_p^n(u_{s,t})|_{s,t=0} = \lim_{n\to\infty} \frac{\partial^2}{\partial t \partial s} E_p^n(u_{s,t})|_{s,t=0}$$

As an application of our first variational formula, and a regularity theorem of Hardt and Lin ([HL] 4.5 and 7) we show

THEOREM (Wei and Yau [WY]). Every p-minimizing, L_1^p map into a manifold with non-positive sectional curvature or into the domain of a strictly convex function is smooth, i.e. has a locally Hölder continuous gradient.

This theorem is on the one hand, closely related to the work of Eells and Sampson [ES] and Hilderbrandt, Kaul and Widman [HKW]; and on the other hand, generalizes a result of Schoen and Uhlenbeck in [SU1] (in which case $p = 2$ and $u(M)$ is contained in a strictly convex ball of N), and augments the following beautiful

THEOREM (Hardt and Lin [HL]). For $p > 1$, the Hausdorff dimension of the singular set of a p-minimizing map $u \varepsilon L_1^p(M^n, N^k)$ in the interior of M cannot exceed $n - [p] - 1$, in general. If $n = [p] + 1$, u has at most isolated singularities. If $n < [p] + 1$ (or if $[p] + 1 \leqq n$, off the singular set) u is locally Hölder continuous up to the boundary and the gradient of u is also locally Hölder continuous in the interior of M.

Generalizing (8) and (7) by the same variational method, we obtain

An average second variational formula on the domain of a p-harmonic map $u : M^n \to N^k$ for p-energy.

$$(21)\ \sum_{j=1}^{q} \frac{d^2}{dt^2} E_p(u_t^{du(V_j^T)})|_{t=0} = \int_M |du|^{p-4}\{(p-2)\sum_{\alpha=1}^{q-n}(\sum_{i=1}^{n} < \widetilde{A^{\nu_\alpha}(e_i)}, \tilde{e}_i >)^2$$
$$+|du|^2 \sum_{i=1}^{n} < \widetilde{Q_i^M(e_i)}, \tilde{e}_i >_N\} dv$$

where we assume that M is isometrically immersed in R^q, with an orthonormal frame $\{V_1, \cdots, V_q\}$

and

an average second variational formula on the target of a map $u : M^n \to N^k$ (not necessarily p-harmonic) for p-energy

$$(22)\ \sum_{j=1}^{q} \frac{d^2}{dt^2}|_{t=0} E_p(u_t^{V_j^T}) = \int_M |du|^{p-4}\{(p-2)|\sum_{i=1}^{n} h(\tilde{e}_i, \tilde{e}_i)|^2 +$$
$$|du|^2 \sum_{i=1}^{n} < Q^N(\tilde{e}_i),\ \tilde{e}_i >_N\} dv$$

where we assume that N is isometrically immersed in R^q with an orthonormal frame $\{V_1, \cdots, V_q\}$.

Thus, via the extrinsic, average variational method and a symmetrizing process, we find p-superstrongly unstable manifolds (for $p \geqq 2$) which fall into SSU manifolds when $p = 2$.

DEFINITION. A complete Riemannian manifold N is said to be p-superstrongly unstable (p-SSU) for $p \geqq 2$, if there exists an isometric immersion in R^q with the second fundamental form h such that the nonlinear functional $F_{p,y}$ defined on the space $\hat{T}_y(N)$ of unit tangent vectors to N at y given by

$$(23)\qquad F_{p,y}(X) = (p-2) < h(X,X), h(X,X) >_{R^q} + < Q_y^N(X), X >_N$$

is *negative* for each $y \varepsilon N$, where Q_y^N is as in (1). That is, there exists a positive function φ_p on N such that

$$(24)\qquad F_{p,y}(X) \leqq -\phi_p(y) \text{ for each } y\varepsilon N \text{ and each } X \varepsilon \hat{T}_y(N)$$

Analogous to SSU manifolds, for every p-SSU manifold N, $\mathrm{Ric}^N > 0$; $\dim N > p$; and $\pi_1(N) = \pi_2(N) = 0$. (Assuming N is compact). Similarly, every complete hypersurface is p-SSU if and only if its principal curvatures satisfy $0 < \lambda_1 \leqq \lambda_2 \leqq \cdots \leqq \lambda_k < \frac{1}{(p-1)}(\lambda_1 + \cdots + \lambda_{k-1})$. For instance, an elliptic paraboloid, the graph of $f(x) = x_1^2 + \cdots + x_k^2$ is p-SSU if and only if $p < k$. The standard sphere S^k is p-SSU if and only if $p < k$.

Expressing the above formulas (21) and (22) in terms of local coordinates and symmetrizing them, we acquire the following generalization of a theorem of Howard and Wei [HW] (in which $p = 2$):

THEOREM. (Wei and Yau [WY]) Every compact p-SSU manifold N is p'-SU for each $p' \leqq p$, (and hence $\pi_1(N) = \pi_2(N) = 0$). Furthermore, the homotopy class of any map into or from a compact p-SSU manifold (from or into an arbitrary compact Riemannian manifold) contains a map of arbitrarily small p'-energy for each $p' \leqq p$.

Proceeding as in [SWW3] and [SWW4] by the same variational method, we obtain the following:

p-Stability Inequality ($p \geqq 2$). Let u be a p-stable map on a complete manifold M into a p-SSU manifold. Then for each $\varphi \varepsilon C_0^\infty(M)$

$$\int_M (\phi_p(u)/(k+p-2))\varphi^2 |du|^p dv \leqq \int_M |\nabla\varphi|^2 du|^{p-2} dv$$

An L^p gradient estimate for a p-stable map into a compact subset N_0 of a p-SSU manifold N. For each $\varphi \varepsilon C_0^\infty(M)$, we have

$$\int_M \varphi^p |du|^p dv \leqq C \int_M |\nabla\varphi|^p dv$$

where C is a constant depending on ϕ_p, k, p and N_0.

We also find p-superstrongly unstable indices associated to p-SSU manifolds.

DEFINITION. The number w_p on a p-SSU manifold N, given by

$$w_p = \inf_{y \varepsilon N} \phi_p(y)/(k+p-2)\mathcal{S}(y) \tag{25}$$

is said to be a p-SSU index where $\mathcal{S}(y)$ is as in (9).

Just as w plays the dominating role in both stable harmonic and energy-minimizing maps into SSU manifolds, so does w_p in p-stable and p-minimizing maps into p-SSU manifolds. Moreover $0 \leqq w_p < 1$. (c.f. $(10\frac{1}{2})$)

Indeed, via the Gauss curvature equation, (23) implies that on a p-SSU manifold N

$$-(k+p-2)\mathcal{S}(y) < F_{p,y} < -\phi_p(y) \text{ at every } y \varepsilon N. \text{ (c.f. (10))}$$

EXAMPLE OF A p-SSU MANIFOLD AND ITS ASSOCIATED p-SSU INDEX. (c.f. [SWW7])

A k-ellipsoid E_a^k is a p-SSU Manifold for $0 < a < \frac{k-1}{p-1}$ and $p+1 \leq k$ and its p-SSU index

$$w_p \geq \begin{cases} \frac{a(k-p)}{k+p-2} & \text{if } 0 < a \leq 1 \\ \frac{k-p}{a(k+p-2)} & \text{if } 1 \leq a \leq \frac{k-p}{p-1} \\ \frac{k-1+(1-p)a}{k+p-2} & \text{if } \frac{k-p}{p-1} \leq a < \frac{k-1}{p-1} \end{cases}$$

Proceeding as in [SWW4], we have

L^q gradient estimates for p-stable maps into p-SSU manifolds. For each $q\varepsilon[p+2, p+2w_p n/(n-1)]$ and for each $\varphi\varepsilon C_0^\infty(M)$, we have

$$\int_M \varphi^q |du|^q dv \leqq C_1 \int_M K\varphi^q dv + C_2 \int_M |\nabla\varphi|^q dv$$

where K is as in (14) and C_1 and C_2 are constants depending on k, p, w_p and K.

As a consequence of these gradient estimates, we have

Liouville Theorem for p-stable Maps into p-SSU manifolds N. Every p-stable map $u : M \to N$ is constant provided (i) M has p-th volume growth i.e., volume $(B_r) = 0(r^p)$ a $r \to \infty$, and $u(M) \subset N_0$, a compact subset of N, or

(ii) $\text{Ricci}^M \geqq 0$, and volume $(B_r) = 0(r^q)$ as $r \to \infty$, where $q\varepsilon[p+2, p+2w_p n/(n-1)]$ and w_p is as in (25).

COROLLARY. (i) Every smooth p-stable map defined on a complete manifold M with p-th volume growth (e.g. R^n, $n \leqq p$) into S^k is constant for $p < k$.

(ii) Every smooth p-stable map on M with volume $(B_r) = 0(r^q)$ as $r \to \infty$ into S^k is constant provided $\text{Ric}^M \geqq 0$ and $q\varepsilon[p+2, p+2(k-p)n/((n-1)(k+p-2))]$ and $p < k$.

Now we set

$$\beta_p = \max\{(2k(p-1)(n-1)-1)/(2p-3), 2-\tfrac{p}{2}\} \tag{26}$$

and

$$w_{p_0} = F_1^{-1}F_2(\beta_p) \tag{27}$$

where $F_1(x) = x + \sqrt{x^2 + (p-1)x}$ and $F_2(x) = x + \sqrt{x^2 + (p-2)x}$.

Furthermore, we set

$$
d(w_p) = \begin{cases}
\begin{cases}
[p+4] \\
[p+3] \\
[p+2] \\
[p+1] \\
[p]
\end{cases}
& \text{if}
\begin{cases}
w_{p_0} \leq w_p & \text{and}
\begin{cases}
\frac{4}{p+3} & < w_p \\
\frac{9}{4(p+2)} & < w_p \leq \frac{4}{p+3} \\
\frac{1}{p+1} & < w_p \leq \frac{9}{4(p+2)} \\
\frac{1}{4p} & < w_p \leq \frac{1}{p+1} \\
& w_p \leq \frac{1}{4p}
\end{cases} \\[4em]
\begin{cases}
6 \\
5 \\
4 \\
3 \\
2
\end{cases}
& w_p < w_{p_0} & \text{and}
\begin{cases}
\frac{4}{5}\beta_p & < w_p \\
\frac{3}{4}\beta_p & < w_p \leq \frac{4}{5}\beta_p \\
\frac{2}{3}\beta_p & < w_p \leq \frac{3}{4}\beta_p \\
\frac{1}{2}\beta_p & < w_p \leq \frac{2}{3}\beta_p \\
& w_p \leq \frac{1}{2}\beta_p
\end{cases}
\end{cases}
\end{cases}
$$

and $\dim(S_p)$ denotes the Hausdorff dimension of the singular set in the interior of M of a p-minimizer u into a p-SSU manifold. Proceeding as in §4, we have

The Liouville Theorem for p-stable tangent maps. Every p-stable tangent map $\bar{u}$ defined on R^n into a p-SSU k-manifold N is constant for $n \leq d(w_p)$.

And hence, via [HL (4.5),p.572],

MAIN REGULARITY THEOREM (Wei and Yau [WY]). Let $u \varepsilon L_1^p(M,N)$ be p-minimizing such that $u(x)\varepsilon N_0$ a.e., for a compact subset N_0 of p-SSU manifold N. Then $\dim(S_{p'}) \leq n-d(w_p)-1$ for each $p' \leq p$. If $n = d(w_p) + 1, S_p$, consists of at most isolated singularities for each $p' \leq p$ and if $n < d(w_p)$, $S_{p'}$ is empty for $p' \leq p$. In general, every p'-minimizer in $l_1^{p'}(M,N)$ is locally Hölder continuous off $S_{p'}$ and its gradient is locally Hölder continuous off $S_{p'}$ and the boundary ∂M for $p' \leq p$.

The above theorem generalizes the work in [SWW4] in which $p = 2$, and reduces $\dim(S_p)$ from $n - [p] - 1$ for general Riemannian manifold ([HL]) to $n - [p] - 5$ for some p-SSU manifolds. The following corollary and Liouville theorem for p-minimizing maps, generalize the work in [GG] [GS] [SU2] and [SWW4] in which $p = 2$.

COROLLARY. If $u \varepsilon L_1^p(M, S^k)$ is p-minimizing, then

$$\dim(S_p) = \begin{cases} n-[p]-5 & \text{if } \frac{p(1+W_{p_0})-2W_{p_0}}{1-W_{p_0}} \leq k \text{ and } p+8 \leq k \\ n-[p]-4 & \text{if } \frac{p(1+W_{p_0})-2W_{p_0}}{1-W_{p_0}} \leq k \text{ and } \frac{4p^2+17p-18}{4p-1} < k \leq p+8 \\ n-[p]-3 & \text{if } \frac{p(1+W_{p_0})-2W_{p_0}}{1-W_{p_0}} \leq k \text{ and } \frac{p^2+2p-2}{p} < k \leq \frac{4p^2+17p-18}{4p-1} \\ n-[p]-2 & \text{if } \frac{p(1+W_{p_0})-2W_{p_0}}{1-W_{p_0}} \leq k \text{ and } \frac{4p^2+p-2}{4p-1} < k \leq \frac{p^2+2p-2}{p} \\ n-[p]-1 & \text{if } \frac{p(1+W_{p_0})-2W_{p_0}}{1-W_{p_0}} \leq k \text{ and } k \leq \frac{4p^2+p-2}{4p-1} \\ n-7 & \text{if } k < \frac{p(1+W_{p_0})-2W_{p_0}}{1-W_{p_0}} \text{ and } \frac{p+\frac{4}{5}(p-2)\beta_p}{1-\frac{4}{5}\beta_p} < k \\ n-6 & \text{if } k < \frac{p(1+W_{p_0})-2W_{p_0}}{1-W_{p_0}} \text{ and } \frac{p+\frac{3}{4}(p-2)\beta_p}{1-\frac{3}{4}\beta_p} < k \leq \frac{p+\frac{4}{5}(p-2)\beta_p}{1-\frac{4}{5}\beta_p} \\ n-5 & \text{if } k < \frac{p(1+W_{p_0})-2W_{p_0}}{1-W_{p_0}} \text{ and } \frac{p+\frac{2}{3}(p-2)\beta_p}{1-\frac{2}{3}\beta_p} < k \leq \frac{p+\frac{3}{4}(p-2)\beta_p}{1-\frac{3}{4}\beta_p} \\ n-4 & \text{if } k < \frac{p(1+W_{p_0})-2W_{p_0}}{1-W_{p_0}} \text{ and } \frac{p+\frac{1}{2}(p-2)\beta_p}{1-\frac{1}{2}\beta_p} < k \leq \frac{p+\frac{2}{3}(p-2)\beta_p}{1-\frac{2}{3}} \\ n-3 & \text{if } k < \frac{p(1+W_{p_0})-2W_{p_0}}{1-W_{p_0}} \text{ and } k \leq \frac{p+\frac{1}{2}(p-2)\beta_p}{1-\frac{1}{2}\beta_p} \end{cases}$$

where w_{p_0} is as in (27) and β_p as in (26).

PROOF OF COROLLARY. This follows from the above regularity theorem and $w_p = (k-2)/(k+p-2)$ on S^k for $k > p$.

A scaling technique similar to [SWW4], yields

Liouville theorem for p-minimizing maps. Every p-minimizing map u defined on R^n into a p-SSU manifold is constant for $n \leqq d(w_p)$.

It should be mentioned that regularity estimates for elliptic systems, in particular the Euler-Lagrange equation for p-energy, were first obtained by K. Uhlenbeck [U] for $2 \leqq p$ and later by Tolksdorf [T] for $1 < p$. The question of minimizing in appropriate homotopy classes has been studied by White [W2].

§7 FURTHER DIMENSION REDUCTION OF S_p. ([SWW7])

This section includes a simple, unified and refined estimate on $\dim(S_p)$. The essence of further dimension reduction of S_p in §6 lies in a delicate balance between two integral estimates: one on the Bochner formula for p-harmonic maps, and the other on a splitting in a p-stability inequality derived from the proposed method (c.f. [SWW7] lemma 1.2 and 1.3). Our technique is sufficiently generally to yield an estimate on $\dim(S_p)$ for p-minimizers into $\overline{S}_+^k$ which will be discussed in the next section. We begin with the following notation.

For each fixed number $0 < \sigma \leq 1$ and each k-SSU index w_p on a p-SSU manifold N, we set a positive integer

$$d(w_p, \sigma) = \begin{cases} 2 + [[p + 2\sqrt{p - \sigma}]] & \text{if } w_p \geq \frac{p+1-\sigma+2\sqrt{p-\sigma}}{p+1+2\sqrt{p-\sigma}} \\ 1 + [\frac{\sigma}{1-w_p}] & \text{otherwise} \end{cases}$$

and we have

THEOREM. S_p is empty for $n < d(w_p, \sigma) + 1$, is a discrete set for $n = d(w_p, \sigma) + 1$ and has $\dim(S_p) \leq n - d(w_p, \sigma) - 1$, in general.

COROLLARY. If $u \in L_1^p(M, S^k)$ is p-minimizing, then

$$\dim(S_p) \leq \begin{cases} n - 3 - [[p + 2\sqrt{p - \sigma}]] & \text{if } (2p^2 - \sigma p - 2(1-\sigma) + 4(p-1)\sqrt{p-\sigma})/\sigma \leq k \\ n - 2 - [\frac{\sigma(k+p-2)}{2p-2}] & \text{if } p < k < (2p^2 - \sigma p - 2(1-\sigma) + 4(p-1)\sqrt{p-\sigma})/\sigma \end{cases}$$
$$\text{for any constant} \quad 0 < \sigma \leq 1.$$

In particular,

$$\dim(S_p) \leq \begin{cases} n - 3 - [[p + 2\sqrt{p - 1}]] & \text{if } 2p^2 - p + 4(p-1)\sqrt{p-1} \leq k \\ n - 2 - [\frac{k+p-2}{2p-2}] & \text{if } p < k < 2p^2 - p + 4(p-1)\sqrt{p-1} \end{cases}$$

COROLLARY. Let $u \in L_1^p(M, E_a^k)$ be p-minimizing, where $0 < a < \frac{k-1}{p-1}$ and $p + 1 \leq k$, then

(i)

$$\dim(S_p) \leq \begin{cases} n - 2 - [\frac{\sigma(k+p-2)}{(1-a)k+(1+a)p-2}] & \text{if } 0 < a \leq 1, \\ & k < \frac{(1+a)p^2+(a-1-\sigma)p+2(\sigma-1)+(2+2a)p\sqrt{p-\sigma}-4\sqrt{p-\sigma}}{(a-1)p+(a-1+\sigma) \ +2(a-1)\sqrt{p-\sigma}} \\ n - 2 - [\frac{a\sigma(k+p-2)}{(a-1)k+(a+1)p-2a}] & \text{if } 1 \leq a \leq \frac{k-p}{p-1}, \\ & k < \frac{(1+a)p^2+(1-a-a\sigma)p+2a(\sigma-1)+(2+2a)p\sqrt{p-\sigma}-4a\sqrt{p-\sigma}}{(1-a)p+(1-a+a\sigma)+(2-2a)\sqrt{p-\sigma}} \\ n - 2 - [\frac{\sigma(k+p-2)}{(a+1)p-(a+1)}] & \text{if } \frac{k-p}{p-1} \leq a < \frac{k-1}{p-1}, \\ & k < ((a+1)p^2 - \sigma p - (1 + a - 2\sigma) + 2(a+1)(p-1)\sqrt{p-\sigma})/\sigma \end{cases}$$

and

(ii)

$$\dim(S_p) \leq n - 3 - [[p + 2\sqrt{p - \sigma}]] \qquad \text{otherwise}$$

LIOUVILLE THEOREM. Let u be a p-minimizing map from a simple Riemannian manifold M^n into a p-SSU manifold with a p-SSU-index w_p.

Then u is constant provided

$$(i) \quad n \leq 2 + [[p + 2\sqrt{p - \sigma}]] \text{ and } w_p \geq \frac{p + 1 - \sigma + 2\sqrt{p - \sigma}}{p + 1 + 2\sqrt{p - \sigma}}$$
$$\text{or } (ii) \quad n \leq 1 + [\frac{\sigma}{1 - w_p}] \text{ and } w_p < \frac{p + 1 - \sigma + 2\sqrt{p - \sigma}}{p + 1 + 2\sqrt{p - \sigma}}$$

for any fixed constant $\quad 0 < \sigma \le 1.$

COROLLARY. Every p-minimizer u from a simple Riemannian manifold M^n into S^k, $p < k$ is constant provided (i) $n \le 2 + [[p + 2\sqrt{p-\sigma}]]$ and

$$(2p^2 - \sigma p - 2(1-\sigma) + 4(p-1)\sqrt{p-\sigma})/\sigma \le k$$

or (ii) $n \le [\frac{\sigma(k+3p-4)}{2p-2}]$ and

$$k < (2p^2 - \sigma p - 2(p-1) + 4(p-1)\sqrt{p-\sigma})/\sigma$$

COROLLARY. Every p-minimizing u from a simple Riemannian manifold M into E^k_a is constant provided

(i) $0 < a \le 1$ and

$$n \le \begin{cases} 2 + [[p + 2\sqrt{p-\sigma}]] & \text{if } \frac{(1+a)p^2 + (a-1-\sigma)p + 2(\sigma-1) + (2+2a)p\sqrt{p-\sigma} - 4\sqrt{p-\sigma}}{(a-1)p + (a-1+\sigma) + 2(a-1)\sqrt{p-\sigma})} \le k \\ 1 + [\frac{\sigma(k+p-2)}{(1-a)k + (1+a)p - 2}] & \text{otherwise} \end{cases}$$

(ii) $1 \le a \le \frac{k-p}{p-1}$ and

$$n \le \begin{cases} 2 + [[p + 2\sqrt{p-\sigma}]] & \text{if } \frac{(1+a)p^2 + (1-a-a\sigma)p + 2a(\sigma-1) + (2+2a)p\sqrt{p-\sigma} - 4a\sqrt{p-\sigma}}{(1-a)p + (1-a+a\sigma) + 2(1-a)\sqrt{p-\sigma}} \le k \\ 1 + [\frac{a\sigma(k+p-2)}{(a-1)k + (a+1)p - 2a}] & \text{otherwise} \end{cases}$$

(iii) $\frac{k-p}{p-1} \le a < \frac{k-1}{p-1}$ and

$$n \le \begin{cases} 2 + [[p + 2\sqrt{p-\sigma}]] & \text{if } ((1+a)p^2 - \sigma p - (1+a-2\sigma) + (2+2a)(p-1)\sqrt{p-\sigma})/\sigma \le k \\ 1 + [\frac{\sigma(k+p-2)}{(1+a)p - (1+a)}] & \text{otherwise} \end{cases}$$

§8. HARMONIC, p-HARMONIC, p-STABLE, p-MINIMIZING AND p-MINIMIZING TANGENT MAPS INTO CLOSED HEMISPHERES.

Applying the Bochner technique, Cheng and the author show:

THEOREM ([CW]). *Every harmonic map* u on a complete n-manifold into a closed k-hemisphere $\overline{S}^k_+$ either lies on the equator or is a constant provided

(i) M is parabolic

or (ii) $\mathrm{Ric}^M \geqq 0$, volume $(B_r) = 0(r^{4+\frac{2}{n-1}})$ as $r \to \infty$.

This theorem is closely related to the work of Choi [CHO], Giaquinta and Souĉek [GS], Karp [K], Schoen and Uhlenbeck [SU2] and Yu [YU]. As an application, Cheng and Wei, via Ruh and Vilm's theorem (on the Gauss map) [RV], gives a simple new proof of

THE CLASSICAL BERNSTEIN THEOREM (c.f. [CW]). Every entire solution to the minimal surface equation, $\operatorname{div}(\nabla u/\sqrt{1+|\nabla u|^2}) = 0$, or to the constant mean curvature equation, $\operatorname{div}(\nabla u/\sqrt{1+|\nabla u|^2}) = nc$, on R^2 is linear.

The following theorems are variations on the idea in [CW]. Assertion (i) is due to Yau and the author:

THEOREM. Every smooth *p-harmonic map* ($p \geqq 2$) on a complete n-manifold into $\overline{S}^k_+$ either lies on the equator or is a constant, provided (i) volume $(B_r) = 0(r^p)$ as $r \to \infty$ or (ii) $\mathrm{Ric}^M \geq 0$, volume $(B_r) = 0(r^q)$ as $r \to \infty$, where $q = p + \frac{2n}{n-1}$.

THEOREM. Every smooth *p-stable map* from a complete n-manifold into $\overline{S}^k_+$ is constant provided (i) M has *p-th* volume growth or (ii) M has nonnegative Ricci curvature tensor and *q-th* volume growth where $q = p + 2n/(n-1)$.

REMARK. The *p-stability* assumption is necessary, as for $p = 2$ one can construct non-constant, unstable harmonic maps from minimal surfaces founded by Lawson [LB] or by Pitts and Rubinstein [PR] in S^3, into $\overline{S}^4_+$ in which S^3 is embedded as a totally geodesic. The same construction indicates that each conclusion in the previous theorem on p-harmonic maps into $\overline{S}^k_+$ does occur.

Similarly one can show for $p + 1 \leq k$ and $0 < a < \frac{k-1}{p-1}$,

COROLLARY. Every smooth p-stable map $u : M \to E^k_a$, is constant if (i) M has *p-th* volume growth or (ii) $\mathrm{Ric}^M \geq 0$ and M has *q-th* volume growth

$$\text{where} \quad q = \begin{cases} p + \frac{a(k-p)}{k+p-2}\frac{2n}{n-1} & \text{if } 0 < a \leq 1 \\ p + \frac{k-p}{a(k+p-2)}\frac{2n}{n-1} & \text{if } 1 < a \leq \frac{k-p}{p-1} \\ p + \frac{k-1+(1-p)a}{k+p-2}\frac{2n}{n-1} & \text{if } \frac{k-p}{p-1} < a < \frac{k-1}{p-1} \end{cases}$$

As discussed in §7 the author has shown

LIOUVILLE THEOREM FOR P-STABLE (RESP. P-MINIMIZING) TANGENT MAPS. *Every p-stable (resp. p-minimizing) tangent map* from R^ℓ into $\overline{S}^k_+$ is constant for $\ell \leq 2 + [p + 2\sqrt{p}]$.

REGULARITY THEOREM [SWW7]. If $u \varepsilon L^p_1(M^n, \overline{S}^k_+)$ is p-minimizing, then u is locally Hölder continuous up to boundary and the gradient of u is also locally Hölder continuous in the interior of M for $n < 3 + [p + 2\sqrt{p}]$, has at most isolated singularities for $n = 3 + [p + 2\sqrt{p}]$ and has a closed singular set of Hausdorff dimensions at most $n - 3 - [p + 2\sqrt{p}]$.

REMARK. If $u \varepsilon L^p_1(M, S^k_+)$ is minimizing, then u is smooth by a theorem of Wei and Yau in §6. Similarly, every p-minimizing L^p_1 map into a complete noncompact manifold with positive section curvature is smooth, due to a theorem of Greene and Wu [GW].

As an application, via a scaling inequality, Wei has proved

THEOREM [SWW7]. Every smooth p-minimizing map u from R^n or a simple Riemannian manifold into $\overline{S}^k_+$ is constant for $n \leqq 2 + [p + 2\sqrt{p}]$.

This generalizes a result of Giaquinta and Giusti [GG], Giaquinta and Souĉek [GS] and Schoen and Uhlenbeck [SU2] in which they treat the case $p = 2$. Their theorem is sharp because of the equator map.

References

[BS] S. Bochner, Vector fields and Ricci curvature, Bull. Amer. Math. Soc., 53 (1947).

[CH] S.Y. Cheng, Liouville theorem for harmonic maps, Proc. Symp. Pure Math, 36 (1980).

[CW] S.Y. Cheng and S.W. Wei, Liouville theorems and Bernstein problems.

[CY] S.Y. Cheng and S.T. Yau, Differential equations on Riemannian manifold and their geometric applications. Comm. Pure Appl. Math. 28, (1975).

[CHO] H. Choi, On the Liouville theorem for harmonic maps, Proc. of A.M.S., 85 No. 1 (1982).

[D] S.K. Donaldson, An application of Gauge theory to the topology of 4-manifolds, J. Diff. Geom. 18(1983).

[DE] E. De Giorgi, Frontiere orientate di misura minima, Sem. Mot. Scuola Normi Sup. Pisa (1961).

[EL] J. Eells and L. Lemaire, Selected Topics in Harmonic Maps, CBMS Regional Conf. Series Number 50.

[ES] J. Eells and J.H. Sampson, Harmonic mappings of Riemannian manifolds, Amer. J. Math. 86 (1964).

[F] H. Federer, The singular sets of area minimizing rectifiable currents with codimension one and of area minimizing flat chains modulo two with arbitrary codimension, Bull A.M.S. 76 (1970).

[FW] W.H. Fleming, On the oriented Plateau problem, Rediconti Circolo Mat., Palerno, Vol. II (1962).

[GG] M. Giaquinta and E. Giusti, The singular set of the minimal of quadratic functionals, Ann. Scuola Norm. Sup. Pisa Ser. IV, Vol 11 (1984).

[GKR] K. Grove, H. Karcher and E.A. Ruh, Group actions and curvature, Inv. Math. 23 (1974).

[GS] M. Giaquinta and J. Souĉek, Harmonic maps into hemisphere, preprint.

[GW] R.E. Greene and H. Wu, Function Theory on manifolds which possess a pole, Lecture Notes in Math, No. 699, springer-Verlag, Berlin-Heidelberg-New York, 1979.

[HS] S. Hilderbrandt, Harmonic mappings of Riemannian manifolds, Spring Lecture Notes 1161, 1985.

[HKW] S. Hilderbrandt, H. Kaul and K.O. Widman, An existence theorem for harmonic mappings of Riemannian manifolds, Acta Math. 138 (1977).

[HL] R. Hardt and F.H. Lin, Mapping minimizing the L^p norm of the gradient, Comm. Pure and Applied Math. Vol. XL (1987).

[H] R. Howard, The non-existence of stable submanifolds, varifolds, and harmonic maps in sufficiently pinched simply connected Riemannian manifolds, Michigan J. 32 (1985).

[HW1] R. Howard and S.W. Wei, Non-existence of stable harmonic maps to and from certain homogeneous spaces and submanifolds of Euclidean space, Trans. A.M.S. 294 No. 1 (1986).

[HW2] R. Howard and S.W. Wei, On the existence and non-existence of stable submanifolds and currents in positively curved manifolds and the topology of submanifolds, preprint.

[HJ] J.E. Humphreys, Introduction to Lie Algebras and Representation Theory, Springer, New York, (1972).

[K] L. Karp, Subharmonic functions on real and complex manifolds, Math. Z 179, (1982).

[KOT] S. Kobayashi, Y. Ohnita, and M. Takeuchi, On instability of Yang-Mills Connections, Math. Z. 193 (1986).

[KW] L. Karp and S.W. Wei, Liouville Theorems on Harmonic Maps.

[LB] H.B. Lawson, *Complete minimal surfaces in* S^3, Ann. of Math. (2) 92 (1970).

[LS] H.B. Lawson and J. Simons, On stable currents and their applications to global problems in real and complex geometry, Ann. of Math. 98 (1973).

[L] P.F. Leung: On the stability of harmonic maps, Lecture Notes in Math, No. 949, Springer-Verlag, Berlin-Heidelberg-New York (1982).

[MO] Min-Oo, *Maps of minimum energy from compact simply-connected Lie groups*, Annals of Global Analysis and Geometry 2 No. 1 (1984).

[M] C.B. Morrey, Jr., *The problem of Plateau in a Riemannian manifold.* Ann. of Math. 49, (1948).

[MJ] J. Moser, A new proof of de Girogi's theorem concerning the regularity problem for elliptic differential equations. Comm. Pure and Applied Math. Vol. (1960).

[MS] S.B. Myers, *Riemann manifolds with positive mean curvature*, Duke Math J., 8 (1941).

[N] T. Nagano, *Stability of harmonic maps between symmetric spaces*, Lecture Notes in Math, No. 949, Berlin-Heidelberg-New York (1982).

[NH] H. Nakajima, *Regularity of minimizing harmonic maps into certain Riemannian manifolds*, preprint.

[O1] Y. Ohnita, Stable minimal submanifolds in compact rank one symmetric spaces, Tohoku Math. J. 39 (1986).

[O2] Y. Ohnita, Stability of Harmonic maps and standard minimal immersions, Tohoku Math. J. 38 (1986).

[OT] T. Okayasu, *Pinching and nonexistence of stable harmonic maps*, (preprint).

[PR] J. Pitts and H. Rubinstein, *Equivariant minimax and minimal surfaces in geomtri 3-manifolds*, Bull. Amer. Math. Soc. (N.S.) 19 (1988).

[RV] E.A. Ruh and J. Vilms, The tension field of the Gauss map, Trans. A.M.S. 149 (1970).

[SU] J. Sacks and K. Uhlenbeck, *The existence of minimal immersions of 2-spheres*, Ann. of Math 113 (1981).

[SSY] R. Schoen, L. Simon and S.T. Yau, *Curvature estimates for minimal hypersurfaces*, Acta Math, 134 (1975).

[SU1] R. Schoen and K. Uhlenbeck, A regularity theory for harmonic maps, J. Diff. Geom. 17, (1982).

[SU2] ______, Regularity of minimizing harmonic maps into the sphere, Invent. Math. 78 (1984).

[SJ] J. Simons, Minimal varieties in Riemannian manifolds, Ann. of Math. 88 (1968).

[S] R.T. Smith, The second variation formula for harmonic mappings, Proc. of AMS 47 (1975).

[SY] J.L. Synge, On the connectivity of spaces of positive curvature, Quart. J. Math. Oxford Ser. 7 (1936).

[TP] P. Tolksdorff, Everywhere regularity for some quasi-linear systems with lack of ellipticity, Ann. Mat. Pura Appl. 134 (1983).

[TT] T. Takahashi, *Minimal immersions of Riemannian manifolds*, J. Math Soc. Japan 18 (1966).

[TW] A. Treibergs and S.W. Wei, Embedded hyperspheres with prescribed mean curvature, J. Differential Geometry 18 (1983).

[SWW1] S.W. Wei, Generalized idea of Synge and its applications to topology and calculus of variations in positively curved manifolds, Proc. Symp. Pure Math. 45 (1986).

[SWW2] ________, An average process in the calculus of variations and the stability of harmonic maps, Bulletin of the Institute of Math Academia Sinica, 11 (1983).

[SWW3] ________, Liouville theorems for stable harmonic maps into either strongly unstable, or δ-pinched, manifolds, Proc. Symp. Pure Math, 44, (1986).

[SWW4] ________, Liouville theorem and regularity of minimizing harmonic maps into super-strongly unstable manifolds, preprint.

[SWW5] ________, On topological vanishing theorems and the stability of Yang-Mills fields, Indiana Univ. Math. J., 33 No. 4 (1984).

[SWW6] ________, The regularity of minimizers, nonlinear analysis and applications, edited by V. Lakshmikantham, lecture notes in pure and applied mathematics, Vol. 109, Marcel Dekker, Inc. (1987).

[SWW7] ________, The minima of the p-energy functional, preprint.

[W1] B. White, Homotopy classes in Sobolev spaces and energy minimizing maps, Bull A.M.S. 13, No. 2, 1985.

[W2] ________, Infima of energy functionals in homotopy classes of mapping, J. Diff. Geometry 23, (1986).

[WY] S.W. Wei and C.M. Yau, Regularity of p-energy minimizing maps and p-superstrongly unstable indices, preprint.

[X] Y.L. Xin, *Some results on stable harmonic maps.* Duke Math. J. 47 (1980).

[YU] Q.H. Yu, Bounded harmonic maps, Acta Math. Sinica 1 (1985).

Department of Mathematics
University of Oklahoma
Norman, OK 73019, USA

Contemporary Mathematics
Volume **101**, 1989

The Dirichlet Problem and Fatou's Theorem for Harmonic Mappings on Regular Domains

Patricio Aviles*

§1. Introduction.

In this note I shall briefly discuss two classical topics of harmonic analysis; the solvability of the Dirichlet problem in regular domains and the Fatou's theorem in the case of harmonic maps. Since every bounded harmonic function can be considered as a harmonic map into $\mathbb{R}$ our analysis below can be considered as a natural and important extension of part of the classical function theory to harmonic maps. There are several new difficulties that we have overcome in the process of doing harmonic analysis for harmonic mappings which can be briefly explained as follows. In the situation at hand we are dealing with a nonlinear elliptic system as opposed to a single equation which furthermore is a degenerated non-linear elliptic system when we are in a complete negatively curved manifold. There are also new difficulties because the harmonic maps may have infinite energy, which, in general, is a difficult problem to overcome to do analysis when dealing with solutions of non-linear elliptic equations or systems. Roughly speaking, this is because the solutions become singular. In our situation this is especially the case in the Fatou theorem.

1980 *Mathematics Subject Classification* (1985 Revision). 58E20.

*"This paper is in final form and no version of it will be submitted for publication elsewhere."

§ 2. Some Definitions and Statements of Results.

We recall the definition of regular domain

Definition. A bounded domain Ω in a complete Riemannian manifold M, is said to be a regular domain for the Laplace–Beltrami operator if at every point ξ of $\partial\Omega$, there exists $\omega_\xi \in C^0(\overline{\Omega})$ so that

(i) ω_ξ is super–harmonic in Ω;

(ii) $\omega_\xi > 0$ in $\overline{\Omega} - \xi$, $\omega_\xi(\xi) = 0$.

It is well known and easy to prove that the classical Dirichlet problem in Ω is solvable for the Laplace–Beltrami operator and for arbitrary continuous boundary values if and only if Ω is a regular domain (c.f. [G–T] Chapter 2).

A more general class of regular domains is obtained by introducing the Martin compactification. We give a brief discussion of the Martin boundary $\mathcal{M}(\Omega)$ of a general domain, that is an open connected set, Ω. We shall assume that Ω is a Greenian domain that is the Green function with pole at $x \in \Omega$ exists for all $x \in \Omega$, i.e. $\Delta_g G = \delta(x)$, $G|_{\partial\Omega} = 0$. We fix an origin $0 \in G$ and for $x,y \in \Omega$ we let

$$h_y(x) = h(y,x) = \frac{G(y,x)}{G(y,0)}$$

denote the normalized Green's function at 0 with pole at y. Let $\{y_i\}$ be a sequence in Ω having no limit points in Ω. The corresponding sequence $\{h_{y_i}\}$ of harmonic functions is uniformly bounded on compact sets by the Harnack inequality. The sequence $\{y_i\}$ is called fundamental if $\{h_{y_i}\}$ converges to a harmonic function h_y on Ω. By Harnack's principle, any sequence $\{y_i\}$ having no limit points in Ω has a fundamental subsequence. Two fundamental sequences are equivalent if the associated limit harmonic functions are equal. An element of the Martin boundary $\mathcal{M}(\Omega)$ is defined to be an equivalence class of fundamental sequences. If $[Y] \in \mathcal{M}(\Omega)$, then

$$h_Y(x) = \lim_{i\to\infty} h_{y_i}(x)$$

where $\{y_i\}$ is a fundamental sequence associated to $[Y]$; thus points $[Y] \in \mathcal{M}(\Omega)$ correspond uniquely to certain positive harmonic functions h_Y on Ω. Let $\hat{\Omega} = \Omega \cup \mathcal{M}(\Omega)$. We define a metric topology on $\hat{\Omega}$ as follows: If $Y, Y' \in \hat{\Omega}$ then

$$\rho(Y,Y') = \sup_{x\in B_1(0)} |h_Y(x) - h_{Y'}(x)|$$

$B_1(0) \subset \Omega$. Then $\hat{\Omega}$ is complete and compact with respect to the topology given by ρ and furthermore the boundary of $\hat{\Omega}$ is $\mathcal{M}(\Omega)$ and the ρ-topology of Ω agrees with its topology as a Riemannian manifold (it is well known that $\mathcal{M}(\Omega)$ is independent of the choice of the origin. We refer to [Db] for more details.)

In Theorem A below we shall understand by a regular domain a domain which is Greenian and every point Y of $\mathcal{M}(\Omega)$ is a regular point, i.e. there exists $\omega_Y \in C^0(\hat{\Omega})$ satisfying (i), (ii). Such domains include for example, bounded Lipschitz domains [H-W], [Da, 1] the non-tangentially accessible domains introduced by Jerison and Kenig [J-K], complete simply connected negatively curved manifolds [Ad], [S], [Ad-Sc] and the complete manifolds studied by Ancona [An].

The image of the mapping in consideration will be required to be in a convex ball $\bar{B}_\tau(p)$ of a complete C^∞ Riemannian manifold N. Consequently, $\bar{B}_\tau(p)$ shall denote the closed geodesic ball of radius τ and center at p in N with

$$\tau < \min\{\frac{\pi}{2\sqrt{K}}, \text{ injectivity radius of } N \text{ at } p\}$$

where $K \geq 0$ is an upper bound for the sectional curvatures of N. We remark that, in general, there are examples that show that the theorems below do not hold without the hypothesis that the image of the mapping lies in a ball as described above (see for example [Hi-K-W] for one of such examples.)

We also recall that for $f \in C^1(M,N)$, the energy density $e(f)$ of f is defined by

$$e(f)(x) = \frac{1}{2} g^{ij}(x) h_{\alpha\beta}(f(x)) \frac{\partial f^\alpha}{\partial x^i} \frac{\partial f^\beta}{\partial x^j}$$

where $x = (x^1,\dots,x^m)$, $Y = (Y^1,\dots,Y^n)$ are local co-ordinates on M and N respectively, $f^\alpha(x) = Y^\alpha(f(x))$, g_{ij}, $dx^i dx^j$ and $h_{\alpha\beta} dY^\alpha dY^\beta$ define the

Riemannian metrics on M and N respectively and the matrix (g^{ij}) is the inverse of (g_{ij}). The energy $E(f)$ of f is defined by $E(f) = \int_M e(f)dV$ where dV = volume element of M. The map $u \in C^1(M,N)$ is said <u>to be harmonic</u>, if it is a critical point of the functional $E : C^1(M,N) \to \mathbb{R}$ with respect to compactly supported variations. A simple calculation shows that the Euler–Lagrange system of equations have to be satisfied by a harmonic map are

$$\Delta u^\alpha(x) + g^{ij}(x)\, \Gamma^\alpha_{\beta\gamma}(u(x))\, \frac{\partial u^\beta}{\partial x^i} \frac{\partial u^\gamma}{\partial x^j} = 0$$

where Δ is the Laplace–Beltrami operator on M and $\Gamma^\alpha_{\beta\gamma}$ are the Christoffel symbols of the metric on N.

Before stating the results I would like to point out that complete proofs of the theorems stated in this note can be found in the joint paper, Aviles–Choi–Micallef [A–C–M].

<u>Theorem</u> A. (Dirichlet Problem). Let Ω be a regular domain for the Dirichlet problem for the Laplace–Beltrami operator Δ. For each $\phi \in C^0(\mathcal{M}(\Omega), \bar{B}_\tau(p))$ there exists a unique $u \in C^0(\hat{\Omega}, \bar{B}_\tau(p)) \cap C^\infty(\Omega, \bar{B}_\tau(p))$ which is a harmonic map on Ω and which equals ϕ on $\mathcal{M}(\Omega)$.

<u>Remarks</u>. (i) In the above theorem we have assumed that the metric in the domain is C^∞ in order to get the C^∞ interior regularity of u. If not then the interior regularity depends on the smoothness of such a metric.

(ii) In the case that Ω is a bounded domain with a smooth boundary, say of class C^2, Theorem A was proved by Hildebrandt–Kaul and Widman [Hi–K–W]. Our methods of proofs in [A–C–M] are very different and mostly unrelated to those in [Hi–K–W].

A classical problem in linear harmonic analysis is to study the boundary regularity of harmonic functions in an arbitrary bounded domain. For the convenience of the reader, we recall the notion of capacity and the Wiener criterion. (We shall assume that the dimension of the domains is greater or equal to three.)

Capacity. Let R be a compact Riemannian manifold with smooth non-empty boundary. We shall let $C^1(R)$ denote the space of functions f for which df exists in the interior of R and extends continuously to all of R. $C_0^1(R)$ will denote those functions in $C^1(R)$ and which vanish on ∂R.

Let K be a compact subset of R such that $\partial K \cap \partial R = \emptyset$. The capacity of K relative to R, $\mathrm{Cap}(K)$, is defined by

$$\mathrm{Cap}(K) = \inf\{\int_{R/K} |df|^2 dv \mid f \in C_0^1(R),\ f = 1 \text{ on } K\}.$$

Wiener condition. Let Ω be a bounded domain which is contained in the interior of a complete Riemannian manifold M. Given a point $x_0 \in \partial\Omega$, let $X(\sigma) = \sigma^{2-n}\,\mathrm{Cap}\{x \in B_\sigma(x_0) \mid x \notin \Omega\}$: the capacity is measured with respect to a fixed smooth compact subset R of M which contains $B_\sigma(x_0)$ in its interior. x_0 is said to satisfy the Wiener condition if $\sum_{j=1}^{\infty} X(\sigma^j)$ diverges for $\sigma \in (0,1)$.

Definition. Let Ω be a bounded domain which is contained in the interior of a complete Riemannian manifold M. We say that $x_0 \in \partial\Omega$ is a regular point for the Laplace-Beltrami operator Δ, if for every $\phi \in L^{\infty}(\partial\Omega,\mathbb{R})$ which is continuous at x_0, the solution of the Dirichlet problem for Δ with boundary values ϕ is continuous at x_0. The following result is well known.

Theorem. (Wiener criterion). x_0 is regular for the Laplace-Beltrami operator Δ iff x_0 satisfies the Wiener condition.

Wiener [Wn] proved the above theorem only for the Euclidean-Laplace operator. Various authors later showed that a boundary point is regular for a large class of elliptic operators (which included the Laplace-Beltrami operator) if and only if it is regular for the Euclidean-Laplace operator (see [L-S-W]).

We now make the following

Definition. Let Ω be a bounded domain which is contained in the interior of a complete Riemannian manifold M. We say that $x_0 \in \partial\Omega$ is regular for the harmonic map system if for all $\phi \in L^{\infty}(\partial\Omega; \overline{B}_{\tau}(p))$ which is continuous at x_0, the solution of the Dirichlet problem for the harmonic map system with boundary values ϕ is continuous at x_0.

We then have

<u>Theorem</u> B. (Wiener criterion). x_0 is a regular point for the harmonic map system iff x_0 satisfies the Wiener condition.

We remark that using totally different methods and ideas a version of the Wiener text for harmonic mappings of finite energy was independently established by Paulik (see [P]).

It is of interest to study the solution of the Dirichlet problem of the harmonic map system with further details. We shall next state a result in which we discuss sharp bounds for such solutions in the important case of simply connected negatively curved manifolds.

If M denote a complete, simply connected Riemannian manifold of dimension m, whose sectional curvature K_M satisfies $-b^2 \leq K_M \leq -a^2 < 0$, the sphere at infinity $S(\infty)$ of M is defined to be the set of asymptotic classes of geodesic rays in M: two rays γ_1 and γ_2 are asymptotic if $\mathrm{dist}(\gamma_1(t), \gamma_2(t))$ is bounded for $t \geq 0$. The cone topology on $\overline{M} = M \cup S(\infty)$ is defined by: let q be a fixed point in M and let γ be a geodesic ray passing through q with tangent vector v at q. The cone $C_q(\gamma, \theta)$ of angle θ about γ is defined by $C_q(\gamma,\theta) = \{x \in M \mid$ angle between v and tangent vector at q of geodesic joining q to x is less than $\theta\}$. Let $T_q(\gamma,\theta,R) = C_q(\gamma,\theta) \mid \overline{B}_R(q)$ denote a truncated cone, then the domains $T_q(\gamma,\theta,R)$ together with the open geodesic balls $B_\delta(x)$, $x \in M$ form a local basis for the cone topology. Let $\xi : [0,1] \to [0,\infty]$ be a fixed homeomorphism which is a diffeomorphism on $[0,1)$. The map $E_\xi(v) = \exp_q(\xi(|v|)v)$ is a diffeomorphism of the open unit ball $B_1(0)$ in $T_q(M)$ onto $\mathbb{M}$; moreover E_ξ extends to a homeomorphism of the sphere $S_1 = \partial B_1(0)$ onto $S(\infty)$. We identify $\overline{M}, M$ and $S(\infty)$ with $\overline{B_1(0)}$, $B_1(0)$ and S_1 respectively. For example, ϕ defined on $S(\infty)$ is C^α, $\alpha \in (0,1]$ it means with respect to the standard smooth structure on S_1.

It was proved by Anderson [Ad], Sullivan [S] and later by Schoen ([Ad–Sc] pp. 435–438]) that every point of $S(\infty)$ is a regular point for the Dirichlet problem with respect to the Laplace–Beltrami operator on M. Furthermore, Anderson and Schoen showed that the Martin boundary of M is homeomorphic to $S(\infty)$. More precisely, they showed that there is a natural map $\Phi : \mathcal{M}(M) \to S(\infty)$ which is a C^α–homeomorphism and $\Phi^{-1} : S(\infty) \to \mathcal{M}(M)$ is also C^α relative to the Martin metric on $\mathcal{M}(M)$, where α depends only on n, a, b (see Theorem 6.3 in [Ad–Sc]). Hence, it follows from Theorem A that for each $\phi \in C^0(S(\infty), \overline{B}_\tau(p))$ there exists a unique harmonic map $u \in C^0(\overline{M}, \overline{B}_\tau(p)) \cap C^\infty(M, \overline{B}_\tau(p))$ which equals ϕ on $S(\infty)$. Furthermore we have

<u>Theorem</u> C. (Bounds for the solution of the Dirichlet problem). Let M be a complete, simply connected Riemannian manifold of dimension m, whose sectional curvatures K_M satisfy $-b^2 \leq K_M \leq -a^2 < 0$ and let $\overline{B}_\tau(p)$ as described before. Given $\phi \in C^\alpha(S(\infty), \overline{B}_\tau(p))$, $\alpha \in (0,1]$, the harmonic map $u : M \to \overline{B}_\tau(p)$, $u|_{S(\infty)} = \phi$ satisfies the following decay estimates:

(i) $\rho(u(x), \phi(x)) \leq c_1 e^{-1/2\ \delta r(x)}$

(ii) for any $\beta \in [0,1)$, $|Du(x)|_{C^{0,\beta}} \leq c_2 e^{-1/2\ \delta r(x)}$

where $r(x)$ = distance of x from some fixed point $q \in M$, $\delta > 0$ is

$$\delta = \begin{cases} \alpha a & \text{if } \alpha < 1 \text{ or } \alpha = 1 \text{ and } m \geq 3 \\ \text{any positive number strictly less than } a & \text{if } \alpha = 1 \text{ and } m = 2, \end{cases}$$

c_1, c_2 depend on the geometry of $\bar{B}_\tau(p)$, β, a, b, ϕ, m except u itself. Note that in (i) ϕ denotes the radial extension of ϕ (see [Ad-Sc]). If only $\phi \in C^0(S(\infty), \bar{B}_\tau(p))$ then the energy density $e(u)(x) \to 0$ as $x \to S(\infty)$.

Remark. It is not possible to obtain decay on higher derivatives of u without assuming global bounds on the derivatives of the curvature of M.

I shall finally discuss the Fatou's theorem for harmonic mappings. This is a very interesting result. Technically speaking, it is interesting because the basic tools that there are used to prove it for bounded harmonic functions, that is, boundary representation, the correct inequality between the non-tangential maximal function and the Hardy-Littlewood maximal function associated to the harmonic function (see for instance the book of Stein [St]) are not available, in general, for harmonic mappings due to the non-linear structure. Geometrically it is interesting because if we combine it with the solution of the Dirichlet problem for L^∞-data we obtain the following: There is a one to one correspondence between $\phi \in L^\infty(\partial\Omega, \bar{B}_\tau(p))$ and harmonic maps $u \in L^\infty(\bar{\Omega}, \bar{B}_\tau(p)) \cap C^\infty(\Omega, \bar{B}_\tau(p))$ so that

$$\lim_{x \to Q} u(x) = \phi(Q),$$

almost everywhere $Q \in \partial\Omega$ with respect to the natural measure associated to $\partial\Omega$ (harmonic measure) and where $x \to Q$ non-tangentially.

Theorem D. (Fatou's theorem). Let $u : \Omega \to \bar{B}_\tau(p)$ be a harmonic map where $\bar{B}_\tau(p)$ is as defined above and Ω is either (i) a bounded Lipschitz domain contains in the interior of a complete Riemannian manifold or (ii) a complete, simply connected Riemannian manifold where sectional curvatures K_Ω satisfy $-b^2 \le K_\Omega \le -a^2 < 0$. Then u has the Fatou property, that is, for almost every $Q \in \partial\Omega$ with respect to the harmonic measure on $\partial\Omega$, $\lim_{x\to Q} u(x)$ exists whenever $x \to Q$ non-tangentially.

We observe that, if Ω is viewed as a finite ball when Ω satisfies condition (ii) in Theorem D, then the system of equations that u satisfies is not uniformly elliptic. In the context of non-uniformly elliptic equations we recall that the Fatou property for bounded solutions of the so-called p-Laplacian, $\operatorname{div}(|\nabla u|^{p-2}\nabla u) = 0$, $p > 2$, in bounded Lipschitz domains with respect to a suitable measure on the boundary has been proved by Fabes, Garafolo, Mortola and Salsa [F-G-M-S]. Earlier Manfredi and Weitsman [M-Wt] proved this theorem in the case that Ω is a ball and T. Wolff [Wo] has established previously that Fatou's theorem for bounded solutions of the p-Laplacian does not hold with respect to Lebesgue measure of the boundary.

§ 3. Sketch of Proofs.

The essential idea to show Theorem A is to compare the harmonic map $u : \Omega \to \overline{B}_{\tau}(p)$, $u|_{\partial\Omega} = \phi$, with the harmonic function h, $h_{\partial\Omega} = \phi$. This is accomplished with the following:

Lemma E. Let Ω be a regular domain, let $\phi \in C^0(\mathcal{M}(\Omega), \overline{B}_{\tau}(p))$ and let $u \in C^0(\hat{\Omega}, \overline{B}_{\tau}(p)) \cap C^{\infty}(\Omega, \overline{B}_{\tau}(p))$ be a harmonic map on Ω which is equal to ϕ on $\mathcal{M}(\Omega)$. With respect to geodesic normal co-ordinates centered at p, ϕ may also be viewed as being $\mathbb{R}^n$ valued. Let $h : \hat{\Omega} \longmapsto \mathbb{R}^n$ be the harmonic extension of ϕ, i.e. $h = (h^1, \ldots, h^n)$ where h^{α} is a harmonic function for each α and $h|_{\partial\Omega} = \phi$. Let $v : \hat{\Omega} \longmapsto \mathbb{R}$ be the harmonic extension of $\frac{1}{2}|\phi|^2 = \frac{1}{2}\sum_{\alpha=1}^{n}(\phi^{\alpha})^2$. Then, there exists a constant $c > 0$, depending only on the geometry of $\overline{B}_{\tau}(p)$ such that

$$\rho(u(x),h(x))^2 \leq c(v(x) - \tfrac{1}{2}|h(x)|^2) \quad \text{for all } x \in \hat{\Omega}.$$

The proof of this lemma can be found in [A–C–M] pp 9–12. It is then easy to show that an exhaustion of Ω by smooth domains, interior a-priori estimates and Lemma E yield the solution of the Dirichlet problem as stated in Theorem A.

I shall not give further details of the proof of Theorems B and C. The reader may consult [A–C–M] pp 14–30.

The proof of Fatou's theorem has two major steps. We first show that the harmonic map $u(x)$ has limits in a certain weak sense (certainly weaker than non-tangential in general) as x approaches the boundary of Ω. The key fact to do this is the following lemma. I first state it in the case of bounded Lipschitz domains in $\mathbb{R}^m$. We consider the uniformly elliptic operator in $\mathbb{R}^m$

$$L = \frac{1}{g^{1/2}} \sum_{i,j=1}^{m} \partial_i (a^{ij}(x) g^{1/2} \partial_j), \quad x \in \mathbb{R}^m.$$

The coefficients $a^{ij}(x)$ are assumed to be defined and measurable on $\mathbb{R}^m$ with (a^{ij}) symmetric, $g = \det(a_{ij})$ with (a_{ij}) being the inverse of (a^{ij}) (a solution of $L\,h = 0$ is a distributional solution and $h \in W^{1,2}_{loc}$). We recall that the L–"harmonic" measure in a bounded regular domain, Ω, is defined as follows. Let h be the unique solution of the Dirichlet problem $L\,h = 0$ in Ω, $h|_{\partial\Omega} = f$ $f \in C^0(\partial\Omega)$. For $x \in \Omega$, the map $f \to h(x)$ is a positive linear functional on $C^0(\partial\Omega)$. Therefore, there exists a Borel measure W_L^x on $\partial\Omega$ so that

$$h(x) = \int_{\partial\Omega} f(Q)\, dW_L^x(Q)$$

W_L^x is called the <u>harmonic measure</u> associated to L and Ω. The Moser's Harnack inequality implies that for any $x_1, x_2 \in \Omega$, $W_L^{x_1}$ and $W_L^{x_2}$ are absolutely continuous to each other. Hence we say that $E \subset \partial\Omega$ is harmonic measure zero if $W_L^0(E) = 0$ for a fixed $0 \in \Omega$.

Let us now assume that Ω is a bounded Lipschitz domain. Let $e : \partial\Omega \to S^{m-1}$ be a non-tangential map in the following sense. For each $Q \in \partial\Omega$ there exists an open truncated cone $\Gamma(Q) \subset \Omega$ with vertex at Q and axis of symmetry e. A map $v : \Omega \to N$ is said to have e–radial limits equal to

$\phi : \partial\Omega \to N$ with respect to a measure ν on $\partial\Omega$ if e is non-tangential and Lipschitz and for almost every $Q \in \partial\Omega$ with respect to ν,

$\lim_{t\downarrow 0} v(Q + te(Q)) = \phi(Q)$.

Lemma F. Let Ω be a bounded Lipschitz domain in $\mathbb{R}^m$ and let $v(x)$ be the L-Green potential

$$v(x) = \int_\Omega G(x,y)|f(y)|dV$$

where $G(x,\bullet)$ is the Green function associated to L with pole at $x \in \Omega$. f is a function defined in Ω so that $f(x)d^2(x,\partial\Omega) \in L^\infty(\Omega)$ where $d(x,\partial\Omega)$ is the distance from $x \in \Omega$ to the boundary of Ω and dV is the volume element corresponding to the metric $a_{ij}(x)dx^idx^j$. If $|v(x)| \not\equiv \infty$, then $v(x)$ has zero e-radial limits with respect to the L-harmonic measure W_L^0.

It is not difficult to see that the lemma above is applicable with $f(x)$ being the density energy of the harmonic map u, $e(u)$, when the density is taken with respect to any Riemannian metric g_{ij} in the domain.

One then shows that there exists a Δ_g-bounded harmonic function h with non-tangential boundary values $\phi \in L^\infty(\partial\Omega; \overline{B}_\tau(p))$ so that the harmonic map u has e-radial limits equal those of h almost everywhere with respect to the natural harmonic measure associated to g_{ij}.

To go from e-radial limits to non-tangential limits (observe that this is, in general, false. That is, there are maps which have e-radial limits without having non-tangential limits) we prove a very general maximum principle and we combine it with Lemma E in a rather delicate manner. I shall refer to [A – C – M] for further details.

Remarks. (i) The discussion above can be easily generalized to bounded Lipschitz domains contained in the interior of a complete Riemannian manifold;

(ii) The existence of zero radial limits for Green's potentials of the form

$$v(x) = \int_{\Omega} G(x,y)d\mu(y)$$

where μ is a non-negative measure and G is the Green function of the Euclidean Laplacian was shown by Littlewood [Li] when Ω is a ball in $\mathbb{R}^2$, by Privalov [Pr] when Ω is a ball in $\mathbb{R}^m$, $m \geq 3$, by Wu [Wu] and in more generality (and independently) by B. Dahlberg [Da, 2] when Ω is a bounded Lipschitz domain in $\mathbb{R}^m$. However, in all these proofs the underlying Euclidean geometry is used in an essential manner. For instance, the relation: For $x_0 \in \partial\Omega$

$$W_L^0(\partial L \cap B(x_0,r)) \cong H^{m-1}(\partial\Omega \cap B(x_0,r)) \cong r^{m-1}, \ r > 0$$

where $B(x_0,r) = \{x : |x - x_0| \leq r\}$ and H^{m-1} is the (m-1)-Hausdorff measure of the boundary is used crucially in some fundamental inequalities in the proofs (see [Li], [Pr], [Wu], [Da, 2]). The above inequality only holds for bounded Lipschitz domain when the harmonic measure W_L^0 is absolutely continuous with respect to the natural (m–1)–Hausdorff measure of the boundary. This is true when L is the Euclidean Laplacian due to a famous result of Dahlberg [Da, 1]. In general, however, this is not even true in smooth domain, see for instance Caffarelli–Fabes and Kenig [C–F–K].

The proof of Lemma F we have given in [A–C–M] is new and it has more geometric flavor than the proofs just mentioned. Furthermore, it is naturally adapted to either:

(a) general uniformly elliptic operators in divergence form in a bounded Lipschitz domain of $\mathbb{R}^m$; or

(b) the Laplace–Beltrami operator in a complete simply connected negatively curved Riemannian manifold.

We close this note by stating the equivalent of Lemma F in a complete simply connected negatively curved manifold.

<u>Lemma</u> G. Let Ω be a complete simply connected Riemannian manifold where the sectional curvatures K_Ω satisfy $-b^2 \le K_\Omega \le -a^2 < 0$. Let v be the Green potential

$$v(x) = \int_\Omega G(x,y)|f(y)|dV$$

where $G(x,\bullet)$ is the Green function associated to the Laplace–Beltrami operator Δ with pole at $x \in \Omega$. f is a function defined in Ω, so that $f \in L^\infty(\Omega)$ and dV is the volume element associated to the metric in Ω. If

$|v(x)| \not\equiv \infty$ then for almost every $Q \in S(\infty)$ with respect to the harmonic measure – W^0 – induced by Δ on $S(\infty)$

$$\lim_{t\to\infty} v(\gamma_Q(t)) = 0$$

where $\gamma_Q(t)$ is the geodesic that joins $Q \in S(\infty)$ and a fixed point 0 in the interior of Ω.

REFERENCES

[An] Ancona, A., Negatively curved manifolds, elliptic operators and the Martin boundary, Ann. of Math. 125(1987), 495–536.

[Ad] Anderson, M. T., The Dirichlet problem at infinity for manifolds of negative curvature, J. Diff. Geom. 18(1983), 701–721.

[Ad-Sc] Anderson, M. T. and Schoen, R. M., Positive harmonic functions on complete manifolds of negative curvature, Ann. of Math. 121(1985), 429–461.

[A-C-M] Aviles, P., Choi, H. and Micallef, M., Boundary behavior of harmonic maps on non-smooth domains and complete negatively curved manifolds, to appear 1988.

[C-F-K] Caffarelli, L., Fabes, E. and Kenig, C., Completely singular elliptic-harmonic measures, Ind. Univ. Math. Journal, Vol. 30 (1981), 917–924.

[Da, 1] Dahlberg, B., On estimates of harmonic measures, Arch. Rational. Mech. Anal. 65(1977), 272–288.

[Da, 2] Dahlberg, B. On the existence of radial boundary values for functions subharmonic in a Lipschitz domain, Ind. Univ. Math. Journal, Vol. 27(1978), 515–526.

[Db] Doob, J. L., Classical Potential Theory and its Probabilistic Counterpart, Grund. der Math. Wiss 262, Springer-Verlag 1984.

[F-G-M-S] Fabes, E., Garafolo, Mortola and Salsa, The Fatou theorem for the p-Laplacian in a Lipschitz domain, to appear 1988.

[G-T] Gilbarg, D. and Trudinger, N. S., Elliptic Partial Differential Equations of Second Order, Second Edition, Grund. der Math. Wiss 224, Springer-Verlag 1983.

[Hi-K-W] Hildebrandt, S., Kaul, H. and Widman, K. O., An existence theorem for harmonic mappings of Riemannian manifolds, Acta. Math. 138(1977), 1–16.

[H-W] Hunt, R. A. and Wheeden, R. L., Positive harmonic functions on Lipschitz domains, Trans. Amer. Math. Soc. 147(1970), 507-527.

[J-K] Jerison, D. and Kenig, C., Boundary behavior of harmonic functions on non-tangentially accessible domains, Adv. in Math. 46(1982), 80-147.

[Li] Littlewood, J. E., On functions subharmonic in a circle. II,, Proc. London Math. Soc. (2) 28(1928), 383-394.

[M-Wt] Manfredi, J. and Weitsman, A., On the Fatou theorem for p-Harmonic function, Comm. in P.D.E. Vol. 13 (6) (1988), 651-668.

[Pr] Privalov, N. Boundary problems of the theory of harmonic and subharmonic functions in space (Russian) Mat. Sbornik 45(1938), 3-25.

[P] Paulik, G., A regularity condition at the boundary for weak solutions of some nonlinear elliptic systems, Proc. Symp. Pure Math. (Amer. Math. Soc.) 44(1986), 353-357.

[St] Stein, E. M., Singular Integrals and Differentiability Properties of Functions, Princeton University Press, Princeton, New Jersey 1970.

[S] Sullivan, D., The Dirichlet problem at infinity for a negatively curved manifold, J. Diff. Geometry 18(1983), 723-732.

[Wn] Wiener, N., The Dirichlet problem, J. Math. and Physics 3(1924), 127-146.

[Wo] Wolff, T., Gap series constructions for the p-Laplacian, Annali Della Scuola Normale Superiore di Pisa, 1987.

[Wu] Wu, J. M., On functions subharmonic in a Lipschitz domain, Proceeding of the American Math. Soc. Vol. 68 (2) (1978), 309-316.

Patricio Aviles
Department of Mathematics
University of Illinois
1409 West Green Street
Urbana, Illinois 61801

June 28, 1988

Contemporary Mathematics
Volume **101**, 1989

New Examples of Singly-Periodic Minimal Surfaces and Their Qualitative Behavior

DAVID A. HOFFMAN

1. Singly-Periodic Minimal Surfaces. I would like to describe some recent research, concerning properly embedded minimal surfaces with periodicity that I have been doing with Michael Callahan and Bill Meeks III. The work includes the construction of new examples and the characterization of the qualitative behavior of all examples in an important class. It is based, in turn, on an analysis of the geometric behavior of such surfaces at infinity. This talk is divided into two parts; the first discusses the examples and qualitative results; and the second gives a feeling for the analytical background.

To begin, it is easy to see that any connected, triply-periodic embedded surface in R^3 must have a single topological end. That is, the part of such a surface outside of any ball in R^3 is a connected set. For any doubly-periodic surface, Scherk's First Surface for example, the same is true, although it is by no means obvious. A proof of this fact will be described below.

For singly-periodic surfaces, things can get more complicated. Many singly-periodic minimal surfaces have a single topological end. (In order to avoid confusion, let me emphasize that we are counting ends of a surface M in R^3, not in R^3 mod T, where T is the cyclic group of symmetries. For example, Scherk's First Surface has *four* topological ends in R^3/T.) Scherk's Second Surface comes to mind. However, this is not always the case. Riemann discovered a 1-parameter family of connected, properly embedded minimal surfaces with an infinite number of flat ends.

The Riemann examples $\mathcal{R}$ possess a quite special set of properties:

(a) They have an infinite number of flat annular ends; (By an *annular end* of a surface we mean an end that has a representative homeomorphic to a punctured disk. Often such a representative is referred to as the end itself. For a properly embedded minimal surface, we may choose this representative to be the image of the punctured unit disk in the complex plane. An annular end is *flat* provided it is asymptotic to a plane in R^3.)

(b) They are invariant under a nontrivial translation T;

1980 *Mathematics subject classifications.* Primary 53C40

The research described in this paper was supported by research grant DE-FG02-86ER250125 of the Applied Mathematical Science subprogram of the Office of Energy Research, U.S. Department of Energy, and National Science Foundation grants DMS-8802858 and DMS-8611574

(c) The surfaces $\mathcal{R}/T$ have genus 1 and 2 ends;

(d) The surfaces $\mathcal{R}/T$ have total curvature equal to -8π.

Inspired by these examples, we established the existence of an infinite family of properly embedded, periodic minimal surfaces, $\mathcal{M}_k$ each with an infinite number of flat annular ends [**1**].

These surfaces have the following properties:

(1) $\mathcal{M}_k$ has an infinite number of annular ends.

(2) $\mathcal{M}_k$ is invariant under the group of translations T generated by $T\colon \vec{x} \mapsto \vec{x} + (0,0,2)$.

(3) $\mathcal{M}_k/T$ has genus $2k+1$ and two ends.

(4) The symmetry group of $\mathcal{M}_k/T$ has order $8(k+1)$.

(5) Reflection in the plane $\{x_3 = n + 1/2\}$, $n \in Z$, is a symmetry of $\mathcal{M}_k$.

(6) $\mathcal{M}_k/T$ has finite total curvature $-4\pi(2k+2)$.

(7) All the ends of $\mathcal{M}_k$ are flat; they are asymptotic to the planes $x_3 = n$, $n \in Z$.

(8) $\mathcal{M}_k \cap \{x_3 = n\}$, $n \in Z$, consists of $k+1$ equally spaced straight lines meeting at $(0,0,n)$.

(9) $\mathcal{M}_k \cap \{x_3 = c\}$, $c \notin Z$, is a simple closed curve.

(10) The subgroup R of the symmetry group of $\mathcal{M}_k$ consisting of rotations about the x_3-axis has order $k+1$ and is generated by rotation by $2\pi/(k+1)$.

(11) $\mathcal{M}_k$ is symmetric under reflection through the $k+1$ vertical planes containing the x_3-axis and bisecting the lines of property 8.

(12) The full symmetry group of $\mathcal{M}_k$ is generated by T, R, one of the reflections in 5, rotation about one of the lines in 8, and reflection through one of the planes in 11.

We have proved that these surfaces are characterized by only a few of their properties. In particular, properties 1–5 above imply properties 6–12. Thus, if properties 1–5 are satisfied on a properly embedded minimal surface with a translational symmetry T and more than one end, the surface $\mathcal{M}_k/T$ must have finite total curvature [**2**].

There is strong computational evidence that the surfaces $\mathcal{M}_k$ are the unique in the sense that they are the only properly embedded minimal surface with a translational symmetry and more than one end on which properties 1–5 are satisfied. For each k, there is a one-parameter family of *immersed* minimal surfaces, which must contain any surface satisfying the conditions above. A surface in this family will be embedded and singly-periodic, and will satisfy these conditions, provided a period vanishes. This period is a smooth function of the parameter describing the family, and in [**2**], we show that this function possesses a zero and is asymptotic to a linear function. Computations indicate that it is also monotonic and hence its zero is unique. Nonetheless, there is the remote possibility that this is not the case and that there is more than one $\mathcal{M}_k$ satisfying the list of conditions.

We point out something quite interesting and (we believe!) nontrivial. Namely, property 9 is forced by properties 1-5. It is quite plausible that this should *not* be the case. Consider the following construction. Take a boundary curve consisting of the union of the positive $x-$ and $y-$ axes, which we shall label L_0, together with L_1, the vertical translation of L_0 by one unit. Suppose $L_0 \cup L_1$ bounds an

FIGURE 1: Scherk's first surface

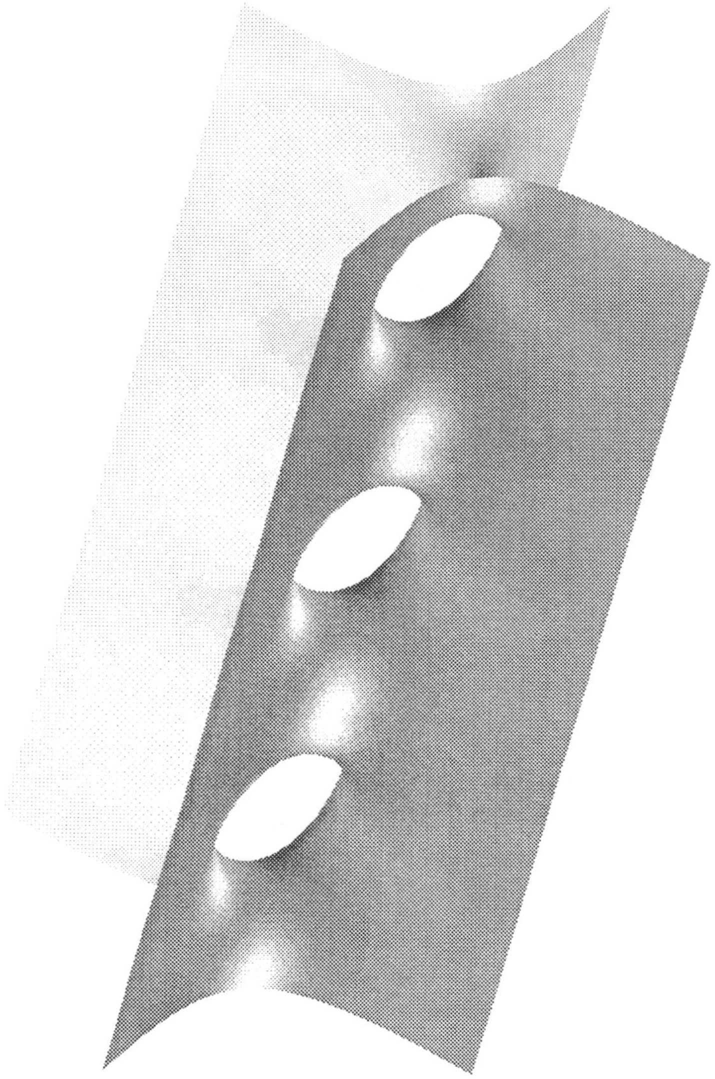

FIGURE 2: Scherk's second surface

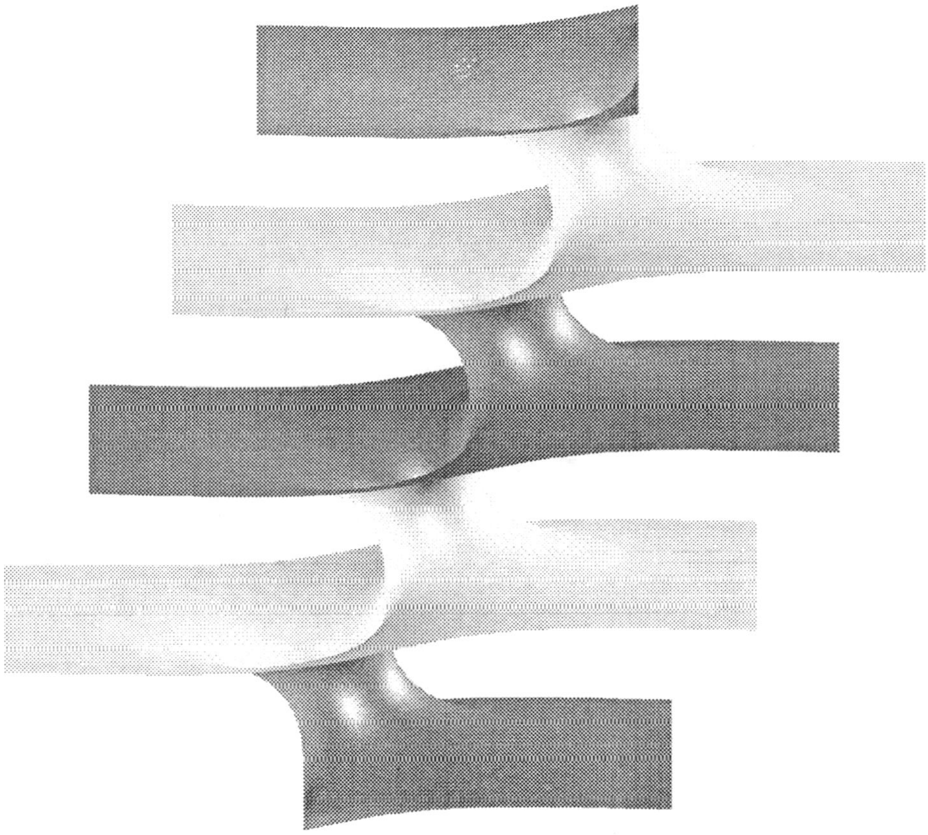

FIGURE 3: Riemann's example

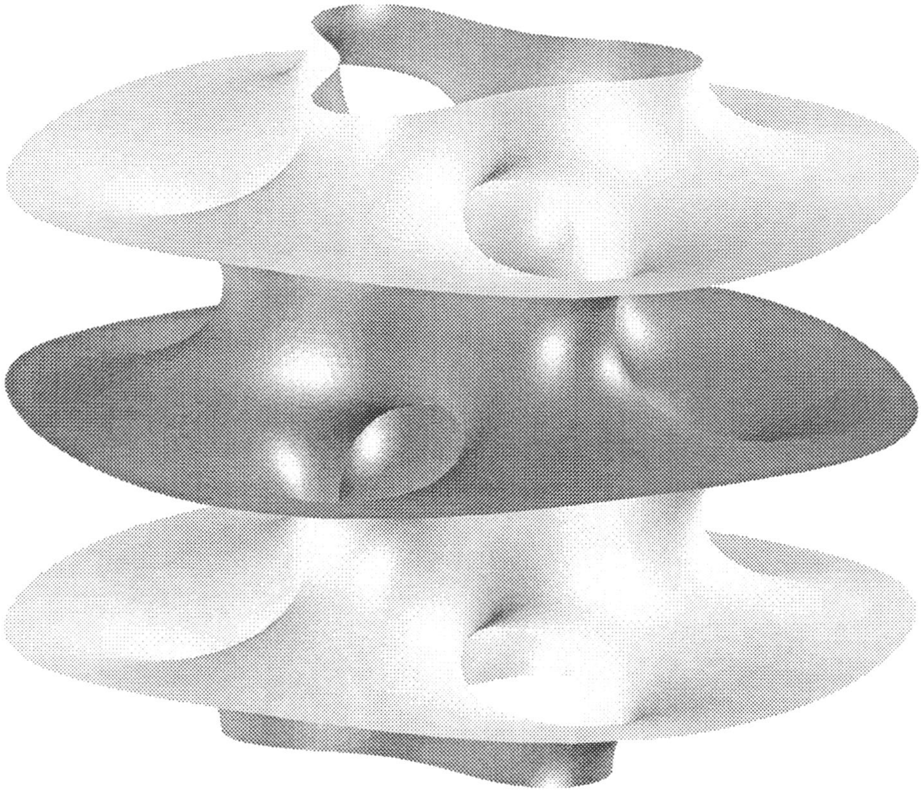

FIGURE 4: The surface M_2

embedded minimal annulus, which lies in the convex hull of this boundary and has all the symmetries of the boundary. Then this surface extends, by Schwarz reflection, to an example with Properties 1-8 and 10, but certainly not Property 9. To establish that Properties 1-5 forced Properties 6-12 we had to prove, among other things, that such a minimal annulus could not exist. We did this using a variant of the Alexandroff reflection principle, as developed for minimal surfaces by Rick Schoen [9] (see Lemma 3 of [2].)

A natural question to ask at this point is whether or not the examples $\mathcal{M}_k$ may be given a helical twist. To make this precise, we will define a *screw motion* of R^3 to be the composition of a nonzero translation, T, and a rotation, R, about the axis defined T. With this definition, we may consider a pure translation to be a (degenerate) screw motion. We were able to show [2] that:

For every positive integer k and angle θ, $0 < |\theta| < \frac{\pi}{k+1}$, there exists a properly embedded minimal surface $\mathcal{M}_{k,\theta}$ that has the following properties.

(i) $\mathcal{M}_{k,\theta}$ has an infinite number of flat annular ends.

(ii) $\mathcal{M}_{k,\theta}$ is invariant under the group of nondegenerate screw motions, $\mathcal{S}$, generated by $(r, \phi, z) \to (r, \phi + 2\theta, z + 2)$.

(iii) Properties 3 and 6–10 above hold for $\mathcal{M}_{k,\theta}$, with T replaced by $\mathcal{S}$.

(iv) The symmetry group of $\mathcal{M}_{k,\theta}/\mathcal{S}$, has order $4(k+1)$.

REMARK 1. *When $\theta = 0$, the construction of $M_{k,\theta}$ yields a surface satisfying all the conditions characterizing the examples $\mathcal{M}_k$. If we knew that the surfaces $\mathcal{M}_k$ and $\mathcal{M}_{k,\theta}$ were unique, then we would have*

CONJECTURE 1. $\{\mathcal{M}_{k,\theta} \mid \theta \in (-\frac{\pi}{k+1}, \frac{\pi}{k+1})\}$ *is a smooth, one-parameter family of embedded minimal surfaces.*

The examples $\mathcal{M}_{k,\theta}$ and the Riemann examples $\mathcal{R}$ comprise all the known properly embedded minimal surfaces with more than one end and infinite symmetry group. They share many geometric properties. We have been able to show that this is not accidental.

THEOREM 1. **(First Structure Theorem)** *Suppose M is a properly embedded minimal surface, with more than one end, whose symmetry group is infinite. Then either M is the catenoid or:*

(a) *M has an infinite number of ends;*

(b) *M is invariant under a screw motion $\mathcal{S}$;*

(c) *all annular ends of M are flat ends;* (d) *the total curvature of $\widetilde{M} = M/\mathcal{S}$ is*

$$C(\widetilde{M}) = 2\pi(\chi(\widetilde{M}) - r(\widetilde{M})),$$

where $r(\widetilde{M})$ is the number of ends of $\widetilde{M}$.

This theorem proved in [2] is a consequence of a more general structure theorem:

THEOREM 2. **(Second Structure Theorem)** *Under the same hypotheses (M is a properly embedded minimal surface, with more than one end, whose symmetry group is infinite) then either M is the catenoid or:*

(a) *There exists a plane parallel to the limit tangent plane, whose intersection with M consists of a finite number of simple closed curves;*

(b) *The symmetry group of M contains an infinite cyclic, normal subgroup, generated by a screw motion $S = T + R$, where T is a translation and the index of this subgroup is finite;*

(c) *If $S = T + R$ and $R \neq 0$, the limit tangent plane is orthogonal to the axis of S.*

In the next section we shall give some indication of the analysis behind the proof of this theorem.

Note, as is evident from the picture of Riemann's example, that the axis of translation is *not* orthogonal to the limit tangent planes. This is not a contradiction to the result above because the Riemann examples are invariant under a degenerate screw-motion; that is, one without a rotational part.

For purposes of comparison, I should describe some very beautiful examples, recently-discovered by H. Karcher and J. Pitts. They are asymptotic to two coaxial helicoids and resemble two such helicoids, one twisted by a fixed angle from the other, with the intersection set replaced by a tower of tunnels resembling the core of Scherk's Second Surface. These surfaces have screw-motion symmetries and would appear to violate the structure theorems above. However, they have *one* topological end.

Using the Second Structure Theorem, we can establish the following topological result: *A doubly-periodic, properly embedded minimal surface in $\mathbf{R}^3$ has one end and infinite genus* (Corollary 2, Section 3 of [2]). The proof is relatively straightforward:

Since the symmetry group of a doubly-periodic minimal surface does not have a cyclic subgroup of finite index, the Second Structure Theorem above implies that a connected doubly-periodic minimal surface has one end. If a doubly-periodic minimal surface with translation group L has genus zero and one end, then, topologically, it is a plane and its quotient in $\mathbf{R}^3/L$ is a torus. Since a closed minimal torus in a flat 3-manifold is totally geodesic (by Gauss-Bonnet), we see that a nonplanar, doubly-periodic minimal surface must have one end and infinite genus.

2. Limit Tangent Planes and Canonical Annular Ends. In this section, I will describe a bit of the underlying theory on which the proofs of the Structure Theorems are based. Recall that the Second Structure Theorem in the previous section uses the existence of a unique limit tangent plane to the surfaces in question. While this is plausible, it is not obvious that such a plane exists in the generality that is required: we make no topological assumptions about the minimal surfaces in question, except that they have more than one end.

Some definitions are necessary. A subsurface $E \subset M$ is an *end-representative* if it is a closed noncompact subset of M with compact boundary. We say that a surface M has more than one end if it contains two or more pairwise-disjoint end representatives. Finally, a surface Σ in a region $R \subset \mathbf{R}^3$ is said to be a surface of least area in R if every compact subdomain $D \subset \Sigma$ has least area among surfaces in R with boundary ∂D.

An end of a complete minimal surface of finite total curvature has a well-defined limit normal vector and limit tangent plane [**2**]. Recall that a properly embedded minimal surface M of finite total curvature has the property that all of its limit tangent planes coincide and thus it makes sense to speak of *the* limit tangent plane to M. We have generalized the notion of limit tangent plane to apply to any properly-embedded minimal surface with more than one end, and have shown that in this case, there is a unique limit tangent plane.

LEMMA 1. *Suppose $M \subset \mathbf{R}^3$ is a properly embedded minimal surface. Let E be an end-representative of M with smooth boundary and having the property that $M - (E)$ is noncompact. Then $S = \partial E$ bounds a smooth, properly-embedded, noncompact, least-area surface Σ of finite total curvature given in Lemma 1 in the closure of one of the components of $\mathbf{R}^3 - M$.*

The limit tangent plane of Σ will be called a *limit tangent plane* to M.

THEOREM 3. *If M is a properly embedded minimal surface in $\mathbf{R}^3$ with more than one end, it has a unique limit tangent plane.*

We will refer to the limit plane defined by Theorem 3 as *the* limit tangent plane to M.

This tangent plane is the one referred to by the Second Structure Theorem. Its existence allows us to begin the analysis of that theorem. Beginning with a properly embedded minimal surface with more than one end and infinite symmetry group,we show first that one of the ends of a least-area surface Σ of finite total curvature in the closure of one of the components of $\mathbf{R}^3 - M$ must have a flat end. The tangent plane P to this flat end is, of course, parallel to the unique limit tangent plane. By using a version of the maximum principle at infinity [**7**] [**8**] we can perturb this tangent plane to the flat end to insure that it intersects M transversally in a compact set. This shows that there exists a plane parallel to the limit tangent plane, whose intersection with M consists of a finite number of simple closed curves.

Without loss of generality, we may assume that P is horizontal and equal to the (x, y)-plane. Let $\mathcal{S}$ be the orientation-preserving symmetries of M whose linear part fixes the vertical vector $(0, 0, 1)$. Because of the existence of a *unique* limit tangent plane, every element of the symmetry group of M either preserves or reverses $(0, 0, 1)$. Observe that $\mathcal{S}$ can contain no horizontal translations because $P \cap M$ must be compact. Therefore $\mathcal{S}$ consists of pure vertical translations, screw motions with vertical axis and rotations with vertical axis. It follows that the index of $\mathcal{S}$ in $Sym(M)$ is either 1, 2 or 4. We now concentrate on the orbit of P under $\mathcal{S}$. Since $P \cap M$ is compact, the end of P is a positive distance from M, which means that the orbit of P under $\mathcal{S}$ consists of a family of horizontal planes, indexed by their third coordinate, z. Since we are assuming that M is not the catenoid, $\mathcal{S}$ must act discretely. Thus $\mathcal{S}$ cannot consist entirely of rotations. Therefore, there exists a minimum positive height, achieved by an element S of $\mathcal{S}$. This is the required screw-motion generator of a cyclic, infinite normal subgroup of isometries. It is evident from the previous paragraph that the cyclic subgroup generated by S has finite index, asserted in the second statement of the Second Structure Theorem.

Let L be the slab in R^3 between the plane P and the plane $\mathcal{S}(P)$. It is clear that $L \cap M$ is not compact; otherwise, M would have one end, a contradiction. It follows that the $\mathcal{S}$ – *orbit* of $L \cap M$ contains an infinite number of distinct ends, as claimed in Statement 1 of the First Structure Theorem.

In order to prove the aforementioned fact that if M has one annular end it has infinitely many, we need to use a technical result about the existence of canonical end representatives:

PROPOSITION 1. (**Canonical Representation of Annular Ends**) *Each annular end of a complete, nonsimply-connected, oriented minimal surface has a unique representative whose boundary is a closed geodesic. These representatives have pairwise-disjoint interiors. If the boundaries of two such annular ends touch, they coincide and M is an annulus.*

The cyclic subgroup generated by $\mathcal{S}$ acts freely on the collection of canonical annular ends of M. Suppose there is one canonical annular end. If there were not an infinite number of distinct canonical annular ends, then at least one of the annular ends would have to be fixed. But then the (compact) boundary of this end would be invariant under an infinite cyclic group of symmetries. It is easy to see that in this circumstance, that this end would be invariant under an action of S^1 by rotation, forcing M to be the catenoid. This proves that if M has one annular end, it has infinitely many, unless M is the catenoid. We can now evoke the Annular End Theorem of [4] which states that any properly embedded minimal surface M can have, at most, two annular ends of infinite total curvature. But the $\mathcal{S}$ – *orbit* of any annular end contains an infinite number of annular ends. Hence every annular end of M must have finite total curvature. An annular end of finite total curvature is asymptotic to either the plane or the catenoid. A straightforward argument using the existence of the plane P will show that the annular ends must be "flat", i.e. asymptotic to planes. This completes the proof of Statement 3 of the First Structure Theorem.

It is hoped that these few details give a feel of the technicalities involved in the proof of the Structure Theorem.

REFERENCES

1. M. Callahan, D. Hoffman, and W. H. Meeks III, *Embedded minimal surfaces with an infinite number of ends*, Inventiones Math.,**96** (1989) 459-505.

2. M. Callahan, D. Hoffman, and W. H. Meeks III, *The structure of singly-periodic minimal surfaces* (to appear in Inventiones Math.).

3. C. Costa, *Example of a complete minimal immersion in of genus one and three embedded ends*, Bull. Soc. Bras. Mat. 15 (1984), 47–54.

4. D. Hoffman and W. H. Meeks III, *The asymptotic behavior of properly embedded minimal surfaces of finite topology* (to appear in J. Amer. Math. Soc., October, 1989.)

5. H. Karcher, *Embedded minimal surfaces derived from Scherk's examples*, Manuscripta Math. **62** (1988) 83-114.

6. H. Karcher, *New examples of periodic minimal surfaces*, Preprint.

7. R. Langevin and H. Rosenberg, *A maximum principle at infinity for minimal surfaces and applications*, Preprint.

8. W. H. Meeks III and H. Rosenberg, *The strong maximum principle for complete minimal surfaces in flat 3-manifolds*, Preprint.

9. R. Schoen, *Uniqueness, symmetry, and embeddedness of minimal surfaces*, J. Diff. Geom., **18** (1983), 791–809.

Contemporary Mathematics
Volume **101**, 1989

THE GAUSS CURVATURE AND MINKOWSKI PROBLEMS IN SPACE FORMS

BY VLADIMIR I. OLIKER[1]

ABSTRACT. We consider here convex hypersurfaces in space forms given either as radial graphs over spheres or as gradient maps constructed from suitably defined support functions. We study questions of existence and uniqueness of such hypersurfaces with prescribed Gauss curvature.

In our previous paper [O] we investigated existence and uniqueness of convex hypersurfaces in R^{n+1} with given Gauss curvature. We considered there radial hypersurfaces, that is, hypersurfaces that can be represented as graphs over a unit sphere S^n. The equation for the Gauss curvature in such circumstances is a special equation of Monge-Ampere type for which in [O] the a priori estimates were derived. We use these estimates here to solve other nonlinear geometric problems connected with the Gauss curvature.

Existence and uniqueness of radial hypersurfaces in R^{n+1} with given mean curvature was established by Bakelman and Kantor [BK] and Treibergs and Wei [TW] and the case of any k-th elementary symmetric function of the principal curvatures was investigated by Caffarelli, Nirenberg and Spruck in [CNS] where actually a more general result is established.

In this paper in §0 we derive the basic differential equation for Gauss curvature of a radial hypersurface in a Riemannian space with a metric of the form $ds^2 = d\rho^2 + f(\rho)e_{ij}du^idu^j$, where $e_{ij}du^idu^j$ is the metric of the Euclidean unit sphere. In §1 we state uniqueness and

----

1980 Mathematics Subject Classification. 53C42, 49F22.

[1] Research supported by NSF Grants MCS-8301904, DMS-8702742 and CNR, Italy.

This paper is in final form and no version of it will be submitted for publication elsewhere.

existence results analogous to those in [O] for hyperbolic and spherical spaces. In §2 we consider a particular class of convex hypersurfaces in hyperbolic and spherical spaces. Hypersurfaces in this class admit a parametrization in terms of the support function and that allows to formulate a version of the classical Minkowski problem. The equations here are more complicated than their analogues in Euclidean space. The corresponding existence and uniqueness results are presented in §3. The Theorem 3.1 and its analogue for spherical space in this section give partial answers to questions posed by Aleksandrov in [A, p.320].

NOTE. The results presented in this paper for the most part are improved versions of results in our unpublished and not widely circulated notes [O1].

§0. DERIVATION OF THE BASIC EQUATION FOR RADIAL HYPERSURFACES.

0.1. In Euclidean space R^{n+1} ($n \geq 2$) consider a unit sphere $S^n (\equiv S)$ centered at some point **O**. Let u denote a point on S, $u^1, \dots, u^n$ smooth local coordinates on S and (u, ρ) the spherical coordinates in R^{n+1}. The standard metric on S induced from R^{n+1} we denote by $e = e_{ij} du^i du^j$, $i,j = 1, \dots, n$. The summation over repeated lower and upper indices is assumed everywhere in this paper.

Let $I = (0,a)$, where $a = \text{const} \leq \infty$, and $f(\rho)$ a positive C^∞ function on I such that $f(0) = 0$. In the ball $B = \{0 \leq \rho < a\}$ introduce a metric

$$h = d\rho^2 + f(\rho) e_{ij} du^i du^j,$$

and so obtained Riemannian space denote by $\tilde{R}^{n+1}$.

In the particular case when $a = \infty$ and $f(\rho) = \rho^2$ we obtain the Euclidean space with the metric written in spherical coordinates. When $a = \pi/2$ and $f(\rho) = \sin^2 \rho$, we get the metric of the hemisphere S_+ of a hypersphere in R^{n+2}. Finally, in case $a = \infty$ and $f(\rho) = \sinh^2 \rho$ we obtain the hyperbolic space H^{n+1}.

0.2. We consider graphs of functions over S, in particular, when they represent embedded hypersurfaces. For a given smooth function $\rho = z(u)$, $u \in S$, we denote by $r(u) = (u, z(u))$ the graph of this function.

Put $X_i = \partial/\partial u^i$, $i = 1, \dots, n$, $R = \partial/\partial \rho$, $z_i = X_i z$ and $r_i = X_i + z_i R$. The latter is a tangent vector to the hypersurface $F = r(S)$. The metric

$g = g_{ij}du^idu^j$ induced on F from $\tilde{R}^{n+1}$ is given by

(0.1) $$g_{ij} = fe_{ij} + z_iz_j, \quad i,j = 1,\dots,n, \text{ and}$$

(0.2) $$\det(g_{ij}) = f^{n-1}[f + |\nabla z|^2]\det(e_{ij}), \text{ where}$$

$$\nabla z = z_ie^{ij}X_j,\ (e^{ij}) = (e_{ij})^{-1},\ |\nabla z|^2 = e^{ij}z_iz_j.$$

It follows from (0.2) that if $z(u) > 0$ on S then $r: S \to \tilde{R}^{n+1}$ is an embedded hypersurface.

0.3. The unit normal vector field on F is given by

(0.3) $$N = \frac{\nabla z - fR}{\sqrt{f^2 + f|\nabla z|^2}}.$$

0.4. The second fundamental form b of F is the normal component of the covariant derivative D_YX where X and Y are tangent vectors to F. Taking $Y = r_i$, $X = r_j$, and putting $z_{ij} = \partial^2z/\partial u^i\partial u^j$, we get

$$D_{r_i}r_j = \tilde{\Gamma}^k_{ij}X_k + \tilde{\Gamma}^R_{ij}R + z_i\tilde{\Gamma}^k_{Rj}X_k + z_{ij}R + z_j\tilde{\Gamma}^k_{iR}X_k,$$

where the indexed $\tilde{\Gamma}$ denote the Christoffel symbols of the second kind of the metric h.

Now for the matrix of the second fundamental form of F we obtain an expression

$$b_{ij} = \frac{f}{\sqrt{f^2 + f|\nabla z|^2}}[\tilde{\Gamma}^k_{ij}z_k + (z_i\tilde{\Gamma}^k_{Rj} + z_j\tilde{\Gamma}^k_{iR})z_k - \tilde{\Gamma}^R_{ij} - z_{ij}].$$

The Christoffel symbols can be represented as follows:

$$\tilde{\Gamma}^k_{ij} = \Gamma^k_{ij},$$

where on the right are the Christoffel symbols of the second kind of the metric e;

$$\tilde{\Gamma}^R_{ij} = -\frac{1}{2}\frac{\partial f}{\partial \rho}e_{ij}, \quad \tilde{\Gamma}^k_{iR} = \frac{\partial \ln\sqrt{f}}{\partial \rho}\delta^k_i,$$

and all other vanish. Hence,

$$(0.4)\qquad b_{ij} = \frac{f}{\sqrt{f^2 + f|\nabla z|^2}}[-\nabla_{ij}z + \frac{\partial \ln f}{\partial \rho}z_i z_j + \frac{1}{2}\frac{\partial f}{\partial \rho}e_{ij}],$$

where $\nabla_{ij} = \frac{\partial^2}{\partial u^i \partial u^j} + \Gamma^k_{ij}\frac{\partial}{\partial u^k}$.

0.5. The (exterior) Gauss curvature of F is defined as

$$K = \frac{\det(b_{ij})}{\det(g_{ij})}.$$

It follows from (0.2) and (0.4) that

$$(0.5)\quad K = f^{-\frac{n}{2}+1}(f + |\nabla z|^2)^{-\frac{n}{2}-1}\frac{\det(-\nabla_{ij}z + \frac{\partial \ln f}{\partial \rho}z_i z_j + \frac{1}{2}\frac{\partial f}{\partial \rho}e_{ij})}{\det(e_{ij})}.$$

0.6. A particular change of the unknown function allows to transform (0.5) to a form more convenient in several instances.

Assume that $z(u) > 0$ on S. For $\rho \in (0,a)$ consider the antiderivative

$$F(\rho) = -\int \frac{d\rho}{f(\rho)}$$

and put

$$p(u) = F(z(u)).$$

Since $f(\rho) > 0$, $F(\rho)$ is a monotonically decreasing function and therefore the functions p and z define each other uniquely. The range for the function $p = F(\rho)$ is $(F(a),F(0))$. We have

$$z_i = -fp_i,$$

$$\nabla_{ij}z = f[-\nabla_{ij}p + \frac{\partial f}{\partial \rho}p_i p_j],$$

where $p_i = \partial p/\partial u^i$, $i = 1,\dots,n$. Then, since $\partial f/\partial \rho = -(1/f)\partial f/\partial p$,

$$(0.6)\qquad -\nabla_{ij}z + \frac{\partial \ln f}{\partial \rho}z_i z_j + \frac{1}{2}\frac{\partial f}{\partial \rho}e_{ij} = f\nabla_{ij}p - \frac{\partial \ln \sqrt{f}}{\partial p}e_{ij},$$

and (0.5) assumes the form

$$(0.7) \qquad K = (1 + f|\nabla p|^2)^{-\frac{n}{2}-1} \frac{\det(\nabla_{ij}p + \frac{1}{2}\frac{\partial(1/f)}{\partial p}e_{ij})}{\det(e_{ij})}.$$

0.7. Remark. If we would have made the change $p = F(\rho)$, $u = u$ in the metric h and then computed the expression for K we would have obtained the expression (0.7) directly.

0.8. Lemma. *Let* $x = (u,\rho)$ *denote a point in* $\tilde{R}^{n+1}$ $(= S \times [0,a))$ *and let* $\phi(x)$ *be a positive continious function defined in the annulus* $T = \{ x \in \tilde{R}^{n+1} \mid R_1 \le \rho \le R_2\}$ where $0 < R_1 < R_2 < a$. *Suppose that for all* $u \in S$

$$(0.8) \qquad \phi(u,\rho) > \left[\frac{\partial \ln\sqrt{f(\rho)}}{\partial\rho}\right]^n \quad \text{when} \quad \rho = R_1$$

$$(0.9) \qquad \phi(u,\rho) < \left[\frac{\partial \ln\sqrt{f(\rho)}}{\partial\rho}\right]^n \quad \text{when} \quad \rho = R_2.$$

Let F *be a hypersurface in* $\tilde{R}^{n+1}$ *given as a graph of the function* $\rho = z(u)$ *over* S, $z \in C^2(S)$, *and the Gauss curvature of* F *at the point* u *is equal to* $\phi(u,z(u))$. *Then* $R_1 < z(u) < R_2$ *for all* $u \in S$. *Furthermore, the gradient* ∇z *admits a uniform estimate depending only on* R_1 *and* R_2.

Proof. Consider a point $\tilde{u} \in S$ such that $\max z = z(\tilde{u})$. Then at the point $\tilde{u}$ we have $z_i = 0$ and Hess $z \le 0$. The equation (0.5) at $\tilde{u}$ takes the form

$$f^{-n} \frac{\det(-\nabla_{ij}z + \frac{1}{2}\frac{\partial f}{\partial\rho}e_{ij})}{\det(e_{ij})} = \phi(\tilde{u},z(\tilde{u})),$$

where the left hand side is evaluated at $\tilde{u}$. Since $-$ Hess $z(\tilde{u}) \ge 0$, we conclude that

$$\phi(\tilde{u},z(\tilde{u})) \ge \left[\frac{\partial \ln\sqrt{f(\rho)}}{\partial\rho}\right]^n \quad \text{at} \quad \rho = z(\tilde{u}).$$

We extend ϕ continiously outside the annulus T so that the inequality in (0.8) holds also for $\rho \le R_1$ and in (0.9) for $\rho \ge R_2$.

Suppose $\max z(\tilde{u}) \geq R_2$. Then the last inequality is in contradiction with the inequality (0.9) (valid now for $\rho \geq R_2$). Hence $z(u) < R_2$ for all $u \in S$. Similarly, with the use of (0.8) one shows that $z(u) > R_1$ for all $u \in S$.

In order to establish the estimate of $|\nabla z|$, we make a change of the unknown function as in section 0.6 and consider the function $P = |\nabla p|^2 + 1/f$. Obviously, the needed estimate of $|\nabla z|$ is implied by an estimate of $|\nabla p|$.

At any critical point of P we have

$$\frac{\partial P}{\partial u^i} = 2(\nabla_{ik} p + \frac{1}{2}\frac{\partial(1/f)}{\partial p} e_{ik}) e^{kj} \frac{\partial p}{\partial u^j} = 0, \quad i = 1,..,n.$$

It follows from (0.7) and our hypotheses that $\det\{\nabla_{ij} p + (1/2)[\partial(1/f)/\partial p] e_{ij}\} \neq 0$. Consequently, grad $p = 0$ at a point of minimum or maximum of P. Then obviously

$$\min 1/f(z(u) \leq P \leq \max 1/f(z(u)), \quad u \in S.$$

Since $|\nabla z| = f\ |\nabla p|$, the desired estimate follows.

§1. RADIAL HYPERSURFACES WITH PRESCRIBED GAUSS CURVATURE IN HYPERBOLIC AND SPHERICAL SPACES.

1.1. Theorem. Let H^{n+1} denote the hyperbolic space with constant sectional curvature -1 and metric $ds^2 = d\rho^2 + \sinh^2\rho\, e_{ij} du^i du^j$. Let $x = (u,\rho)$ denote a point in H^{n+1} and $\phi: H^{n+1}\setminus\{0\} \dashrightarrow R^+ = (0,\infty)$. Suppose ϕ satisfies the following conditions:

a) $\phi \in C^k(H^{n+1}\setminus\{0\})$, $k \geq 3$;

b) there exists two numbers R_1 and R_2, $0 < R_1 \leq 1 \leq R_2 < \infty$ such that

$$\phi(u,\rho) \geq \coth^n\rho \text{ when } \rho = R_1 \text{ and}$$

$$\phi(u,\rho) \leq \coth^n\rho \text{ when } \rho = R_2 \text{ for all } u \in S;$$

c) $\frac{\partial}{\partial\rho}(\phi(u,\rho)\sinh^n\rho\, \cosh^n\rho) \leq 0$ for all $u \in S$ and $\rho \in [R_1, R_2]$.

Then there exists a unique closed convex hypersurface F in H^{n+1} such that

i) F is a graph of a radial function $\rho = z(u) > 0$ over S;

ii) $z(u) \in C^{k+1,\alpha}(S)$ for any $\alpha \in (0,1)$ and if $\phi(x)$ is analytic then $z(u)$ is analytic;

iii) the Gauss curvature of F is given by $\phi(u,z(u))$;

iv) F is located between two concentric spheres of radii R_1 and R_2 centered at $\mathbf{O}$.

1.2. The proof of this theorem is analogous to the proof of the corresponding theorem in Euclidean space [O]. It is based on the continuation method. But before applying the method we observe that the substitution $y(u) = \tanh z(u)$ transforms the equation (0.5) to the form

$$M(y) \equiv (1-y^2)^{\frac{n}{2}+1} y^{2-2n} [y^2(1-y^2) + |\nabla y|^2]^{-\frac{n}{2}-1} \times \frac{\det(-y\nabla_{ij}y + 2y_i y_j + y^2 e_{ij})}{\det(e_{ij})} = \phi \text{ on } S,\ 0 < y < 1, \tag{1.1}$$

where $y_i = \partial y/\partial u^i$.

The continuation method involves three steps (cf. [N]). Consider the family of functions

$$\phi_t(u,y) = t\phi(u,y) + \frac{a^\varepsilon(1-t)}{y^{n+\varepsilon}},\ y \in (0,1),\ t \in [0,1],$$

where a and ε are positive constants to be specified later, and a family of equations

$$M(y(u)) = \phi_t(u,y(u)),\quad u \in S. \tag{1.1$_t$}$$

Put $\Lambda = \{t \in [0,1]$ for which (1.1_t) admits a solution $y_t \in C^{2,\alpha}(S)$, $0 < \alpha < 1$, and such that $y_1 = \tanh R_1 \le y_t \le \tanh R_2 = y_2\}$.

Now, if it is shown that $\Lambda \ne \emptyset$, and Λ is open and closed on $[0,1]$ then $\Lambda = [0,1]$ and the equation (1.1) is solved in $C^{2,\alpha}(S)$.

In order to show that Λ is not empty we take $t = 0$ and $y = a$, where a is any fixed constant in (y_1, y_2). Clearly, such y is a unique positive solution of (1.1_0). Hence, $\Lambda \neq \emptyset$.

The openness of Λ is shown by using the implicit function theorem. Here one needs to prove that the kernel of the equation obtained by linearization of (1.1_t) is trivial. Using the same arguments as in [O], §6, and making the appropriate substitutions connected with the change $y = \tanh \rho$, it is shown that the latter is true if

$$\frac{\partial}{\partial y}\left[\left(\frac{y}{1-y^2}\right)^n \phi_t(u,y)\right] < 0 \text{ for any } t \in [0,1),\ y(u) \in [y_1,y_2],\ \ u \in S.$$

Taking into account condition c), we see (omitting simple manipulations) that the last inequality is satisfied if

$$2ny^2(u) - \varepsilon(1-y^2(u)) < 0 \quad \text{for } u \in S.$$

Since $0 < y \leq y_2 < 1$, we can choose ε, independent on the particular y, so that this inequality holds.

In order to prove that Λ is closed one needs to show that the $C^{2,\alpha}$- norm of a solution $y_t \in \Lambda$ can be bounded independent of t. This is done first for functions ϕ_δ, approximating ϕ in C^3-norm, satisfying all hypotheses in the theorem, and for which strict inequalities in b) hold. Then the C^0- and C^1- esimates follow from Lemma 0.8. The bounds for the second derivatives are given in [O], sections 4.2-4.4 and 5.2. In [O] the estimate of the third derivatives is based on an estimate of Calabi [C]. The same arguments apply to the equation (1.1_t). Alternatively one could use here estimates of the Hölder norm of the second derivatives as in [GT] and then apply the arguments in section 5.2 in [O].

The estimates in Lemma 0.8 are independent of δ. The same is true for the C^2- and $C^{2,\alpha}$- estimates for all sufficiently small δ. Hence, one can construct a sequence of solutions of $M(y_\delta) = \phi_\delta$ converging to a $C^{2,\alpha}$- solution of $M(y) = \phi$. Further smoothness of y is established by arguments standard in the theory of elliptic equations.

The uniqueness of the solution follows essentially from the maximum principle as in [A1]. For $n = 2$ the uniqueness was considered in [F] where the condition c) is given in a slightly different form.

1.3. The scheme described above is followed with minor variations in all of the existence proofs in this paper. For this reason, below, we concentrate on the geometric side of the results.

We note that one could treat in the same manner the general equation (0.5), and it is only the estimates of the second derivatives that are still missing. It would be interesting to find out if one needs to impose some further restrictions on the function f in addition to those in Lemma 0.8 in order to establish these estimates.

1.4. A similar theorem can be stated for the spherical space interpreted as a hemisphere of the unit radius in R^{n+2} with the metric given by $ds^2 = d\rho^2 + \sin^2\rho e_{ij}du^i du^j$. All the hypotheses remain the same except for b) and c) which should be replaced by the following:

b′) there exist two numbers R_1 and R_2, $0 < R_1 \le R_2 < \pi/2$ such that

$$\phi(u,\rho) \ge \cot^n\rho \quad \text{when} \quad \rho = R_1 \quad \text{and}$$

$$\phi(u,\rho) \le \cot^n\rho \quad \text{when} \quad \rho = R_2, \quad u \in S;$$

c′) $\frac{\partial}{\partial\rho}(\phi(u,\rho)\tan^n\rho) \le 0$ for all $u \in S$ and $\rho \in [R_1, R_2]$ *).

In this case the equation (0.5) after substitution $y(u) = \tan z(u)$ becomes

$$(1.2) \quad (1+y^2)^{\frac{n}{2}+1} y^{2-2n}[y^2(1+y^2) + |\nabla y|^2]^{-\frac{n}{2}-1} \times$$

$$\frac{\det(-y\nabla_{ij}y + 2y_iy_j + y^2e_{ij})}{\det(e_{ij})} = \phi \text{ on } S, \quad y > 0.$$

§2. BASIC EQUATIONS FOR HYPERSURFACES ADMITTING REPRESENTATION IN TERMS OF THE SUPPORT FUNCTION

2.1. We will be working with the projective models of the hyperbolic and spherical spaces. First we consider the hyperbolic space.

---

*) It suffices to require that

$$\frac{\partial}{\partial\rho}(\ln\phi(u,\rho)) \le \tan^{-1}\rho[(n+2)\sin^2\rho - n] \text{ for all } u \in S \text{ and } \rho \in [R_1, R_2].$$

In addition to the spherical coordinates (u,ρ) of a point x in R^{n+1} and in H^{n+1}, that were used in previous sections, we now consider a system of Cartesian coordinates $x_1, x_2, \ldots, x_{n+1}$ in the unit (Euclidean) open ball B. The boundary of B is denoted by S. The standard Euclidean inner product we denote by $\langle\ ,\ \rangle$.

Let λ denote the metric of the hyperbolic space. Thus, $\lambda \equiv ds^2 = d\rho^2 + \sinh^2\rho\, e$ where $0 \le \rho < \infty$ and e the standard (induced from R^{n+1}) metric on S. The metric of the projective model of H^{n+1} is obtained by the change $|x| = \tanh \rho$, where $|x| = \langle x,x\rangle^{1/2}$. The metric λ assumes the form

$$\lambda = \frac{(d|x|)^2}{(1-x^2)^2} + \frac{x^2}{1-x^2} e_{ij} du^i du^j,$$

where $x^2 = \langle x,x\rangle$. On the other hand for the Euclidean metric we have $\langle dx,dx\rangle = (d|x|)^2 + x^2 e_{ij} du^i du^j$. Then, since $(d|x|)^2 = \langle x,dx\rangle^2/x^2$, we obtain after elementary calculations

$$\lambda = \sum_{\beta,\gamma=1}^{n+1} \lambda_{\beta\gamma} dx_\beta dx_\gamma, \quad \text{where } \lambda_{\beta\gamma} = \frac{\delta_{\beta\gamma}(1-x^2) + x_\beta x_\gamma}{(1-x^2)^2}$$

The Riemannian space $H^{n+1} = (B,\lambda)$ is the projective or the Cayley-Klein model of the hyperbolic space. The inner product in the metric λ we denote by $\langle,\rangle^H$.

2.2. Let F be a piece of a smooth hypersurface embedded in the ball B and r is the embedding map. On F we have two metrics: one induced from R^{n+1} which we denote by g and the other one induced from H^{n+1} which we denote by g^H. It is not difficult to see that the coefficients of the two metrics on F are related as follows:

$$g^H_{ij} = \frac{g_{ij}}{1-r^2} + \frac{\langle r,r_i\rangle\langle r,r_j\rangle}{(1-r^2)^2}, \quad i,j = 1,\ldots,n, \quad r^2 = \langle r,r\rangle,$$

where $r_i = \partial_i r$ and ∂_i denoting differentiation in some local coordinates on F.

2.3. Let N be the Euclidean unit normal vector field on F and N^H the unit normal vector field on F in H^{n+1}. The following relations can be verified directly

$$N^H = \sqrt{\frac{1-r^2}{1-p^2}}\,(N + pr), \text{ where } p = -\langle r,N\rangle. \text{ Also } \langle N^H,N^H\rangle^H = 1.$$

Let b be the second fundamental form of F in R^{n+1}. Then the second fundamental form of F in H^{n+1} is given by

$$b^H = \frac{b}{\sqrt{1-p^2}\,\sqrt{1-r^2}}. \tag{2.1}$$

The last formula implies the following geometric property. Suppose F is a compact hypersurface without boundary and its second fundamental form b is definite everywhere. Then F is a boundary of a strictly convex finite body (see [KN, p. 41]). On the other hand in the projective model of H^{n+1} the geodesics are represented by segments of Euclidean straight lines. Hence, the body bounded by F is convex in H^{n+1} if and only if it is convex in Euclidean sense [A, p. 83].

2.4. The third fundamental form e^H of F has coefficients

$$e^H_{ij} = \langle \partial_i N^H, \partial_j N^H\rangle^H = \frac{e_{ij}}{1-p^2} - \frac{p_i p_j}{(1-p^2)^2}, \tag{2.2}$$

where $e_{ij} = \langle \partial_i N, \partial_j N\rangle$, $p_i = \partial_i p$, $i,j = 1,\ldots,n$. Note that the form e with coefficients e_{ij} is the Euclidean third fundamental form.

2.5. We recall here the setting of the classical Minkowski problem. Let again F be a closed hypersurface with positive Gauss curvature K. Then F is strictly convex and the Gauss map υ: $F \dashrightarrow S$ $(=\partial B)$ is a diffeomorphism [KN, p. 41]. Then the position vector r of F can be identified with $\upsilon^{-1} = \operatorname{grad} p + pu$: $S \dashrightarrow R^{n+1}$, where $u \in S$ and $p = -\langle r,N\rangle$ is the support function of F transplanted to S via υ. In fact, $N = u$ and the matrix of the second fundamental form b of F is given by

$$(b_{ij}) = \operatorname{Hess} p + pe, \tag{2.3}$$

where Hess p denotes the matrix of second covariant derivatives in the standard metric e of S; the third fundamental form is precisely the metric e (see, for example, [HW]).

The principal radii of curvature of F are the eigenvalues of b relative to e and $K = (\det b/\det e)^{-1}$ is the Gauss curvature of F given as a function of the unit normal field on F, that is, as a function on S.

The classical Minkowski problem consists in finding a closed convex hypersurface in Euclidean space for which 1/K evaluated as a function of the unit normal vector field on F is a prescribed in advance function on S. The solution can be found in [N] (n = 2), [P], and [CY].

2.6. In the hyperbolic space the classical Gauss map is not available and the third fundamental form is related to the metric of S in a more complicated way than in R^{n+1}; see formula (2.2).

However, one can use the following construction. Let $q: S \to R$ be a smooth function such that $q^2 + |\nabla q|^2 < 1$. Then the map

$$-v = \text{grad } q + qu: S \to B \tag{2.4}$$

is defined and, furthermore, one can define e^H by (2.2) and b^H by (2.1) and (2.3) with p replaced by q and r by v. In addition, since

$$\det(e^H_{ij}) = (1 - q^2)^{-n-1}(1 - q^2 - |\nabla q|^2)\det(e_{ij}), \tag{2.5}$$

the form e^H is nondegenerate.

Further, if $u^1,\dots,u^n$ are smooth local coordinates on S then

$$(dv^2)^H = (\nabla_{mi}q + qe_{mi})(e^{mk} + q^m q^k)(\nabla_{kj}q + qe_{kj})du^i du^j,$$

where $(e^{mk}) = (e_{ij})^{-1}$ and $q^m = e^{mi}\partial_i q$. Since $\det(e^{mk} + q^m q^k) \neq 0$, it is clear that v is an immersion in H^{n+1}, as well as in R^{n+1}, if and only if

$$\det(\text{Hess } q + qe) \neq 0. \tag{2.6}$$

Consider now the vector field

$$u^H = \sqrt{\frac{1-v^2}{1-q^2}}\,(u + qv), \quad u \in S.$$

We have $\langle \text{grad } v, u^H\rangle^H = \langle \text{grad } v, u\rangle = 0$, $\langle u^H, u^H\rangle^H = 1$. Thus, u^H is automatically the unit "normal" vector field on F in H^{n+1}, and u is the Euclidean normal vector on F. Note that the orthogonality holds even if the differential dv is degenerate.

Finally, since the form e^H is nondegenerate, one may consider the eigenvalues of b^H relative to e^H. Those are the principal radii of curvature of $F = v(S)$.

2.7. The Minkowski problem in H^{n+1} is now formulated as follows. Let $\phi(u)$ be a smooth positive function on S. Is it possible to find a closed strictly convex hypersurface F in H^{n+1} such that at the point of the hypersurface with the tangent hyperplane orthogonal to the direction **O**u ($u \in S$) the exterior Gauss curvature K as a function of u satisfies the equation

$$1/K(u) = \phi(u), \quad u \in S. \tag{2.7}$$

It follows from preceeding arguments that analytically the problem reduces to solving for a positive q the equation

$$\frac{\det(\text{Hess}\, q + qe)}{\det(e)} = \phi\left(\frac{1 - q^2 - |\nabla q|^2}{1-q^2}\right)^{\frac{n}{2}+1} \quad \text{on } S \tag{2.8}$$

under the condition

$$q^2 + |\nabla q|^2 < 1 \text{ on } S. \tag{2.9}$$

Note, that in (2.8) we identified the form e with its coefficient matrix.

2.8. For the spherical space S_+ everything works out similarly. For that reason we just record here the needed formulas.

The ball B is replaced by the Euclidean space R^{n+1}. The metric of S_+ is given by

$$\theta = \sum_{\beta,\gamma=1}^{n+1} \theta_{\beta\gamma} dx_\beta dx_\gamma, \quad \text{where } \theta_{\beta\gamma} = \frac{\delta_{\beta\gamma}(1+x^2) - x_\beta x_\gamma}{(1+x^2)^2}.$$

The space $S_+ = (R^{n+1}, \theta)$ is the projective model of a hemisphere of a unit hypersphere in R^{n+1}. The inner product in S_+ we denote by $<,>^\theta$. For a smooth hypersurface F in R^{n+1} the metric induced from $<,>$ is denoted by g and the metric induced from S_+ by g^θ. We have

$$g^\theta_{ij} = \frac{g_{ij}}{1+r^2} - \frac{<r,r_i><r,r_j>}{(1+r^2)^2}, \quad i,j = 1,\dots,n, \quad r^2 = <r,r>.$$

The normal vector field on F in S_+ is

$$N^\theta = \sqrt{\frac{1+r^2}{1+p^2}}\,(N - pr), \text{ where } p = -\langle r,N\rangle. \text{ Also } \langle N^\theta, N^\theta\rangle^\theta = 1.$$

The second fundamental forms are related as follows:

$$b^\theta = \frac{b}{\sqrt{1+p^2}\,\sqrt{1+r^2}}.$$

For the coefficients of the third fundamental form we get

$$e^\theta_{ij} = \langle \partial_i N^\theta, \partial_j N^\theta\rangle^\theta = \frac{e_{ij}}{1+p^2} + \frac{p_i p_j}{(1+p^2)^2}.$$

Let, as before, S be a unit n-dimensional sphere in R^{n+1} centered at the origin **O**. Then repeating the arguments in sections 2.6-2.7 we arrive at the following analogue of (2.8); that is, from the analytical point of view the Minkowski problem in S_+ consists in solving for positive q the equation

$$\text{(2.10)} \qquad \frac{\det(\text{Hess } q + qe)}{\det(e)} = \phi\left(\frac{1 + q^2 + |\nabla q|^2}{1+q^2}\right)^{\frac{n}{2}+1} \quad \text{on } S.$$

Here ϕ is a given positive function on S.

2.9. Remark. Formula (2.1) in case n = 2 is given in [KI]. It is derived there with the use of properties of vector products.

§3. EXISTENCE AND UNIQUENESS OF A SOLUTION TO (2.8).

3.1. The necessary conditions for solvability of (2.8) are not known. In the following result we give a set of conditions sufficient for its solvability.

Theorem 3.1. *Consider the projective model of* H^{n+1} *as in §2 and let* $\phi(u,q)$: $S \times [0,1) \to R^+ = (0,\infty)$. *Suppose that* ϕ *satisfies the following conditions:*

a) $\phi(u,q) \in C^k(S \times [0,1))$, $k \geq 3$;

b) <u>there exist two numbers</u> R_1 <u>and</u> R_2, $0 < R_1 \le R_2 < 1$, <u>such that for all</u> $u \in S$

$$\phi(u,q) \le q^n \quad \text{when } q = R_1 \text{ and}$$

$$\phi(u,q) \ge q^n \quad \text{when } q = R_2;$$

c) $$\frac{\partial \phi}{\partial q} \ge \frac{\phi}{q}\left[n + (n+2)\frac{R_2^2 - R_1^2}{(1-R_2^2)^2}\right] \quad \text{for } q \in [R_1,R_2],\ u \in S.$$

<u>Then there exists in</u> H^{n+1} <u>a unique convex hypersurface</u> F <u>such that</u>

i) <u>F is given by the embedding</u> $-r = \text{grad } q + qu$, $u \in S$, <u>where</u> q <u>is a function of class</u> $C^{k+1,\alpha}(S)$, $\alpha \in (0,1)$, <u>and if</u> ϕ <u>is analytic then</u> q <u>is analytic;</u>

ii) <u>the Gauss curvature of</u> F <u>satisfies the relation</u>

$$K^{-1}(u) = \phi(u,q(u));$$

iii) <u>the hypersurface</u> F <u>is located between two concentric n-spheres of radii</u> R_1 <u>and</u> R_2 <u>centered at</u> **O**.

3.2. The proof uses the continuation method in the same way as described in 1.2. The condition b) implies uniform C^0- and C^1-estimates as in Lemma 0.8 (for the C^1 -estimates consider the maximum of the function $w \equiv q^2 + |\nabla q|^2$ and then one can see that $w \le \max q^2 \le R_2^2 < 1$, and consequently, $|\nabla q|^2 \le R_2^2 - q^2$). The C^2- and C^3- estimates are essentially the same as in [0].

The condition c) is needed in order to show that the kernel of the operator obtained by linearization of (2.8) is trivial. This condition can be replaced by a slightly weaker one, namely,

c') $$\frac{\partial \phi}{\partial q} \ge \frac{\phi}{q}\left[n + (n+2)\frac{R_2^2 - q^2}{(1-q^2)(1-R_2^2)}\right] \quad \text{for } q \in [R_1,R_2],\ u \in S.$$

3.3. In the case of the spherical space S_+ a theorem similar to the Theorem 3.1 can be also formulated. The changes that should be made are as follows: in condition b) the restriction $R_2 < 1$ is replaced by $R_2 < \infty$; the condition c) is replaced by $\partial(\phi q^{-n})/\partial q \ge 0$ for $q \in [R_1, R_2]$, $u \in S$.

NOTE. We take the opportunity to point out the following misprints in [O1].

p. 19, condition b). $\cosh^n \rho$ should be replaced by $\coth^n \rho$.

p. 33, condition c). The factor ϕq should be replaced by ϕ/q.

REFERENCES

[A] A.D. Aleksandrov, Convex Polyhedra, GIITL, M.-L.,1950 (in Russian); German transl., Akademie-Verlag, Berlin, 1958.

[A1] ---------------, Uniqueness theorems for surfaces in the large, III, Vestnik Leningrad University, 7(1958), 15-44 (in Russian); Engl. transl., Amer. Math. Soc. Transl., Ser. 2, 21(1962), 341-416.

[BK] I. Bakelman and B. Kantor, Estimates for solutions of quasilinear elliptic equations connected with problems of geometry "in the large", Mat. Sbornik 91(133) (1973), 336-349 (in Russian); Engl. transl., Math. USSR Sbornik, no. 3, 20(1973), 348-363.

[CNS] L. Caffarelli, L. Nirenberg, J. Spruck, Nonlinear second order elliptic equations IV. Starshaped compact Weingarten hypersurfaces, Current Topics in PDE's, edited by Y. Ohya, K. Kasahara, N. Shimakura, Kinokunia Company LTD, Tokyo 1986, 1-26.

[C] E. Calabi, Improper affine hyperspheres of convex type and a generalization of a theorem by K. Jorgens, Mich. Math. J. 5(1958), 105-126.

[CY] S. Y. Cheng and S.T. Yau, On the regularity of the solution of the n-dimensional Minkowski problem, Comm. Pure Appl. Math. 29(1976), 105-126.

[F] A.P. Filimonova, Estimates in the C^2-metric and uniqueness of a convex surface with given Gauss curvature in H^3, Geometry, 4(1975) 64-68 (in Russian).

[GT] D. Gilbarg and N.S. Trudinger, Elliptic Partial Differential Equations of Second Order, Second edition, Springer-Verlag, Berlin-Heidelberg-New York, 1983.

[HW] P. Hartman and A. Wintner, On the third fundamental form of a surface, Amer. J. Math., 75(1953), 298-334.

[IK] L. Il'ina and A. Kagan, Surfaces in Lobachevski space in Cayley-Klein model and infinitesimal deformations, Vestnik Leningrad Univ., no. 19(1972), 21-28.

[KN] S. Kobayashi and K. Nomidzu, Foundations of Differential Geometry, v. II, New York, 1969.

[N] L. Nirenberg, The Weyl and Minkowski problems in differential geometry in the large, Comm. Pure Appl. Math. 6(1953), 337-394.

[O] V.I. Oliker, Hypersurfaces in R^{n+1} with prescribed Gaussian curvature and related equation of Monge-Ampere type, Comm. in PDE's, 9(8)(1984), 807-838.

[O1] ----------, Existence and uniqueness of convex hypersurfaces with prescribed Gaussian curvature in spaces of constant curvature, Seminari dell' Istituto di Matematica Applicata "Giovanni Sansone", Universita degli studi di Firenze, 1983.

[P] A.V. Pogorelov, The Minkowski Multidimensional Problem, Nauka, Moscow, 1975 (in Russian); Engl. transl., John Wiley and Sons, New York, 1978.

[TW] A. E. Treibergs and S.W. Wei, Imbedded hypersurfaces with prescribed mean curvature, J. Diff. Geometry, 18(1983), 513-521.

DEPARTMENT OF MATHEMATICS AND COMPUTER SCIENCE
EMORY UNIVERSITY
ATLANTA, GEORGIA, 30322

Contemporary Mathematics
Volume **101**, 1989

SELF-DUALITY, TWISTOR THEORY, ITS GENERALIZATION AND APPLICATION

Y. Sun Poon

§1 Twistor theory has its root in mathematical physics [P]. In 1978, Atiyah et. al. [A] developed this theory on 4-dimensional Riemannian self-dual manifolds. Then Salamon realized that quaternionic structure on 4n dimensional manifold can be considered as a high dimensional analogue of the (anti-)self-dual conformal structure on a 4-fold [S_1]. We shall discuss some results related to the author's work in this area.

A four dimensional oriented Riemannian manifold is self-dual [A] if the Weyl tensor of the Levi-Civita connection is invariant under the action of the Hodge $*$-operator. As the $*$-operator on 2-forms over 4-fold and the Weyl tensor are conformal invariants, a conformal change of self-dual metric is again self-dual. Fundamental nontrivial examples are the Euclidean 4-sphere, the complex projective plane with Fubini-Study metric.

1980 *Mathematics Subject Classification* (1985 Revision). 32C10, 53C25.

More exotic example is the K3 surface with Calabi-Yau metric and orientation opposite to the canonical one [B].

Twistor space is a complex manifold associated to a self-dual manifold naturally as follow: the holonomy of an oriented 4-dimensional Riemannian manifold, X, is contained in the group SO(4), i.e. $SU(2)xSU(2)/\mathbb{Z}_2$. Each copy of SU(2) has an irreducible representation of dimension 3. Associated to the principal bundle of frame on X are two rank 3 bundles. They are the bundles of self-dual and anti-self-dual 2-forms respectively. The total space, Z, of the unit sphere bundle of anti-self-dual 2-forms is the twistor space. It can also be considered as the associated fibre bundle induced by the action of SU(2) acting on the projective space of the 2-dimensional representation of SU(2). As one can use the metric to identify 2-forms with skew-symmetric endomorphism on tangent bundle, points on twistor space are almost complex structures on X. Using the induced connection of the twistor space to split the tangent space into horizontal and vertical parts, one can choose the complex structure at a point J on the twistor space to be the standard Riemann sphere structure on the vertical part and the J on the horizontal part. On a self-dual manifold, the obstruction to the integrability vanishes, and therefore Z is a complex manifold [A].

Moreover, the fibres of the projection, π, from Z onto X are holomorphic rational curves. This fibration is called the twistor fibration and the fibres are called the twistor lines.

Since the Hodge $*$-operator on 2-forms over a four dimensional manifold is conformally invariant, the holomorphic geometry depends only on the conformal geometry on X. On the other hand, given a twistor space, i.e. a complex manifold foliated by appropriate family of rational curves, one can always find a self-dual conformal class on a 4-fold [B]. This one-to-one correspondence is usually called the twistor correspondence or Penrose correspondence.

We shall discuss how the twistor theory and its generalization can be applied to construct or classify self-dual manifolds and its higher dimensional analogue.

§2 As the twistor correspondence is one-to-one, one may try to construct self-dual manifolds by producing a twistor space. For example, $\mathbb{CP}^3$ is the twistor space over S^4, the projection is simply the Hopf fibration. On the complex projective plane $\mathbb{CP}^2$, the twistor space is the flag F(1,2) of lines and planes in $\mathbb{C}^3$. The projection from F(1,2) onto $\mathbb{CP}^2$ is to take the intersection of a given plane and the orthogonal complement of a given line in the given plane. With this approach, new examples of self-dual structures were found on the connected-sum of projective planes [DF, P_1]. It was shown in [P_1] that a small resolution of the intersection of two quadrics with four nodes in $\mathbb{CP}^5$ is the twistor spaces associated to the connected-sum of two projective planes. The fibres of the projection from the small resolution to the connected-sum are conics. Algebraically, the intersection is given by:

$$Q_0 = \{z \in \mathbb{CP}^5: \alpha z_0^2 + \alpha z_1^2 + \beta z_2^2 + \gamma z_3^2 + \lambda z_4^2 + \lambda z_5^2 = 0\};$$

and $$Q_\infty = \{z \in \mathbb{CP}^5: z_0^2 + z_1^2 + z_2^2 + z_3^2 + z_4^2 + z_5^2 = 0\},$$

where $\alpha > \beta > \gamma > \lambda$. There are two interesting limits in this family of intersection of quadrics described by Donaldson and Kronheimer [D,K] respectively as follows:

1) when $\beta = \gamma$, and α approaches to λ by passing through the point at infinity, the above intersection is deformed to the intersection of

$$Q_0 = \{z \in \mathbb{CP}^5: z_2^2 + z_3^2 = 0\};$$

and $$Q_\infty = \{z \in \mathbb{CP}^5: z_0^2 + z_1^2 + z_2^2 + z_3^2 + z_4^2 + z_5^2 = 0\}.$$

This intersection can be considered as gluing the two quadrics

$$F_1 = \{z \in \mathbb{CP}^5 : z_0^2 + z_1^2 + z_4^2 + z_5^2 = 0,\ z_2 - iz_3 = 0\}$$

and $$F_2 = \{z \in \mathbb{CP}^5 : z_0^2 + z_1^2 + z_4^2 + z_5^2 = 0,\ z_2 + iz_3 = 0\}$$

along the quadric surfaces

$$Q = \{z \in \mathbb{CP}^5 : z_0^2 + z_1^2 + z_4^2 + z_5^2 = 0,\ z_2 = 0,\ z_3 = 0\}.$$

Note that F_1 and F_2 can be considered as the blowing-up of the twistor space F(1,2) along a twistor line whose exceptional divisor is exactly the quadric surface Q.

In general, Donaldson and Friedman [DF] recently found a general construction of self-dual structure on the connected-sum of two self-dual manifolds. In short, they blow up a twistor line from the twistor space of each manifold. The exceptional divisors are quadric surface because the normal bundle of a twistor line is $\mathbf{H} \oplus \mathbf{H}$, where $\mathbf{H}$ is the hyperplane bundle on a complex projective curve. When there is no obstruction, the complex space obtained by gluing the two twistor spaces along the quadric surfaces can be deformed to a twistor space. Then the Penrose correspondence provides the last step to produce a self-dual structure. Examples on which their construction can be carried out are the connected-sums of the complex projective planes.

2) when $\alpha=\beta$, and $\gamma=\lambda$, the intersection is deformed to be

$$Q_0 = \{z \in \mathbb{CP}^5 : z_0^2 + z_1^2 + z_2^2 = 0\}$$

and $Q_\infty = \{z \in \mathbb{CP}^5 : z_3^2 + z_4^2 + z_5^2 = 0\}$.

This is the image of $\mathbb{CP}^3$ in $\mathbb{CP}^5$ via the map

$$[z_0 : z_1 : z_2 : z_3] \to [z_0^2 : z_0 z_1 : z_1^2 : z_2^2 : z_2 z_3 : z_3^2].$$

i.e. the singular space $\mathbb{CP}^3/\mathbb{Z}_2$. Although this space is not smooth, it is the twistor space of the orbifold $S^4/\mathbb{Z}_2$. This orbifold is not only self-dual but also Einstein. But this orbifold can be considered as trivial in the sense that it is the quotient of a smooth manifold by finite group of automorphism.

On the other hand, Galicki and Lawson found a family of nontrivial self-dual Einstein orbifold [GL]. From a generalization of the symplectic reduction and hyper-Kähler reduction [HK], they developed the so-called quaternionic reduction on quaternionic Kähler manifolds and found new examples of compact self-dual Einstein orbifolds and quaternionic Kähler orbifolds with positive scalar curvature. By definition, a quaternionic Kähler manifold is a Riemannian manifold whose holonomy is in the group $Sp(n)Sp(1)$. It is known for some time that quaternionic Kähler manifolds are Einstein space [B]. However, Salamon [S_1,S_2] found that quaternionic Kähler manifold can be considered as a high dimensional generalization of (anti-)self-dual Einstein manifold in the sense that the associated fibre bundle of the action of Sp(1) on $\mathbb{CP}^1$ is a manifold with a natural integrable complex structure. As in the four dimensional case, the total space of this fibre bundle is called the twistor space associated to the quaternionic Kähler manifold. Moreover, the Levi-Civita connection induces a horizontal distribution on the twistor space. When the scalar curvature is not zero, this distribution is the kernel of a complex contact structure [S_2]. Basic example of quaternionic Kähler manifold is projective quaternionic $\mathbb{HP}^n$. Its twistor space is $\mathbb{CP}^{2n+1}$. He further generalized to define a quaternionic manifold as a 4n-dimensional manifold, $n \geq 2$, with a $GL(n,\mathbb{H})Sp(1)$-structure admitting a torsion-free connection. The definition is tailored so that the associated twistor space has a natural integrable complex structure. This type of geometry should be considered as a high dimensional version of a (anti-)self-dual conformal structure on a 4-fold. To complete the Penrose correspondence, the author and Pedersen [PP_2] recently proved that a twistor space determines a quaternionic structure on a manifold.

This observation enables us to find new examples of quaternionic manifolds. We can also mimic the four dimensional case and prove that a twistor space with appropriate contact structure will generate a quaternionic Kähler manifold, a large family of new examples construced by this twistorial method is found by LeBrun [L_1] who also independently proved this result. As the constructions of LeBrun and Galicki and Lawon are already discussed in another survey article [S_4], we skip it here.

§3 Since the holomorphic data on a twistor space determines a quaternionic structure, it can not only help to construct a self-dual or quaternionic structure but also provide a way to classification. The first such theorem is Hitchin's result [H_1] that a compact Kählerian twistor space on a self-dual manifold is necessarily the twistor space of the Euclidean sphere S^4 and the twistor space of the projective plane with the Fubini-Study metric. As the twistor space of a self-dual Einstein manifold with positive scalar curvature is Kählerian, this theorem implies that the Riemannian manifolds are the only examples of compact self-dual Einstein manifolds with positive scalar curvature, a result also proved by Friedrich and Kurke [FK]. The outline of Hitchin's method is the following:

on a twistor space, there is always an anti-holomorphic involution σ induced by antipodal map on the twistor lines. It is usually called a real structure. If $[\omega]$ is a Kähler class, then $[\omega]-\sigma^*[\omega]$ is a σ-invariant class represented by positive form. As σ-invariant class is generated by the first Chern class of the twistor space, the anticanonical bundle is positive. In particular, the Chern number c_1^3 is positive. With the Hodge symmetry, it imposed topological constraints on the twistor space and hence on the self-dual manifold. To be precise, the signature of X must be 0, 1, 2, or 3. Moreover, from the tautalogical construction of the twistor space, one can find a holomomrphic line bundle $\mathbf{K}^{-1/2}$ such that $(\mathbf{K}^{-1/2})^2=\mathbf{K}^{-1}$. This is also a positive line bundle. With the Riemann-Roch formula and the given topological constraints, one can prove that the Euler number of the bundle $\mathbf{K}^{-1/2}$ is positive. Then an application of the vanishing theorem of Kodaira will implies that this bundle has global nontrivial sections. The main body of the work is to show that the associated map of this line bundle is an embedding. Finally, when the signature is equal to 2 and 3, he identified the image of the embedding as the complete intersection of two smooth quadrics in $\mathbb{CP}^5$ and the double covering of $\mathbb{CP}^3$ branched along a smooth quartic surface. However, they cannot be a twistor space because the topological data are not compatible. When the signature is equal to 0 or 1, he identified the twistor space as stated by computing the group of conformal transformation of the self-dual manifold. This computation is possible because the twistor correspondence identifies this group as the real form of the group of holomorphic transformation on the twistor space.

This entire programme is adapted by the author and Salamon [PS] to show that any 8-dimensional compact quaternionic Kähler manifold with positive scalar curvature is symmetry. In fact, with the assumption that the scalar curvature is positive, one can construct a canonical Einstein Kähler metric with positive scalar curvature on the twistor space. Then a natural third root of the anticanonical bundle, $\mathbf{K}^{-1/3}$, is positive. Except when the twistor space is the complex projective space, this bundle is the generator of the Picard group on the twistor space [S2]. In other words, the twistor space is a 5-dimensional Fano-manifold with index 3. In this case, it is not hard to apply the Riemann-Roch formula and the Serre duality to deduce that the bundle $\mathbf{K}^{-1/3}$ has global sections. The work again is to show that the associated map of this bundle is an embedding. On the other hand, the algebra of Killing vector field on X is the real form of the space of sections of $\mathbf{K}^{-1/3}$. When this space is sufficiently large, then the manifold is homogeneous and hence symmetric [Al]. When it is not large, we are able to identify the image of the embedding and hence determine the twistor spaces of all 8-dimensional quaternionic Kähler manifolds with positive scalar curvature.

It remains to see whether there are non-symmetric compact quaternionic Kähler manifolds with positive scalar curvature in higher dimension. However, LeBrun [L_2] shows that any compact quaternionic Kähler manifold with positive scalar curvature is rigid in the sense that such geometrical structure has no deformation on the given differentiable manifold. This is the consequence of his result that there is no contact deformation on the twistor space when the scalar curvature on X is positive.

The idea of Hitchin can also be further developed in the four dimensional case [P_3, P_4, V]. In fact, when a compact twistor space is Moishezon, the self-dual conformal class must contain a metric with constant positive scalar curvature. The basic reason is that the meromorphic function field on such a complex manifold is always generated by the sections of a certain holomorhic line bundle **L**. Then the sections of the real holomorphic bundle $\mathbf{L}\otimes\sigma^*\bar{\bar{\mathbf{L}}}$ also generates the function field. As real holomorphic line bundles on such a twistor space can be proved to be generated by the anticanonical bundle, $\mathbf{L}\otimes\sigma^*\bar{\mathbf{L}}$ is a positive power of the anticanonical bundle. However, when the scalar curvature is negative, a Bochner type argument, using the twistor operator, shows that any positive power of the anticanonical bundle has no nontrivial sections. When the scalar curvature is equal to zero, this argument is refined to show that sections of any power of the anticanonical bundle cannot generate the function field. Examples of Moishezon twistor spaces are the twistor spaces of a self-dual structure on the connected-sum of two and three copies of the complex projective plane [P_1, P_2]. They are the small resolution of the intersection of two quadrics in $\mathbb{CP}^5$ with four nodes and the double covering of $\mathbb{CP}^3$ branched along a quartic with 13 nodes respectively. More generally, when the 4-fold is simply connected, if the algebraic dimension of a compact twistor space is positive, one can prove that the self-dual 4-fold is either the K3-surface with Calabi-Yau metric and opposite orientation or homeomorphic to the connected-sum of complex projective planes with a self-dual conformal class containing metrics with positive scalar curvature [P_4, V]. The motivation of this generalization is that the construction of Donaldson and Friedman shows that the connected-sum of projective planes always admits such a family of self-dual conformal class. One might want to know if there are special elements in this family. As the conformal structure is encoded in the holomorphic structure of the twistor space, one may want to compute a certain algebraic invariant on the twistor space. Therefore, the algebraic dimension is one of the invariant that we want to know. For example, a generic twistor space over the connected-sum of four $\mathbb{CP}^2$ has algebraic dimension one [DF]. Can it jump or not?

§4 In [S_1], Salamon showed that there is a generalization of the Ward correspondence of instantons and holomorphic bundles on the twistor spaces. Essentially, the Ward correspondence states that there is a one-to-one correspondence between SU(2) bundles $\mathbf{E}$ with self-dual connection on a self-dual manifolds and real holomorphic rank 2 bundle $\pi^*\mathbf{E}$ on the twistor space that is trivial on every real twistor line. A self-dual SU(2) connection is usually called an instanton. It is well-known that instanton is the minimum of the Yang-Mills functional:

$$\mathcal{YM}(\nabla) = \int_X \| F_\nabla \|^2 \, vol$$

$$= -\int_X Tr(F_\nabla \wedge * F_\nabla) \quad ,$$

where F_∇ is the curvature of the connection of ∇. In fact, the curvature can be splitted into a direct summands:

$$F_\nabla = F^+ + F^-,$$

where F^+ and F^- satisfy the self-dual and anti-self-dual equations respectivley:

$$*F^+ = F^+ , \quad *F^- = -F^- .$$

Then $\mathcal{YM}(\nabla) = \| F^+ \|^2 + \| F^- \|^2$.

This functional is bounded below by $8\pi^2|k|$ where $-k$ is the second Chern number of the SU(2) bundle. And

$$8\pi^2 k = \| F^+ \|^2 - \| F^- \|^2.$$

Salamon generalizes the Ward correspondence algebraically in the following way: by definition, the principal bundle of frame on a quaternionic manifold X can be reduced to $GL(n,\mathbb{H})Sp(1)$. If E is the complex rank 2n dimension representation of $GL(n, \mathbb{H})$ given by the inclusion in $GL(2n, \mathbb{C})$ and H is the irreducible rank 2 representation of Sp(1), then the bundle of 2-forms on X is associated to the frame bundle by the representation $\Lambda^2(E \otimes H)$ [S_2]. It has a decomposition:

$$\Lambda^2(E\otimes H) \cong S^2E\otimes\Lambda^2H \oplus \Lambda^2E\otimes S^2H$$
$$\cong S^2E \oplus \Lambda^2E\otimes S^2H. \quad (4.1)$$

If the manifold is actually a quaternionic Kähler manifold, then E has an invariant symplectic form and $\Lambda^2(E\otimes H)$ is further decomposed into

$$\Lambda^2(E\otimes H) \cong S^2E \oplus S^2H \oplus \Lambda_0^2E\otimes S^2H, \quad (4.2)$$

where Λ_0^2E is the orthogonal compliment of the symplectic form on E in Λ^2E. Then the bundle of 2-forms has a corresponding decomposition. Of course, when n=1, Λ^2E is one dimensional and the above sum is simply the sum of the complexification of self-dual and anti-self-dual 2-forms. Salamon called a vector bundle **V** on a quaternionic manifold quaternionic if it admits a GL(p,$\mathbb{H}$) structure and a connection whose curvature 2-form is in the component S^2E. And he found that for such bundle, as in the Ward cooresponderice, π^***V** on the twistor space is a holomorphic bundle. Examples are the tangent bundles on quaternionic Kähler manifolds [S_1].

On the other hand, the Yang-Mills functional on quaternionic Kähler manifold can be modified [GP] as follows:

$$\mathcal{YM}_c(\nabla) \equiv \tfrac{1}{2}\int_X \left[\, \| F_\nabla \|^2 + c^2 \| F_\nabla \wedge \Omega^{n-1} \|^2 \,\right] \quad ,$$

where c is a real number and

$$\Omega \equiv \omega_1 \wedge \omega_1 + \omega_2 \wedge \omega_2 + \omega_3 \wedge \omega_3 ,$$

when $\omega_1, \omega_2, \omega_3$ forms an orthogonal frame with length $\sqrt{2}n$ on the bundle associated to the representation S^2H. Part of the reasons that we choose this modification is that this functional has a lower bound that depends only on the topology of the bundle **V** and the geometry of X [GP]:

$$c\,(8\pi^2 \int_X p_1(\mathbf{V}) \wedge \Omega^{n-1}) \le \mathcal{YM}_c(\nabla)$$

with equality if and only if

$$*F_\nabla = c\, F_\nabla \wedge \Omega^{n-1}. \tag{4.3}$$

When n=1, this equation is reduced to

$$*F_\nabla = c\, F_\nabla .$$

And because the $*$-operator on 2-forms over four dimensional manifold is an involution, c must be either 1 or -1. Equation (4.3) is reduced to the self-dual and anti-self-dual equations (4.1). Note that although the Yang-Mills functional is modified so that solutions to equation (4.3) are absolute minima or maxima, the solutions are also critical points of the standard Yang-Mills functional because the Euler-Lagrange equations of the Yang-Mills functional are $d^\nabla F_\nabla = 0$

and $d^{\nabla} * F_{\nabla} = 0$. And these equations are satisfied because the fundamental 4-form Ω is parallel and F_{∇} satisfied the Bianchi identity.

Now the obvious question is how the Salamon's notion of quaternionic bundle is related to this generalization of Yang-Mills functional. It turns out that if F is a 2-form such that

$$* F = - \frac{1}{(2n-1)!} F \wedge \Omega^{n-1} \tag{4.4}$$

then F is in the component S^2E. To express this equation pointwisely, one may choose an orthonormal basis $\{\omega_0^j, \omega_1^j, \omega_2^j, \omega_3^j;\ j=1,...,n.\}$ on the cotangent bundle over a point on X. If

$$\omega_1 \equiv \Sigma_1^n (\omega_0^j \wedge \omega_1^j + \omega_2^j \wedge \omega_3^j),$$

$$\omega_2 \equiv \Sigma_1^n (\omega_0^j \wedge \omega_2^j - \omega_1^j \wedge \omega_3^j),$$

$$\omega_3 \equiv \Sigma_1^n (\omega_0^j \wedge \omega_3^j + \omega_1^j \wedge \omega_2^j),$$

then $\{ \omega_1, \omega_2, \omega_3 \}$ is an orthogonal frame on S^2H with lenght $\sqrt{2n}$.

and $\Omega \equiv \omega_1 \wedge \omega_1 + \omega_2 \wedge \omega_2 + \omega_3 \wedge \omega_3$. When $F\binom{i}{\alpha}\binom{j}{\beta}$ is the coefficient of a 2-form F with respect to the frame

$$\{\ \omega^i_\alpha \wedge \omega^j_\beta\ :\alpha,\ \beta=0,\ 1,\ 2,\ 3;\ \ i,\ j=1,...,n\ \},$$

then F satisfies equation (4.4) if and only if

$$F_{\binom{i}{0}\binom{j}{1}} = -F_{\binom{i}{2}\binom{j}{3}}, \qquad F_{\binom{i}{0}\binom{j}{2}} = F_{\binom{i}{1}\binom{j}{3}}, \qquad F_{\binom{i}{0}\binom{j}{3}} = -\ F_{\binom{i}{1}\binom{j}{2}},$$

$$F_{\binom{i}{\alpha}\binom{j}{\alpha}} = F_{\binom{i}{\beta}\binom{j}{\beta}}, \qquad F_{\binom{i}{\alpha}\binom{j}{\beta}} = F_{\binom{j}{\alpha}\binom{i}{\beta}}, \qquad \forall\ \alpha,\ \beta,\ i,\ j. \tag{4.5}$$

Going through the identification:

$$\Lambda^2T^* \cong \Lambda^2(E\otimes H) \cong S^2E \oplus S^2H \oplus \Lambda^2_0E\otimes S^2H,$$

one can check that such an F is in the component S^2E. Counting the degree of freedom, one can conclude that any F in S^2E satisfies these equations. In particalur, the connection defining a quaternionic vector bundle is a minimun of the generalized Yang-Mills functional. This observation justifies Salamon's approach to generalize the concept of instantons on quaternionic Kähler manifold.

Moreover, the construction of S^2E does not depend on the choice of a quaternionic Kähler metric within a given quaternionic structure, equation (4.4) is independent of the choice of metric so long as there is one quaternionin Kähler metric within the given structure. This is the analogue in the four dimensional case that the self-dual and anti-self-dual equations are conformally invariant. Similarly, one can show that a 2-form F is in the component S^2H if and only if it satisfies the following generalization of self-dual equation:

$$* F = \frac{6n}{(2n+1)!} F \wedge \Omega^{n-1}$$

However, except when n=1, this equation depends on the choice of a metric as the decomposition (4.2) does.

To find examples of solution to this pair of equations and see the relation between equation (4.4) and other topics. Let's consider the projection [H_2,H_3]

$$p: \mathbb{R}^4 \otimes \mathbb{R}^n \to \mathbb{R}^3 \otimes \mathbb{R}^n$$

given by

$$(x_0^i, x_1^i, x_2^i, x_3^i) \mapsto (x_1^i, x_2^i, x_3^i),$$

where $1 \le i \le n$. Suppose that **V** is a vector bundle on $\mathbb{R}^4 \otimes \mathbb{R}^n$ with connection ∇' invariant under the translation in the coordinates x_0^i, then $\mathbf{V} = p^* \widetilde{\mathbf{V}}$ for a bundle $\widetilde{\mathbf{V}}$ on $\mathbb{R}^3 \otimes \mathbb{R}^n$ and

$$\nabla' = p^* \nabla + \Sigma_i \Phi^i dx_0^i,$$

where Φ^i is a section of the adjoint bundle of $\widetilde{\mathbf{V}}$. One can check that when the curvature F of ∇' satisfies the equation (4.4), or equivalently equation (4.5), then

$$F_{X^i_\alpha X^j_\beta} = \Sigma_\gamma \varepsilon_{\alpha\beta\gamma} \nabla_{X^i_\gamma} \Phi^j + \tfrac{1}{2}\delta_{\alpha\beta}[\Phi^i,\Phi^j]$$

and $\nabla_{X^i_\alpha}\Phi^j = \nabla_{X^j_\beta}\Phi^i$, where $\alpha,\beta=1,2,3$; $i,j=1,\ldots n$.

This is the generalization of Bogomolny equation found in [PP_1]. If the adjoint bundle of $\widetilde{\mathbf{V}}$ is abelien, this pair of equation is reduced to

$$F_{X^i_\alpha X^j_\beta} = \Sigma_\gamma \varepsilon_{\alpha\beta\gamma} \nabla_{X^i_\gamma} \Phi^j \text{ and } \nabla_{X^i_\alpha}\Phi^j = \nabla_{X^j_\beta}\Phi^i.$$

It was shown [HK, PP_1] that coefficients of a hyperkähler metric on a 4n dimensional space with n commuting Killing field preserving the hyperkähler structure will determine a solution to this pair of equation.

REFERENCE

[A] M. F. Atiyah, N, J, Hitchin and I. M. Singer, *Self-duality in four dimensional Riemannian geometry*, Proc. roy. Soc. London Ser. A**362**(1978)425-461.

[Al] D. V. Alekseevskii, *Compact quaternion spaces,* Functional Anal. Appl. **2** (1968) 106-114.

[B] A. Besse, *Einstein manifolds*, Spring-Verlag, (1987).

[D] S. K. Donaldson, private communication.

[DF] S. K. Donaldson, R. Friedman, *Connected sums of self-dual manifolds and deformations of singular spaces,* preprint.

[FK] T. Friedrich, H. Kurke, *Compact four-dimensional self-dual Einstein manifolds with positive scalar curvature,*

[GL] K. Galicki, H. B. Lawson Jr., *Quaternionic reduction and quaternionic orbifolds,* to appear in Math. Ann..

[GP] K. Galicki, Y. S. Poon, manuscript.

[H_1] N. J. Hitchin, *Kählerian twistor spaces* , Proc. London Math. Soc. (3) **43** (1981) 133-150.

[H_2] N. J. Hitchin, *Monopoles and geodesics,* Commun. Math. Phys. **83** (1982) 579-602.

[H_3] N. J. Hitchin, *On the construction of monopoles,* Commun. Math. Phys. **89** (1983) 145-190.

[HK] N. J. Hitchin, A. Karlhede, U. Lindström, M. Roček, *Hyperkähler metrics and supersymmetry,* Commun. Math. Phys. **108** (1987) 537-589.

[K] P. Kronheimer, private communication.

[L_1] C. LeBrun, *A rigidity theorem for quaternionic Kähler manifolds,* preprint.

[L_2] C. LeBrun, *Quaternionic Kähler manifolds and conformal geometry,* preprint.

[P] R. Penrose, *Twistor theory, its aims and achievements,* in Quantum Gravity, C. J. Isham, R. Penrose and D. W. Sciama, eds., Clarendon Press, Oxford (1975) 268-407.

[PP_1] H. Pedersen and Y. S. Poon, *Hyper Kähler metrics and a generalization of the Bogomolny equations,* to appear in Comm. Math. Phy.

[PP_2] H. Pedersen and Y. S. Poon, *Twistorial construction of quaternionic manifolds,* preprint (being revised).

[P_1] Y. S. Poon, *Compact self-dual manifolds with positive scalar curvature*, J. Differential Geo. **24** (1986), 97-132.

[P_2] Y. S. Poon, *Small resolution of double solids as twistor spaces*, preprint (being revised)

[P_3] Y. S. Poon, *Algebraic dimensions of twistor spaces*, to appear in Math. Ann.

[P_4] Y. S. Poon, *Twistor spaces with meromorphic functions,* preprint.

[PS] Y. S. Poon and S. M. Salamon, *Eight-dimensional quaternionic Kähler manifolds with positive scalar curvature,* preprint.

[S_1] S. M. Salamon, *Quaternionic Manifold,* Symp. Math., **26** (1982) 139-151.

[S_2] S. M. Salamon, *Quaternionic Kähler Manifolds,* Invent. Math., **67**(1982) 143-171.

[S_3] S. M. Salamon, *Differential geometry of Quaternionic manifolds,* Ann. scient. Éc. norm. sup., 4^c série, t.19 (1986) 31-55.

[S_4] S. M. Salamon, *Quaternionic Kähler geometry,* in preparation.

[V] M. Ville, *On twistor spaces with positive algebraic dimension*, preprint.

Department of Mathematics
Rice University
P.O.Box 1892
Houston, TX77251

June, 1988

Contemporary Mathematics
Volume **101**, 1989

Spectral geometry of Riemannian manifolds

Peter B. Gilkey *

Let M be a compact Riemannian manifold without boundary of dimension m and let V be a smooth vector bundle over M. Let $D:C^\infty(V)\rightarrow C^\infty(V)$ be a self-adjoint elliptic second order partial differential operator. We assume the leading symbol $\sigma_L(x,\xi)=|\xi|^2\cdot I$ is scalar so D is a generalized Laplacian. Let $\{\lambda_\nu\}$ be the eigenvalues of D where each eigenvalue is repeated according to its multiplicity; only a finite number of the λ_ν are negative and e^{-tD} is trace class on $L^2(V)$. As $t\rightarrow 0^+$, there is an asymptotic series of the form:

$$\mathrm{Tr}_{L^2}(e^{-tD})=\Sigma_\nu\, e^{-t\lambda_\nu}\sim(4\pi t)^{-m/2}\cdot\Sigma_{n\geq 0}\, t^n a_n(D).$$

The $\{a_n(D)\}$ are spectral invariants of D which are locally computable. There are local invariants $a_n(x,D)$ involving the jets of the total symbol of D so that $a_n(D)=\int_M a_n(x,D)dvol(x)$. Although the eigenvalues of D reflect global information, the $a_n(x,D)$ are locally computable and give information regarding the spectrum.

Let Δ be the Laplacian δd on functions. In his famous paper "Can one hear the shape of a drum", M.Kac [10] asked the question of whether the spectrum determines the geometry of M up to isometry. Kac originally asked the question for planar domains but it has been generalized to the setting of compact manifolds and is in fact more tractable there. The first result along these lines is due to Milnor [11] who showed there existed two 16 dimensional tori which had the same

1980 Mathematics Subject Classification (1985 Revision). 58G25
*Research partially supported by NSF grant DMS8614715

spectrum but which were not isometric.

The invariants $a_n(\Delta)$ are given by integrating complicated expressions in the curvature tensor and are known explicitly for n=0,1,2,3. Let R_{ijkl} be the components of the curvature tensor with the sign convention that $R_{1212}=-1$ on the standard sphere S^2. Adopt the convention of summing over repeated indices with respect to a local orthonormal frame for the tangent space T(M). Let $\rho_{ij}=-R_{ijkj}$ and $\tau=\rho_{ii}$ be the Ricci tensor and its trace; τ is twice the scalar curvature. Then

$$a_0(\Delta)=\mathrm{vol}(M),\ a_1(\Delta)=\int_M \tau/6\ ,\ \text{and}$$

$$a_2(D)=\int_M\{5\tau^2-2|\rho|^2+2|R|^2\}/360.$$

The formula for a_3 is more complicated; see Gilkey [5]. If m=2, the Gauss-Bonnet formula implies $\chi(M)$ is a multiple of $a_1(\Delta)$ so the genus of M is a spectral invariant of Δ; consequently the topological type of a Riemann surface is determined by the spectrum of Δ. Using the asymptotics of the heat equation, a number of positive isospectral results have been proved; for example it is known that the standard sphere in low dimensions is characterized by the spectrum of Δ [2]. Patodi [15] showed by taking into consideration the spectrum of the Laplacian on differential forms, it was possible to determine if a manifold has constant sectional curvature. There are many other reuslts in this direction.

Milnor's isospectral tori have the same underlying topology. It was very surprising when M.Vigneras [21] constructed examples of 3 dimensional manifolds with constant curvature -1 which were isospectral but which had different fundamental groups and which consequently had different topologies. Vigneras also constructed isospectral but not isometric Riemann surfaces with constant curvature -1. Since then, many other such examples have been found. Ikeda [9] constructed examples of spherical space forms (constant curvature +1) with metacyclic fundamental groups which were isospectral but not isometric; this implies they are not diffeomorphic and in fact they are not equivariantly cobordant [6]. Urakawa [20] constructed examples of bounded domains in $\mathbf{R}^n$ which are not congruent but so that the spectrum of the Laplacian with both Neumann and Dirichlet boundary conditions coincides. Sunada [19] has given an approach which unifies these examples in a single conceptual framework.

The examples cited above are all discrete. It was therefore most surprising when Gordon and Wilson [8] constructed continuous families of isospectral nonisometric manifolds; there examples were of the form $(G/\Gamma,g)$ where G is an exponential solvable Lie group, Γ is a uniform discrete subgroup, and g is a right invariant metric on G. Gordon has also constructed manifolds which are isospectral on functions but not on 1-forms. The examples of Gordon and Wilson can all be compactified in a suitable sense. This is leads one to ask whether in general isospectral metrics modulo the action of Diff(M) form a compact subset. This question has been answered in the affirmative by Osgood, Phillips, and Sarnak [13] for compact Riemann surfaces and for planar domains. Brooks, Perry, and Yang [3] and Chang and Yang [4] have similar compactness theorems for isospectral metrics within a fixed conformal class for 3 dimensional manifolds..

These compactness theorems are an interesting mixture of global analysis and local analysis. If M has constant sectional curvature, then essentially the only local invariant is the volume of M. Consequently the invariants of the heat equation can not serve to distinguish non isospectral metrics and some global information is required. Let $\zeta(s)=\Sigma_\nu \lambda_\nu^{-s}$ be the zeta function where the sum ranges over the non-zero eigenvalues of the Laplacian. The Mellin transform relates the zeta function and the heat equations and shows ζ has a meromorphic extension to **C** with isolated simple poles on the real axis. 0 is a regular value and $\zeta(0)=a_m(\Delta)$. $\zeta'(0)$ is a global invariant of the spectrum and this plays a crucial role the compactness theorems cited above. There is a natural interpretation of log det $\Delta=e^{-\zeta'(0)}$ so $\zeta'(0)=-\log\det(\Delta)$ is closely related to the functional determinant.

We describe the proof of Osgood et al as follow. Let M be a compact Riemann surface; suppose the genus is non-negative for simplicity; the argument for the sphere is different in flavor owing to the conformal group. Since $a_0(\Delta)$ is the normalized volume of M, it is a spectral metric. We normalize the metrics hence forth to have constant volume $-2\pi\cdot\chi(M)$. Such a metric g determines a conformal class which contains a unique metric $\sigma(g)$ of constant curvature -1. Osgood et al [14] showed $\zeta'(0,\sigma(g))\le\zeta'(0,g)$ with equality if and only if $\sigma(g)=g$; ζ' attains its minimum within a conformal class on the metric of constant curvature -1. Let $L(\sigma)$ be the length of the shortest closed geodesic of the metric σ. Wolpert [22] showed there exists $\epsilon, c>0$ depending only on the genus of M so $e^{-\zeta'(0,\sigma)}\le c/L\cdot e^{-\epsilon/L}$. Let $\{g_n\}$ be an isospectral family, let $\sigma_n=\sigma(g_n)$, and let $L_n=L(\sigma_n)$. Since $\zeta'(0,g_n)$ is a spectral invariant, it is constant. This provides an upper bound for $\zeta'(0,\sigma_n)$ and hence a positive lower bound for the L_n. It now follows by a theorem of Mumford [12] that by passing to a subsequence, one may assume the σ_n converge to a constant curvature metric σ. This controls variation across conformal classes and gives a lower bound for $\zeta'(0,\sigma_n)$. Let $g_n=e^{2\phi_n}\sigma_n$. Polyakov [16] has proved:

$$12\pi\cdot\zeta'(0,g_n)-12\pi\cdot\zeta'(0,\sigma_n)=\int_M|\nabla_{\sigma_n}\phi_n|^2 d\mathrm{vol}(\sigma_n)$$
$$+4\pi\chi(M)\int_M\phi_n d\mathrm{vol}(\sigma_n).$$

This leads to uniform estimates for the Sobolev 1-norm $|\phi_n|_1$. Using $a_2(D)$, one obtains uniform estimates for $|\phi_n|_2$. At this point, the higher order asymptotics of the heat equation enter. Using the Selberg-trace formula, Osgood et al show

$$a_n(\Delta)=c(n)\cdot\int_M|\nabla^{n-2}\tau\cdot\nabla^{n-2}\tau|+\text{lower order terms}$$

where c(n) is a non-zero constant and where the lower order terms involve lower order jets of the metric. Consequently

$$a_n(\Delta(g_n))=c(n)\cdot\int_M|\nabla^n\phi_n\cdot\nabla^n\phi_n|+\text{lower order terms.}$$

This controls the higher order norms from which the compactness theorem follows.

We have generalized the result of Osgood et al to compute the leading terms in the asymptotics of the heat equation in arbitrary dimensions. We have shown (see [7] for details).

Theorem 1: Let $n\ge3$ and let $c(n)=(-1)^n/2^{n+1}\cdot1\cdot3\cdots(2n+1)$.

$$a_n(P)=c(n)\cdot\int_M(n^2-n-1)\nabla^{n-2}\tau\cdot\nabla^{n-2}\tau+2\nabla^{n-2}\rho\cdot\nabla^{n-2}\rho+\dots$$

where the omitted terms involve lower order derivatives of the metric.

Let (M,g_0) be a three dimensional Riemannian manifold and let $g_n=e^{2\phi_n}\cdot g_0$ be a family of isospectral metrics within the conformal class defined by g_0. By the solution of the Yamabe problem [1,17], g_0 can be chosen to be a metric of constant scalar curvature. Brooks et al [3] have shown that if the scalar curvature of g_0 is negative, then the g_n form a compact family. Theorem 1 gives bounds for the Sobolev norms of ϕ_n modulo lower order terms; these together with other suitable estimates using the asymptotics of the heat equation a_0,a_1, and a_2 complete the proof. If (M,g_0) is the standard metric on S^3, the conformal group is infinite dimensional and the situation is more complicated. Chang et al [4] establish the same compactness result once gage equivalence under the conformal group is factored out; again the proof relies heavily upon Theorem 1 to estimate the higher order Sobolev norms.

Theorem 1 can be generalized to the vector valued case. Let V be a smooth vector bundle over M and let

$$D=-\{\Sigma_{i,j}\, g^{ij}\partial^2/\partial x_i\partial/x_j+\Sigma_k\, p^k\partial/\partial x_k + q\}$$

be an elliptic operator on $C^\infty(V)$. If ∇ is a connection on V, let $D_\nabla=-g^{ij}\nabla_i\nabla_j$ be the reduced or Bochner Laplacian. There exists a unique connection ∇ on V and a unique endomorphism E of V so $D=D_\nabla-E$; the local invariants of the operator D can be expressed in terms the data $\{R_{ijkl;\dots},\Omega_{ij;\dots},E_{;\dots}\}$ where ";..." denotes covariant differentiation and where Ω is the curvature of the operator on E.

Theorem 2: Let $n\geq 3$ and let $c(n)=(-1)^n/2^{n+1}\cdot 1\cdot 3\cdots(2n+1)$.

$$\begin{aligned}a_n(D)=c_n\cdot\int_M \mathrm{Tr}\{&(n^2-n-1)\nabla^{n-2}\tau\cdot\nabla^{n-2}\tau+2\nabla^{n-2}\rho\cdot 2\nabla^{n-2}\rho\\ &+4(2n+1)(n-1)\nabla^{n-2}\tau\cdot\nabla^{n-2}E+2(2n+1)\nabla^{n-2}\Omega\cdot\nabla^{n-2}\Omega\\ &+4(2n+1)(2n-1)\nabla^{n-2}E\cdot\nabla^{n-2}E+\text{lower order terms}\}.\end{aligned}$$

The proof is quite calculational and uses the calculus of pseudo-differential operators depending on a complex parameter developed by Seeley [18]. We refer to Gilkey [6] for details.

References

[1] J.P.Aubin, Nonlinear analysis on manifolds, Monge-Ampere equations, Springer-Verlag.

[2] M.Berger, Eigenvalues of the Laplacian, Proc. Symp. in Pure Math (AMS) VXVI (1968), 121-126.

[3] R.Brooks, P.Perry, and P.Yang, Isospectral sets of conformally equivalent metrics (preprint)

[4] S.A.Chang and P.Yang, Compactness of isospectral conformal metrics on S^3. (preprint)

[5] P.Gilkey, The spectral geometry of a Riemannian manifold, J.Diff.Geo. V10 (1975), 601-618.

[6] P.Gilkey, On spherical space forms with meta-cyclic fundamental group which are isospectral but not equivariant cobordant. Compositio Math V 56 (1985), 171-201.

[7] Leading terms in the asymptotics of the heat equation (to appear).

[8] C.Gordon and E.Wilson, Isospectral deformations of compact solvmanifolds, J.Diff.Geo. 19 (1984), 241-256.

[9] A.Ikeda, On spherical space forms which are isospectral but not isometric, J.Math.Soc. Japan V35 (1983), 437-444.

[10] M.Kac, Can one hear the shape of a drum?, American Math Monthly, V73 (1966), 1-23.

[11] J.Milnor, Eigenvalues of the Laplace operator on certain manifolds, Proc Nat Acad Sci USA 51 (1964), 542.

[12] D.Mumford, A remark on Mahler's compactness theorem, Proc Amer Math Soc 28 (1972) 289-294.

[13] B.Osgood, R.Phillips, P.Sarnak, Compact isospectral sets of surfaces (preprint).

[14] B.Osgood, R.Phillips, R.Sarnak, Extremals of determinants of Laplacians (preprint).

[15] V.K.Patodi, Curvature and the fundamental solution of the heat operator, J.Indian Math Soc 34 (1970), 269-285.

[16] A.Polyakov, Quantum geometry of bosonic strings", Physics letters 103B (1981) 207-210.

[17] R.Schoen, Conformal deformations of a Riemannian metric to constant scalar curvature, J.Diff.Geo 20 (1984), 479-496.

[18] R.T.Seeley, Complex powers of an elliptic operator, Proc. Symp. Pure Math V 10 (AMS) (1967), 288-307.

[19] T.Sunada, Riemannian coverings and isospectral manifolds, Ann Math 121 (1985), 169-186.

[20] H.Urakawa, Bounded domains which are isospectral but not congruent, Ann. scient. Ec. Norm. Sup. 15 (1982), 441-456.

[21] M.Vigneras, Variete Riemanniennes Isospectrales et non-isometriques, Annals Math 112 (1980), 21-32.

[22] S.Wolpert, Asymptotics of the Selberg zeta function for degenerating Riemann surfaces (preprint).

Peter B Gilkey

Mathematics Department

University of Oregon

Eugene Oregon 97403 USA.

Contemporary Mathematics
Volume **101**, 1989

EIGENVALUE ASYMPTOTICS AND THEIR GEOMETRIC APPLICATIONS

Mark A. Pinsky[1]

1. INTRODUCTION. Let (M_1,g_1) and (M_2,g_2) be Riemannian manifolds. The principal Dirichlet eigenvalue of the Laplace-Beltrami operator Δ on the ball of radius r is denoted $\lambda(B_i(O_i;r))$ $(i=1,2)$. In 1975 S. Y. Cheng [1] obtained comparison theorems stating that if the metric g_2 has constant sectional curvature k and sect $(g_1) \leqslant k$ then $\lambda(B_1(O_1;r)) \geqslant \lambda(B_2(O_2;r))$ for r sufficiently small. Theorems of this type provide valuable geometric information, but do not enable us to infer eigenvalue comparisons from corresponding estimates on the scalar curvature alone, for example. In addition, they do not permit one to obtain isometry of the metrics from equality of the two eigenvalues for a sufficiently large class of balls.

For these purposes, we propose to develop asymptotic expansions of the principal Dirichlet eigenvalue when $r \downarrow 0$. Corresponding expansions for the volume of small balls and mean exit time of Brownian motion (the solution of $\Delta f = -1$) are already well-known ([2], [3]). Many geometric studies have been devoted to asymptotic expansions for the partition function $\Sigma e^{-\lambda_j t}$ which depends on all of the eigenvalues. To the best of our knowledge, the only previous work dealing with asymptotics of the principal eigenvalue alone is the paper of Levy-Bruhl [6]. The work presented here is joint with Leon Karp ([4], [5]).

The methods used to handle the above case of the ball also generalize to study the principal Dirichlet eigenvalue of a tubular neighborhood of a compact hypersurface in a Riemannian manifold. In both cases we develop a suitable perturbation theory for the Laplacian as an asymptotic power series in the radius. The following results are obtained.

THEOREM 1. *When $r \downarrow 0$ we have the expansion*

$$\lambda(B(m;r)) = c_0/r^2 + c_1\tau_m + r^2 c_2[|R|^2 - |\rho|^2 + 6\Delta\tau]_m + O(r^4)$$

c_0, c_1, c_2 *depend only on the dimension* n *and satisfy* $c_0>0$, $c_1<0$, $c_2<0$; τ *is the*

1980 Mathematics Subject Classification (1985 Revision). 58G25, 53B20.

[1]Supported by National Science Foundation Grant DMS 8803154.

scalar curvature, ρ *is the Ricci tensor and* R *is the Riemann curvature tensor*.

COROLLARY 1. *If* (M_1,g_1) *and* (M_2,g_2) *are Riemannian manifolds whose scalar curvatures satisfy* $\tau_1(O_1)<\tau_2(O_2)$ *for some* $O_1 \in M_1$, $O_2 \in M_2$, *then there exists* $r_0>0$ *such that* $\lambda(B_1(O_1;r)) > \lambda(B_2(O_2;r))$ *for all* $r<r_0$.

COROLLARY 2. *Let* $n < 6$. *If for every* $m \in M$ *we have* $\lambda(B(m;r))=c_0/r^2+O(r^2)$, *then* (M,g) *is locally isometric to* R^n *with the Euclidean metric. More generally, if* (M_2,g_2) *is a rank-one symmetric space and for every* $O_1 \in M_1$, $O_2 \in M_2$, *we have* $\lambda(B_1(O_1;r)) = \lambda(B_2(O_2;r))+O(r^2)$, *then* (M_1,g_1) *is isometric to* (M_2,g_2).

We remark that the condition $n < 6$ is also necessary. The manifold $M^6 = S^3 \times H^3$ is not flat, but $\lambda(B(m;r)) = c_0/r^2 + O(r^2)$.

To formulate the next result, consider the case of an imbedded hypersurface $P \subset M^n$ which may be assumed to be compact. The tubular neighborhood $N_\varepsilon(P)$ is defined as $N_\varepsilon(P) = \{x: \text{dist}(x,P) < \varepsilon\}$. The eigenvalue problem $\Delta f+\lambda f = 0$ with zero boundary conditions on $\partial N_\varepsilon(P)$ has a well-defined positive solution with principal eigenvalue $\lambda(\varepsilon;P)$. The following result gives the first-order asymptotics in case of $M = R^n$.

THEOREM 2. *When* $\varepsilon \downarrow 0$ *we have the asymptotic result*

$$\lambda(\varepsilon;P) = \pi^2/4\varepsilon^2 + (4V)^{-1}\int_P\{(\Sigma k_i)^2 - 2\Sigma k_i^2\}d\text{ vol} + O(\varepsilon)$$

where V = *volume of* M, *and* k_i *is the* i^{th} *principal curvature*.

It is interesting to note two important special cases of the above expansion.

Case i: P is a sphere of radius R imbedded in R^n. In this case $k_i = 1/R$ and the quantity $2\Sigma k_i^2 - (\Sigma k_i)^2 = (n-1)(3-n)/R^2$.

Case ii: P is a compact surface in R^3. In this case $2\Sigma k_i^2 - (\Sigma k_i)^2 = \frac{1}{4}(k_1-k_2)^2 = K - H^2$ where $K = k_1k_2$ is the gaussian curvature and $H = \frac{1}{2}(k_1 + k_2)$ is the mean curvature. $K-H^2$ is non-negative and zero only for umbilic surfaces. Thus we have the following.

COROLLARY 3. *Let* $P \subset R^3$ *be a compact surface for which* $\lambda(\varepsilon;P)=\pi^2/4\varepsilon^2+O(\varepsilon)$. *Then* P *is a round sphere*.

METHOD OF PROOF. Let M_m be the tangent space and $\exp_m$ be the exponential mapping of the Riemannian manifold at $m \in M$. Define a mapping $\Phi_\varepsilon:C^\infty(M_m) \to C^\infty(M)$ by $(\Phi_\varepsilon f)(\exp_m x) = f(x/\varepsilon)$. From our 1983 paper with Alfred Gray [3], we quote the asymptotic expansion

$$\Phi_\varepsilon^{-1}\Delta\Phi_\varepsilon f = (\varepsilon^{-2}\Delta_{-2} + \Delta_0 + \varepsilon\Delta_1 + \dots + \varepsilon^N\Delta_N)f + O(\varepsilon^{N+1})$$

Here $f \in C^\infty(M_m)$, N is a natural number and $\Delta_{-2},\Delta_0,\dots,\Delta_N$ are second order differential operators on the tangent space M_m with polynomial coefficients, such that Δ_j decreases the degree of a polynomial by j; in detail the first two are

$$\Delta_{-2} = \Sigma_i \partial^2/\partial x_i^2$$

$$\Delta_0 = (1/3)\Sigma_{iajb} R_{iajb} x_a x_b \partial^2/\partial x_i \partial x_j - (2/3)\Sigma_{ia} \rho_{ia} x_a \partial/\partial x_i$$

All sums run from 1 to n and (R_{iajb}) is the Riemann tensor at m while $\rho_{ij} = \Sigma_a R_{iaja}$ is the Ricci tensor, all expressed with respect to an orthonormal basis of M_m.

This asymptotic development of the Riemannian Laplacian allows us to develop a perturbation analysis on the unit ball of R^n. Let H be the Hilbert space of functions on the unit ball which are square-integrable with respect to Lebesgue measure $dx = dx_1 \ldots dx_n$, with the inner product $\langle f,g \rangle = \int f(x)g(x)\,dx$. The Euclidean Laplacian Δ_{-2} with zero boundary conditions is a self-adjoint operator with a simple principal eigenvalue λ_{-2} and a compact resolvent operator $(\lambda - \Delta_{-2})^{-1}$ for $\lambda > 0$.

Let $(\lambda_{(\varepsilon)}, f_{(\varepsilon)})$ be the principal eigenvalue and corresponding eigenfunction of the geodesic ball; these satisfy $(\Phi_\varepsilon^{1} \Delta \Phi_\varepsilon + \lambda_{(\varepsilon)}) f_{(\varepsilon)} = 0$, $f_{(\varepsilon)} = 0$ on the boundary and $\lim_{\varepsilon \downarrow 0} \varepsilon^2 \lambda_\varepsilon = \lambda_{-2}$. The C^∞ perturbation theory guarantees that these are C^∞ functions at $\varepsilon = 0$. The eigenfunctions $f_{(\varepsilon)}$ are further normalized by $\langle f_{(\varepsilon)}, f_0 \rangle = 1$ for $\varepsilon > 0$. Thus

$$f_{(\varepsilon)} = f_0 + \varepsilon f_1 + \varepsilon^2 f_2 + \varepsilon^3 f_3 + \varepsilon^4 f_4 + O(\varepsilon^5)$$

$$\varepsilon^2 \lambda_{(\varepsilon)} = \lambda_{-2} + \varepsilon \lambda_{-1} + \varepsilon^2 \lambda_0 + \varepsilon^3 \lambda_1 + \varepsilon^4 \lambda_4 + O(\varepsilon^5)$$

Substituting in the Laplacian and equating like powers of ε results in the "perturbation equations" $\Sigma_{j=0}^{k+2} (\Delta_{k-j} + \lambda_{k-j}) f_j = 0$ for $k = 2,-1,0,1,2$, with the boundary condition that $f_n = 0$ on the boundary and the normalizations $\langle f_n, f_0 \rangle = 0$ for $n > 0$ while $\langle f_0, f_0 \rangle = 1$.

Elementary manipulations of the first few of these yield immediately that $\lambda_{-1} = 0 = f_1$; f_0 is a standard Bessel function and λ_0 is obtained as $\lambda_0 = -\langle \Delta_0 f_0, f_0 \rangle = -\tau_m/6$.

Further analysis shows that $\lambda_1 = 0$, while f_2 is expressed in terms of a quadratic polynomial and the Bessel function f_0. The detailed work is in the evaluation of $\lambda_2 = -\langle f_0, \Delta_0 f_2 \rangle - \langle f_0, \Delta_2 f_0 \rangle$. The operator Δ_2 contains a fourth degree polynomial whose coefficients depend on the curvature tensor and its covariant derivatives of order less than or equal to two. The first inner product will also involve integrals of fourth degree terms. Using the system of invariants developed in [3], we obtain after a lengthy computation the result $\lambda_2 = (|R|^2 - |\rho|^2 + 6\Delta\tau)\omega_n \int_0^1 f_0 f_0' r^{n+2} dr/30n(n+2)$ where ω_n is the surface area of the n-dimensional sphere.

Having established theorem 1, we can immediately obtain the corollaries which characterize R^n by the principal eigenvalue. For this purpose it helps to note the identity $|R|^2 - |\rho|^2 = |W|^2 + ((6-n)/(n-2))|\rho|^2$ valid for $n>2$, where W is Weyl's conformal tensor. Now if $\lambda_{(\varepsilon)} = c_0/\varepsilon^2$ for all $\varepsilon>0$ and all $m \in M$ we see immediately from the theorem that $\tau_m = 0$ and the quadratic term is also zero. Hence $\Delta\tau_m = 0$ and from the preceding remark we see that if $2<n<6$, both the Ricci tensor ρ and the conformal tensor W are both zero. This means that the curvature tensor is identically zero.

To obtain the expansion in theorem 2, we introduce Fermi coordinates (p,t) where $p \in P$ and $|t| < \varepsilon$. In these coordinates the Laplacian takes the form $\Delta = \partial^2/\partial t^2 + (\theta_t/\theta)\partial/\partial t + \Delta_P\hat{}$, where $\theta = \theta(p,t)$ is the change-of volume factor induced by the normal exponential mapping and $\Delta_P\hat{}$ is a smooth extension of the Laplacian of P, involving only differentiation in the p-variables. From the standard estimates, $\theta(t,p) = 1 - ht + \frac{1}{2}\tau t^2 + O(t^3)$, $t\downarrow 0$ hence $\theta_t/\theta = -h + (\tau - h^2)t + O(t^2) \doteq \alpha + \beta t - O(t^2)$. To develop a suitable perturbation theory, we define a mapping of functions by $(\Phi_\varepsilon f)(p,t) = f(p,t/\varepsilon)$ and look at the transformed Laplacian $L_\varepsilon = \varepsilon^2\Phi_\varepsilon\Delta\Phi_\varepsilon = \Delta_{-2} + \varepsilon\Delta_{-1} + \varepsilon^2\Delta_0 + O(\varepsilon)$. In terms of the above notations we have

$$\Delta_{-2} = \partial^2/\partial t^2 \qquad \Delta_{-1} = \alpha\partial/\partial t \qquad \Delta_0 = \beta t\partial/\partial t + \Delta_P\hat{}$$

The eigenvalue and eigenfunction are expanded in the forms

$$\lambda = \varepsilon^{-2}\lambda_{-2} + \varepsilon\lambda_{-1} + \varepsilon^2\lambda_0 + O(\varepsilon^3)$$

$$f = f_0 + \varepsilon f_1 + \varepsilon^2 f_2 + \varepsilon^3 f_3 + O(\varepsilon^3)$$

Substituting in the eigenvalue problem and equating like powers of ε yields

$$(\Delta_{-2} + \lambda_{-2})f_0 = 0 \text{ in } P_1$$

$$(\Delta_{-2} + \lambda_{-2})f_1 + (\Delta_{-1} + \lambda_{-1})f_0 = 0 \text{ in } P_1$$

$$(\Delta_{-2} + \lambda_{-2})f_2 + (\Delta_{-1} + \lambda_{-1})f_1 + (\Delta_0 + \lambda_0)f_0 = 0 \text{ in } P_1$$

In addition we have the boundary conditions that $f_0 = 0$, $f_1 = 0$ and $f_2 = 0$ on ∂P together with the normalizations $\langle f_0, f_0\rangle = 1$, $\langle f_1, f_0\rangle = 0$, $\langle f_0, f_2\rangle = 0$.

We have immediately that $f_0 = V^{-\frac{1}{2}}\cos(\frac{1}{2}\pi t)$ and $\lambda_{-2} = \frac{1}{4}\pi^2$ where $V = \text{vol } P$. From this it follows that $\lambda_{-1} = -\langle\Delta_{-1}f_0, f_0\rangle = 0$ and $f_1 = \frac{1}{2}t\,\alpha V^{-\frac{1}{2}}\cos(\frac{1}{2}\pi t)$. Finally $\lambda_0 = -\langle\Delta_0 f_0, f_0\rangle - \langle\Delta_{-1}f_1, f_0\rangle$ which, after doing some integrations leads to the final formula $\lambda_0 = (4V)^{-1}\int(\alpha^2 + 2\beta)$.

REFERENCES

1. S. Y. Cheng, Eigenvalue comparison theorems and its geometric applications, Math. Z., 143 (1975), 289-297.

2. A. Gray, L. Karp and M. Pinsky, The Mean Exit Time from a tube in a Riemannian manifold, in Probability and Harmonic Analysis, J. Chao and W. Woyczynski editors, Marcel Dekker, New York, 1986, 113-117.

3. A. Gray and M. Pinsky, The mean exit time from a small geodesic ball in a Riemannian manifold, Bulletin des Sciences Mathematiques, 107(1983), 345-370.

4. L. Karp and M. Pinsky, The first eigenvalue of a small geodesic ball in a Riemannian manifold, Bulletin des Sciences Mathematiques, 111(1987), 229-239.

5. L. Karp and M. Pinsky, First order asymptotics of the principal eigenvalue of tubular neighborhoods, in Geometry of Random Motion, AMS Series "Contemporary Mathematics", 1988.

6. A. Levy-Bruhl, Invariants infinitesimaux, Comptes Rendus Acad. Sci. Paris 279 (1974), 197-200.

MATHEMATICS DEPARTMENT
NORTHWESTERN UNIVERSITY
EVANSTON, IL 60208

Contemporary Mathematics
Volume **101**, 1989

A LOWER BOUND FOR THE NUMBER OF ISOSPECTRAL SURFACES

Richard M. Tse*

ABSTRACT. Two surfaces are said to be isospectral if the spectrum of the Laplacian on the two surfaces are the same. We give in this note a lower bound for the number of isospectral Riemann surfaces of arbitrarily large genus g.

Let M be a smooth compact surface. The Laplacian Δ acts on the C^∞ functions of M by

$$\Delta f = -div(grad)f$$

It is well known that the set of eigenvalues (counting multiplicities) of the Laplacian Δ of M consists of an infinite sequence

$$0 = \lambda_0 < \lambda_1 \leq \lambda_2 \leq \lambda_3 \leq \ldots \uparrow \infty.$$

We call such sequence $\{\lambda_i\}$ the spectrum of M. It is also known that the spectrum of M itself determines, through the Selberg trace formula, a lot of geometric data such as the area, the genus and the lengths of the closed geodesics of M.

It is then a natural question to ask whether or not there exist any isospectral surfaces that are not isometric. Vigneras [3] answered this question affirmatively in 1980 by constructing explicitly a pair of isospectral but nonisometric surfaces of genus 24 through some consideration of the arithmetic of quaternion algebras. In 1985, Sunada [2] gave a general algebraic criterion for when surfaces are isospectral. His work transforms the above geometric question into one of group theoretic nature. Brooks and Tse [1] then constructed examples of isospectral but nonisometric surfaces of small genus based on Sunada's work. As regard to the number of possible isospectral surfaces of a given genus, Buser gave the following upper bound.

* AMS (MOS) Classification 58G99

Theorem 1: *There exists a function $k(g)$ such that there are at most $k(g)$ number of mutually isospectral Riemann surfaces of genus g that are not isometric.*

In this note we shall give a sketch of the proof of the following lower bound.

Theorem 2: *For an arbitrarily large genus g, there exist at least $c\sqrt{g}$ number of mutually isospectral Riemann surfaces of genus g that are not isometric.*

Again, the proof of the above theorem is based on the work of Sunada. Before we state Sunada's theorem, we need a definition.

Definition: Let G be a finite group and H_1 and H_2 be two subgroups of G satisfying the following condition

i) $|H_1 \cap [g]| = |H_2 \cap [g]|$ for every conjugacy class $[g]$ in G,

ii) H_1 is not conjugate to H_2 in G.

Then the triple (G, H_1, H_2) is said to satisfy the conjugacy condition.

Examples of groups that satisfy the conjugacy condition are plentiful. We shall give two of them here without proof.

Example 1: $G = SL(n,p)$, where p is a prime,

$$H_1 = \begin{bmatrix} * & * & * & * \\ 0 & * & * & * \\ 0 & * & * & * \\ 0 & * & * & * \end{bmatrix} \text{ and } H_2 = \begin{bmatrix} * & 0 & 0 & 0 \\ * & * & * & * \\ * & * & * & * \\ * & * & * & * \end{bmatrix}$$

Example 2: $G = Z_8 * \rtimes Z_8$,

$$H_1 = (1,0),(3,0),(5,0),(7,0) \text{ and}$$

$$H_2 = (1,0),(3,4),(5,4),(7,0).$$

Theorem (Sunada): Suppose (G, H_1, H_2) satisfies the conjugacy condition and M be a compact manifold with a surjective homomorphism : $\pi_1(M) \to G$. Let M_1, M_2 be two Riemannian coverings with $\pi_1(M_i) = f^{-1}(H_i), i = 1,2$, then M_1 is isospectral to M_2 for any metric on M. Furthermore, for a generic metric on M, M_1 is not isometric to M_2.

In view of Sunada's theorem, in order to give a lower bound on the number of isospectral surfaces we need a group G with a lot of subgroups H_i satisfying the conjugacy condition. A good candidate for such a group is $A \rtimes H$ where H is a finite group and A is an H-module. There is a close relation between the first cohomology group of a group H relative to A and the complements of A satisfying the conjugacy condition.

Lemma 1: Let $G = A \rtimes H$ as above. Then the conjugacy classes of complements of A are in $1-1$ correspondence with the elements of $H^1(H, A)$.

Lemma 2: Let G as in lemma 1, and K be a complement of A corresponds to $\beta \in H^1(H, A)$ and $\beta = 0$ in $H^1(< h >, A)$ for every cyclic subgroup $< h >$ of H. Then $|[g] \cap H| = |[g] \cap K|$ for every conjugacy class $[g]$ in G.

The above two lemmas give us a convenient method of checking when $G = A \rtimes H$ has a lot of subgroups satisfying the conjugacy condition. Now let $H = SL(2, q)$ with $q = 2^n, n > 1$, and A be the natural module for H. Then with the help of the above two lemmas, we can show

Lemma 3: With H and A as stated above, $|H^1(H, A)| = q$ and each $\beta \in H^1(H, A)$, $\beta = 0$ in $H^1(< h >), A)$ for every cyclic subgroup of H. We are now in a position to give a proof of theorem 2.

Proof (of Theorem 2): Let $G = A \rtimes H$ where $H = SL(2, q)$ as above. Lemma 1,2, and 3 tell us that there are exactly q distinct subgroups of G that satisfy the conjugacy conditions. Given a compact Riemann surface M, Sunada's theorem tells us that there are q isospectral coverings of M. By the Gauss-Bonnet theorem, the genus g of these coverings is given by

$$\begin{aligned} g &= \frac{(\text{index of})(\text{Area of})}{4\pi} + 1 \\ &= \frac{q^2(\text{Area of})}{4\pi} + 1 \end{aligned}$$

Thus $q^{\sim}\sqrt{g}$ as desired. Q.E.D.

BIBLIOGRAPHY

[1] R. Brooks and R. Tse, Isospectral Surfaces of Small Genus, Nagoya Math. J., 107 (1987), 13-24.

[2] T. Sunada, Riemannian Covering and Isospectral Manifolds, Ann. Math., 121 (1985), 169-186.

[3] M. T. Vigneras, Varietes Riemanniennes Isospectrales et Non Isometriques, Ann. Math., 112 (1980), 21-32.

DEPARTMENT OF MATHEAMTICS
UNIVERSITY OF SOUTHERN CALIFORNIA
LOS ANGELES, CA. 90089-1113

Current Address:
Department of Mathematics
Texas Tech University
Lubbock, Texas 79409

Contemporary Mathematics
Volume **101**, 1989

THE CONFORMAL DEFORMATION EQUATION AND ISOSPECTRAL SET OF CONFORMAL METRICS

Sun-Yung A. Chang[1] and Paul C. Yang[2]

ABSTRACT In this survey article we discuss the conformal deformation equation on a compact Riemannian manifold. The problem of prescribing scalar curvature and the problem of compactness of a set of isospectral comformal metrics both involve the analysis of concentration phenonmenon. A central part of the analysis is to show that non-compactness in the problem is caused by the non-compactness of the group of conformal transformations.

§0 Introduction.

Given a compact Riemannian manifold M with metric g, a conformally related metric on M is given by $g' = \rho g$ where ρ is a positive function on M. In order to avoid terms involving gradient of the conformal factor in the differential equation relating the scalar curvatures, the conformal factor is expressed as $\rho = e^{2u}$ when the dimension of M is two, and $\rho = u^{4/n-2}$ when the dimension of M is at least three. The scalar curvature of the new metric R' is then related to the scalar curvature R of the original metric g according to

$$(0.1) \qquad \Delta u + R'e^{2u} = R, \qquad \dim M = 2;$$

$$(0.1') \qquad 4\frac{n-1}{n-2}\Delta u + R'u^{\frac{n+2}{n-2}} = Ru, \qquad \dim M \geq 3.$$

We will discuss two problems concerned with this equation. The first deals with the question of solvability of the equation when R' is a prescribed function, generally known as the problem of prescribing scalar curvature. The second problem is to describe a set of isospectral conformal metrics on a given manifold. The results may be considered as applications of recent advances in

1980 Mathematics Subject Classification: 53C20, 58C40

[1]Research supported by NSF Grant DMS 88-01776

[2]Research supported by NSF Grant DMS 87-02871

the analysis of the nonlinearity of these differential equations.

The basic analytic difficulty relating to the equation (0.1) (0.1') is most apparent in the Yamabe problem where R' is a constant. The equations (0.1) (0.1') are the Euler equations for the variational functionals

(0.2) $$F_g[u] = \log \fint R'e^{2u} - \fint (|\nabla u|^2 + 2u), \qquad n = 2;$$

(0.2') $$Q_g[u] = \left(\int R' u^{\frac{2n}{n-2}}\right)^{-\frac{n-2}{n}} \left(\int 4\frac{n-1}{n-2}|\nabla u|^2 + Ru^2\right), \quad n \geq 3.$$

In the variational approach to the existence problem, it is necessary to prove the compactness of a sequence $\{u_j\}$ approaching an extremal of the functionals F or Q. In case where the manifold M possesses a noncompact group of conformal transformations (such manifold must be conformally equivalent to S^n according to the theorem of Lelong-Ferrand ([L])), compactness becomes a problem. A conformal transformation $\varphi: M \to M$ induces a transformation on the conformal factor: $u \to T(\varphi)(u)$ where

$$\exp(2T(\varphi)(u))g = \varphi^*(\exp 2u \cdot g), \qquad \dim M = 2;$$

$$(T(\varphi)(u))^{4/n-2}\, g = \varphi^*(u^{4/n-2}\, g), \qquad \dim M \geq 3.$$

It is clear that the functionals F,Q remain invariant under conformal transformation:

$$F_{\varphi^*(g)}[T(\varphi)(u)] = F_g[u], \qquad \dim M = 2;$$

$$Q_{\varphi^*(g)}[T(\varphi)(u)] = Q_g[u], \qquad \dim M \geq 3.$$

So when the underlying manifold is S^n with the standard metric, there is no compactness. The basic philosophy in the analysis is to show the converse statement that noncompactness can occur only in the presence of large conformal groups.

§1. Prescribing scalar curvature.

We recall the situation where R' is constant. In dimension two the uniformization theorem implies that for any metric g on a surface M, there exists a conformal factor e^{2u} so that the Gauss curvature of $e^{2u}g$ is a constant whose sign is determined by the Euler characteristic χ of M. In terms of a direct analysis solving the differential equation (0.1), we have

Theorem 1 (Poincare [P], Berger [Be], Hamilton [Ha]) On a compact Riemann surface with hermitian metric g, the equation (0.1) with R' a constant, has a solution.

Poincare gave in the case $\chi < 0$ a solution by iterating a sequence of approximate solutions to (0.1) to show that the limit is a solution of (0.1). Melvyn Berger gave a variational argument for solutions of (0.1) in cases $\chi \leq 0$. In this situation, there is compactness in a sequence of functions approaching an extremal value for the functional F. When M is topologically the two-sphere, Hamilton introduced a heat equation associated to the variational functional F and was able to demonstrate long time existence and convergence of the solution metrics to a solution of the equation (0.1) with $R' \equiv 1$. It would still be of interest to find a direct variational argument for the existence question.

In dimensions greater than two the problem when R' is a constant is known as the Yamabe problem.

Theorem 2. (Yamabe [Y], Trudinger [T], Aubin [A1], Schoen [S1]). For any compact Riemannian manifold of dimensions ≥ 3, the equation (0.1') with R' a constant has a solution.

Yamabe considered the functional Q[u] and showed that it is bounded from below. Denoting by $Q(M) = \inf \{Q[u] \mid u > 0, u \in H_1\}$, it is easy to see that a minimizing sequence $\{u_j\}$ for the functional must be bounded in H_1. It is not clear that the weak limit is a strictly positive function. Trudinger showed that when $Q(M) < 0$, the weak limit is non-zero, hence a solution to the Yamabe problem. Aubin established that we always have $Q(M) \leq Q(S^n)$ and that when $Q(M) < Q(S^n)$ the weak limit is non-zero and was able to verify this inequality by a local construction when $\dim M \geq 6$ and M is not locally conformally flat. The remaining cases require global considerations, and Schoen showed using the positive mass theorem ([SY1, SY2]) that except when M is conformally equivalent to S^n, $Q(M) < Q(S^n)$, in this way the full Yamabe problem is solved. The reader is referred to the survey of Lee-Parker [LP] for the detailed exposition of this development. The Yamabe equation also has important applications to understanding certain Kleinian groups as developed in Schoen-Yau ([SY3]). Recently Schoen ([S2]) has constructed solutions of the Yamabe equation with prescribed singular behavior at a thin set, for example, a finite number of points on the sphere S^n.

To discuss the problem (0.1) when R' is a variable function it is convenient to normalize the background metric g to have constant scalar curvature R. In dimension two this problem was studied by Kazdan-Warner. In the cases where $\chi(M) \leq 0$, there is a reasonably complete picture.

Theorem 3.(Kazdan-Warner [KW1]) When $\chi(M) < 0$, $R \equiv -1$, equation (0.1) has a solution if and only if equation (0.1) has a supersolution u_+ and a subsolution u_- with $u_- \leq u_+$. When $\chi(M)=0$, $R \equiv 0$, equation (0.1) has a solution if and only if R' changes sign, and $\int R' \leq 0$.

The situation when M is the 2-sphere is more complicated. There is an implicit obstruction observed by Kazdan-Warner. Consider S^2 as the standard unit sphere in $\mathbb{R}^3$. The equation (0.1) implies the following condition:

(1.1) $$\int < \nabla R', \nabla x_i > e^{2u} = 0, \qquad \text{for } i = 1, 2, 3.$$

Thus for the functions $R' = 1 + \varepsilon x_i$ which can be arbitrarily close to the constant function 1, equation (0.1) has no solutions. The condition (1.1) expresses the fact that u is a critical point with respect to one-parameter variations of u induced by the conformal vector fields ∇x_i. In fact, more generally for any one-parameter-group of conformal transformations ϕ_t we have

$$\frac{d}{dt} F[T(\phi_t)(u)] \,|_{t=0} = 0.$$

In this way Bounguignon-Ezin ([BE]) gave further implicit conditions for the solvability of the equations (0.1) and (0.1').

To analyze the existence question on the standard 2-sphere we examine the variational functional F[u] in some detail:

$$F[u] = \log ⨍ R'e^{2u} - S[u], \quad S[u] = ⨍ (|\nabla u|^2 + 2u).$$

This functional has two natural symmetries:

i) $F[u + c] = F[u]$;

ii) $S[T(\varphi)(u)] = S[u]$, for any conformal transformation φ.

For later exposition it is convenient to point out the intrinsic significance of S[u] as the regularized determinant of the Laplacian associated to the metric $e^{2u}g$. More precisely, setting

$$\zeta(s) = \sum \lambda_i^{-s},$$

where λ_i are the eigenvalues of the Laplacian associated to $e^{2u}g$, we have ([Pol])

$$\zeta'(0) = -(1/3)S[u] + C,$$

where C is a universal constant independent of u.

The first important progress on the equation for S^2 was due to Moser:

Theorem 4. (Moser [Mo1]). If R' is an even function on the 2-sphere, positive somewhere, the equation (0.1) with $R \equiv 1$ has an even solution.

Moser's approach to this problem is to derive a sharp form of an inequality ([Mo2]) originally discovered by Trudinger: There is a constant c_0 such that for $u \in \mathcal{C}^1(S^2)$, $\int u = 0$ and $\int |\nabla u|^2 = 1$, we have

(1.2) $$\int \exp(4\pi u^2) \le c_0 .$$

(It is curious to note that there is a extremal function u_0 in H_1 realizing the supremum of this inequality for the corresponding Dirichlet problem on the ball in $\mathbb{R}^n$ (Carleson-Chang [CC]).) This inequality gives immediately the boundedness of the functional F[u] for all $u \in \mathcal{C}^1(S^2)$. However, a maximizing sequence of function $\{u_j\}$ will not converge unless R' is identically constant. The reason is that if u_0 is a local maximum of the functional F, a second variation computation shows that there is a lower bound for the first eigenvalue λ_1 of the Laplacian associated to the metric $R'e^{2u}g : \lambda_1 \geq 2$. This would contradict the sharp upper bound for λ_1 due to Hersch: $\lambda_1 \geq 2$ with equality holding if and only if $R'e^{2u}g$ has constant curvature 1. In Moser's case, when u is restricted to be an even function, a better inequality is available: in place of (1.2), we have

$$\int \exp(8\pi u^2) \leq c_0 .$$

It follows easily from this inequality that for a maximizing sequence $\{u_j\}$ of even function for F[u], a uniform bound on $\int|\nabla u_j|$ is available. This yields the convergence of a subsequence of $\{u_j\}$ to an even solution for (0.1).

There are two ways to understand the role evenness plays in this inequality. The first returns to Moser's proof of the inequality by a symmetrization arguement. It turns out that the optimal coefficient for the quadratic term u^2 in the exponential is related to the isoperimetric constant associated to the high level sets of u^2. To be precise, let $\Omega_c = \{x \in S^2 \mid u^2(x) \geq c\}$, A_c = area of Ω_c, L_c = length of $\partial\Omega_c$. Then the optimal coefficient is the number given by

$$\beta = \lim_{c \to \sup u^2} (L_c^2/A_c).$$

In view of the isoperimetric inequality for regions on the two sphere $L^2 \geq A(4\pi\text{-}A)$, we see that β can be taken to be 4π. In the case of even functions β can be taken to be 8π. There is a generalization of this inequality for domains ([CY2]): if D is a piecewise $\mathcal{C}^2$ bounded domain in $\mathbb{R}^2$ with a finite number of vertices, let θ be the minimum interior angle at the vertices; there exists a constant c_D such that for all $u \in \mathcal{C}^1(S^2)$ with $\int_D u = 0$, $\int|\nabla u|^2 = 1$ that

$$(1.2') \qquad \int \exp(2\theta u^2) \leq C_D .$$

In particular when D is smooth, the best exponent is 2π.

A second way to relax the evenness assumption is through the observation that for even functions u, the center of mass of the measure $e^{2u}g$ is at the origin:

$$(1.3) \qquad \int x_i\, e^{2u} = 0 , \qquad i = 1,2,3.$$

Let $\mathcal{S}$ denote the more general class of functions u satisfying (1.3). Aubin observed the following improvement in the linear exponential inequality ([A2]): Given $\varepsilon > 0$, there exists $C_\varepsilon > 0$ so that for $u \in \mathcal{S}$, we have

$$(1.4) \qquad ⨍\exp(2u) \leq C_\varepsilon \exp [\, ((1/2)+ \varepsilon)\, ⨍|\nabla u|^2 + 2\, ⨍u \,].$$

As a consequence, there is compactness in a sequence of function $\{u_j\} \subset \mathcal{S}$ on which S[u] remain bounded. Aubin proved that the equation (0.1) can be solved modulo a Lagrange multiplier term:

$$\Delta u + R'e^{2u} = 1 + \sum_{i=1}^{3} \lambda_i\, x_i\, e^{2u} .$$

Also as a consequence, Onofri ([On], see also [Ho], [OPS1]) obtained a sharp form of the linear exponential inequality: for all $u \in \mathcal{C}^1(S^2)$, we have

(1.5) $\int e^{2u} \le \exp(S[u]);$

and the equality is achieved achieved only by conformal factors $e^{2u} = \det(d\varphi)$, where φ is a conformal transformation. Onofri's argument uses also the fact that given any mass distribution e^{2u}, there is a conformal transformation φ of S^2 with the property that $T(\varphi)(u) \in \mathcal{S}$. In fact there is a conformal transformation of the form $\varphi_{Q,t}$ described in terms of the complex stereographic coordinates $z = (x_1 + ix_2)(1 - x_3)^{-1}$ obtained by projecting from Q as the north pole (0,0,1) to the x_1-x_2 plane:

(1.6) $\varphi_{Q,t}(z) = t\,z\,.$

It is convenient for subsequent discussion to introduce the class $\mathcal{S}_{Q,t} = \{u \mid T(\varphi_{Q,t})(u) \in \mathcal{S}\}$. The (Q,t) parameter measures the extent of concentration of the mass distribution.

It is possible to sharpen Onofri's inequality somewhat for functions $u \in \mathcal{S}$ ([CY2]): There exists a constant $a < 1$ such that for all $u \in \mathcal{S}$ we have

(1.7) $⨍ e^{2u} \le \exp(a ⨍ |\nabla u|^2 + 2 ⨍ u).$

In view of (1.3) it would be of interest to know if a can be taken to be 1/2 in (1.7).

The inequality (1.7) and the centering technique allow us to gain control on a sequence of function $\{u_j\}$ on which $F[u_j]$ tend to a critical value.

Concentration Lemma: If $\{u_j\}$ is a sequence of $\mathcal{C}^1$ function with $⨍ \exp(2u_j) = 1$ and $S[u_j] \le C$, then either

i) $\int |\nabla u_j|^2 \le c'$ or

ii) a subsequence $\exp(2u_j)$ concentrates at some point $Q \in S^2$, i.e.

$$\lim_{R\to 0} \lim_{j\to\infty} \int_{B(Q,R)} \exp(2u_j) = 1.$$

For concentrated masses $\exp(2u)$, it turns out that if $S[u]$ is small enough, we have very accurate control of the measure $\exp(2u)g$:

Asymptotic Formula. Suppose $u \in \mathcal{S}_{Q,t}$ with $S[u] = O(t^{-\alpha})$ for some $\alpha > 0$ and t sufficiently large, then for every $\mathcal{C}^2$ function f we have:

$$⨍ f\, e^{2u} = f(Q) + 2\Delta f(Q) t^{-2} \log t + O(t^{-2}) + O(|\nabla f(Q)| (t^{-2}\log t)^{1/2} (S[u]^{1/2})).$$

The concentration lemma and asymptotic formula are used to construct variations of test functions in a couple of variational schemes involving max-min procedures to prove the following criteria for existence of (0.1):

Theorem 5 ([CY2]). On S^2, let $R' > 0$ be a smooth function with non-degenerate critical points, and in addition $\Delta R'(Q) \neq 0$ at any critical point Q of R'. Suppose R' has p+1 local maxima and q saddle points where $\Delta R' < 0$. Then if $p \neq q$, the equation (0.1) has a solution.

A modified form of the concentration lemma for domains together with the inequality (1.2') give a generalization of Moser's theorem for functions R' with reflection symmetries:

Theorem 6 ([CY2]). Suppose D is a smooth domain of S^2 and R' a smooth function. Then for the Neumann boundary value problem corresponding to (0.1) to have a solution: we have
a) When area (D) < 2π it is sufficient that R' be positive somewhere;
b) When D = hemisphere, it is sufficient that

$$(\text{area}(D))^{-1}\int_D R' > \max\left(\max\{R'(x) \mid x \in \partial D\}, 0\right).$$

It is possible to obtain other existence results by varying conditions on the critical points of R', for example Chen-Ding ([CD], also [2]) gave an existence argument when R' has two global maxima and a subsidiary condition on the value of other saddle points. Hong ([Ho]) gave an existence theorem when R' has symmetry with respect to a finite group. Ko ([Ko]) has an existence result as a special case of Theorem 6.

In dimension greater than two, similar non-compactness phenomenon occurs for equation (0.1'). In dimension 3 an analogue of the concentration lemma holds. However in dimensions higher than three, more complicated concentration can occur; in particular concentration can occur at several points. As an analogue of Moser's theorem, we have:

Theorem 7 (Escobar-Schoen ([ES]). Suppose M is a locally conformally flat manifold with positive scalar curvature which is not simply connected, and R' is a smooth function on M which is somewhere positive and satisfies the condition: there is a maximum point Q of R' at which all partial derivatives of R' of order less than or equal to n-2 vanish. Then equation (0.1') has a solution.

For the three sphere, Bahri- Coron ([BC]) have announced the exact analogue of Theorem 5:
Theorem 8 ([BC]). On the standard S^3 if $R' > 0$ is a smooth function with only non-degenerate critical points and, in addition, $\Delta R'(Q) \neq 0$ at all critical points of R'. Denote by k_i the Morse index of the critical point Q_i. If $\Sigma (-1)^{k_i} \neq -1$ where the sum is taken over those critical points where $\Delta R'(Q_i) < 0$, then equation (0.1') has a solution.

So far all of these existence results impose primarily topological conditions on the function R', it should be noted however that no simple topological condition on R' will give necessary and sufficient condition. This observation follows from a different type of existence assertion due to

Kazdan-Warner ([KW2]): For any function R' positive somewhere, there is a diffeomorphism φ of S^2 so that the equation (0.1) is solvable with $R'\circ\varphi$ in place of R'. To make further progress on the equation (0.1) it would be important to describe analytic condition on a function R' with only one maximum and one minimum making the equation (0.1) solvable.

§2 Isospectral conformal metrics.

Given a compact Riemannian manifold with $g = \Sigma\, g_{ij}dx^i dx^j$, the Laplacian operator on scalar functions is defined by

$$\Delta u \;=\; g^{-1/2}\,\partial_i(g^{ij}\,g^{1/2}\partial_j u).$$

The eigenvalues λ_i paired with eigenfunctions φ_i, $\Delta\varphi_i + \lambda_i\varphi_i = 0$, form a discrete set $0 < \lambda_1 \le \lambda_2 \le \cdots$. It is by now a well known problem to study the extent to which the spectrum determines the metric. There is a vast literature and several excellent surveys on this problem ([BGM], [B]). In the following we will be concerned only with development resulting from the recent understanding of the conformal deformation equation.

In two dimension it is natural to separate the discussion according to the sign of the Euler characteristic of the surface. The most important isospectral invariant turns out to be the determinant of the Laplacian $\zeta'(0)$ where $\zeta(s)$ was defined in the previous section. it is again convenient to regard all metrics as conformal deformations $e^{2u}g$ of a fixed background metric g of constant curvature ± 1 or 0. Polyakov ([Pol]) pointed out that in that case the difference between $\zeta'(0)$ for the two metrics $g' = e^{2u}g$ and g is expressible as an integral over M involving u:

(2.1) $$\zeta'_{g'}(0) - \zeta'_g(0) = -\,1/3\ S[u] = -1/3\ \int(|\nabla u|^2 + 2Ru)$$

where the integration is taken with respect to the area induced from g, and R is the scalar curvature of g.

The better known heat invariants can also be expressed in terms of integrals of local expressions in the metric. Recall that the heat kernel H(x,y,t) is given in L^2 by

$$\Sigma\, \exp(-\lambda_i t)\varphi_i(x)\varphi_i(y).$$

The trace of the heat kernel has an asymptotic expansion ([MS]):

$$\int_M H(x,x,t)\ \sim\ (4\pi t)^{-n/2}(a_0 + a_1 t + a_2 t^2 + \cdots),$$

which is valid for $t \to 0$. The heat invariants a_k can clearly be recovered from knowledge of $\{\lambda_i\}$. While the expression for a_k are known for small values of k, we can only hope to know the dominant terms in the general case when k is large. To formulate these expansions we need some additional notations. For a Riemannian metric $g = \Sigma\, g_{ij}dx^i dx^j$, $\rho = \rho_{ij}$ will denote the Ricci tensor, R the scalar curvative, dv the volume form. For any tensor T, ∇T will denote the covariant derivative, and $|\nabla T|^2$ the pointwise norm $|\nabla T|^2 = \Sigma|\nabla_i T|^2$ where $\{e_i\}$ form an orthonormal frame. Similarly, $|\nabla^k T|^2 = \Sigma|\nabla_{i_1} \nabla_{i_2} \cdots \nabla_{i_k} T|^2$.

For the low order heat invariants we have ([BGM]):

$$a_0 = c_0 \int dv$$
$$a_1 = c_1 \int R\, dv$$
$$a_2 = \int A_2 R^2 + B_2 |\rho|^2 dv, \qquad \text{for } \dim M \le 4,\ A_2, B_2 > 0 .$$

And for $k \ge 3$ (Gilkey [G]):

(2.2) $$a_k = \int A_k |\nabla^k R|^2 + B_k |\nabla^k \rho|^2 + E_k\, dv,$$

where A_k and B_k have the same sign and E_k is a polynomial contraction of lower order covariant derivatives of the curvature tensor of weight 2k+4. The weight of $\nabla_{i_1} \cdots \nabla_{i_m} R_{ijk\ell}$ is m+2, and the weight of a monomial is the sum of the weights of its factors.

Although there is a large number of examples exhibiting families of isospectral but not isometric metrics ([Mi], [V], [Su], [BT], [Bu], [GW]), it is probably true that such families are compact. In dimension two this is indeed so.

Theorem 9 (Osgood, Phillips and Sarnak ([OPS2]). For a compact surface M a set of isospectral metrics on M is compact in the $\mathfrak{C}^\infty$ topology.

The set of metrics on a given manifold naturally separate into conformal classes. In dimension two, theory of conformal moduli gives detailed knowledge of the set of conformal structures in the cases where $\chi(M) \le 0$. The compactness of the set of underlying conformal structures of an isospectral family of metrics follows immediately from Wolpert's calculation for the asymptotics of the corresponding determinants of Laplacian ([W]). The question then reduces to showing the compactness of a family of isospectral conformal metrics. In this case R is either 0 or -1, and (2.1) gives immediately bounds for $\int|\nabla u|^2$. In case M is the two sphere, $R \equiv 1$, and Aubin's inequality (1.4) gives bounds for $\int|\nabla u|^2$ as well when u is normalized by an appropriate conformal transformation. So the determinant of the Laplacian controls the first order derivatives of the conformal factors u in a family of isospectral conformal metrics. Bounds on the higher order derivative of u are derived from an inductive argument using the ellipticity of the equation (0.1) and the boundedness of the heat invariants a_k.

In higher dimensions there seem to be serious difficulty in carrying through the same line of reasoning. First of all there is lacking at this moment a theory of moduli for the set of conformal structures on a given manifold. Secondly even within a fixed conformal structure, an analogue of Polyakov's formula for the determinants of Laplacians is not available. Nevertheless, for conformally equivalent metrics in dimension three, we can manage to get by with only the heat invariants, by exploiting fully the differential equation (0.1').

Theorem 10 ([CY3] [CY4] and [BPY]). For a compact manifold of dimension three, a set of isospectral conformal metrics form a compact set in the C^∞ topology.

The proof of this result divides naturally into three cases according to the sign of the underlying constant scalar curvature R of the Yamabe metric and in the case $R > 0$, whether the manifold is conformally equivalent to the standard S^3.

The proof proceeds in two stages. The first part is concerned with bounds for the conformal factor $\rho = u^4$ up to second order derivatives. For a family of isospectral conformal metrics $\{u_j^4 g\}$ the ratio of the first two heat invariants a_1/a_0 is up to a factor the functional Q[u] displayed in (0.2'). Thus we get immediately the bounds for $\int |\nabla u_j|^2$, and a subsequence converges weakly to a conformal factor u. In contrast to the Yamabe problem the limit doesn't need to satisfy a differential equation, thus it will not be sufficient to show u is not identically zero. The additional information on the a_2 term and the first eigenvalue λ_1 of the Laplacian of the metric $u_j^4 g$ are then used to obtain bounds on the low order derivatives:

Proposition 2.1. Suppose M^3 is a compact Riemannian manifold with given $a_k \le \alpha_k$, $k = 0,1,2$ and $\lambda_1 \ge \Lambda > 0$, then writing the metric $g' = u^4 g_0$ where g_0 is the Yamabe metric, except when M is the standard S^3, there are constants $c_1, c_2 > 0$ such that

i) $\quad 0 < c_1 \le u \le c_1^{-1}$,

ii) $\quad \int |\nabla u_j|^2 , \int |\nabla\nabla u_j|^2 \le c_2$;

in the case $M = S^3$, there exists a conformal map φ of S^3, so that (i) and (ii) hold for $u' = T(\varphi)(u)$.

The idea is to reduce i) and ii) to the condition that there is an $\varepsilon > 0$ such that $\int u^{6+\varepsilon} dv$ has an a priori bound in terms of the data. This bound is available when there exists $\delta > 0$, $L > 0$ so that

(2.3) $\quad \text{meas}\,\{x \in M \mid u(x) > \delta\} \ge L.$

When the background Yamabe metric has negative scalar curvature, there is in fact a positive lower bound for u ([BPY]). When the background Yamabe metric has nonnegative scalar curvature, if the condition (2.3) is not satisfied, there will be a sequence of conformal factors $\{u_j\}$ satisfying the bounds in the statement of the proposition, but $u_j \to 0$ almost everywhere. We then show that after suitably rescaling $v_j = c_j u_j$ with $c_j \to \infty$ that there is a point $Q \in M$ and a subsequence of $\{v_j\}$ that converges strongly in H_1 on compact sets off Q to a positive solution of the conformal Laplacian equation $8\Delta v - R_0 v = 0$. In case $R_0 = 0$, v is necessarily a constant; and in the case when $R_0 > 0$, v is in fact the Green's function G of the conformal Laplacian with pole at Q. The original bounds for a_2, after rescaling, implies that $v^4 g$ is a Ricci flat metric on $M-\{Q\}$. In dimension three Ricci flat metrics are flat. Thus in case $R_0 = 0$, this means that the original metric on M is flat. In case $R_0 > 0$, this means that $(M - \{Q\}, G^4 g)$ is isometric to $\mathbb{R}^3$. So unless M is conformally equivalent to S^3 or a flat manifold, we recover the statement (2.3). When M is in fact S^3, it is possible to transform the conformal factor u by a conformal transformation to bring the center of mass of the measure $u^{6+\varepsilon}dv$ on S^3 in $\mathbb{R}^4$ to the origin. In this normalized situation, the lower bounds on the λ_1 of the metric $u^4 g$ give directly the required bounds for $\int u^{6+\varepsilon}dv$ ([CY3]). When M is a flat manifold with such a sequence $\{u_j\}$ of conformal factors concentrating at a point Q, we show that we can cap off a small ball B around Q to obtain a sequence of similarly concentrating conformal metrics on the sphere S^3. The compactness result for the sphere then yield a blow up rate for the sequence $\{u_j\}$near Q in contradiction to the convergence of $\{u_j\}$ to constant established earlier.

The second stage of the argument is the following:

Proposition 2.2 For a compact manifold (M^3, g) the set of conformal metrics $g' = u^4 g$ which satisfy i) and ii) in proposition (2.1), and the condition $a_k(g) \leq \alpha_k$ for $k = 3,4,5,\ldots$, form a compact set in the $\mathcal{C}^\infty$ topology.

The proof consists in exploiting the ellipticity of the equation (0.1') to inductively control the L^2 norm of the k-th order derivatives of u. To bound the k-th derivatives of u, we use the bound on a_k, observing that the inductive bounds on the lower order derivative of u control the corresponding lower order terms in a_k. A crucial point is that, thanks to Gilkey's calculations, the two dominant terms in a_k have the same sign, and thus cannot cancell each other.

For dimensions higher than three, similar arguments for Proposition 1 seem to break down at almost every stage. It is very likely that higher order heat invariants are needed to derive an initial bounds on the low order derivatives. This is an indication that probably much more is going on in dimensions four and up. However at least in even dimensions it is possible that the determinant of the Laplacians can play the crucial role here as in the case of a surface.

References

[A1] Aubin, T., Equations differentielles nonlineaires et Problem de Yamabe concernant la courbure scalaire. J. Math. Pure Appl. 55 (1976), 269-296.

[A2] ______, Meilleures constantes dans le theoreme d'inclusion de Sobolev et un theoreme de Fredholm non lineaire par la transformation conforme de la courbure scalaire, J. Funct. Anal. 32 (1979), 148-174.

[B] Berard, P.; Spectral geometry: direct and inverse problems. Lecture Notes in Math. 1207, Springer Verlag.

[Be] Berger, Melvyn; On Riemannian structure of prescribed Gaussian curvature for compact 2-manifolds, Jour. Diff. Geom. 5(1971), 325-332.

[BC] Bahri, A., and Coron, J.M., Une theorie des points critiques a l'infini pour l'equation de Yamabe et le problem de Kazdan-Warner, C. R. Acad. Sci. Paris Math. 15 (1985), 513-516.

[BE] Bourguignon, J.P., anf Ezin, J.P.; Scalar curvarture functions in a conformal class of metrics and conformal transformations, Trans. AMS 301(1987), 723-730.

[BGM] Berger, M; Gauduchon, P; and Mazet, E; Le Spectre d'une Varieté Riemannienne, Lecture Notes in Mathematics, 194, Springer-Verlag.

[BPY] Brooks, R; Perry, P.; and Yang, P.; Isospectal sets of conformally equivalent metrics, to appear in Duke Math. Jour.

[BT] Brooks, R., and Tse, R.; Isospectral surfaces of small genus, Nagoya Math. Jour. 107(1987), 13-24.

[Bu] Buser, P.; Isospectral Riemann surfaces, preprint.

[CC] Chang, S.Y.; and Carleson, L.; On the existence of an extremal function for an inequality of J. Moser, Bull. SC. Math. 2^e serie, 110(1986), 113-127.

[CD] Chen, W.-X., and Ding, W.-Y.; Scalar curvature on S^2, Trans. AMS 303(1987), 369-382.

[CY1] Chang, S.Y., and Yang, P.C.; Prescribing Gaussian curvature on S^2, Acta. Math. 159 (1987), 215-259.

[CY2] _____; Conformal deformation of metrics on S^2, J.D.G. 27(1988), 259-296.

[CY3] _____; Compactness of isospectral conformal metrics on S^3, to appear in Comm. Math. Helv.

[CY4] _____; A compactness theorem for conformal metrics on 3-manifolds and application to isospectral metrics, preprint.

[ES] Escobar, J., and Schoen, R.; Conformal metrics with prescribed scalar curvature, Invent. Math.86(1986), 243-254.

[Gi] Gilkey, P; Leading terms in the asymptotics of the heat equation, preprint.

[GW] Gordon, C., and Wilson, E.; Isospectral deformation of compact solvemanifolds, JDG 19(1984), 241-256.

[Ha] Hamilton, R.; Lecture at AMS meeting Hawaii, 1987.

[Ho] Hong, C.; A best constant and the Gaussian curvature, Proc. AMS 97(1986), 737-747.

[KW1] Kazdan,J.; and Warner F.; Curvature functions for compact 2-manifolds, Ann. Math. 99(1974), 14-47.

[KW2] _____; Existence and conformal deformaion of metrics with prescribed Gaussian and scalar curvatures, Ann. of Math. 101(1975)317-331.

[Ko] Ko, K.-S.; Ph.D. Thesis. USC, 1986.

[LP] Lee, J., and Parker, T.; The Yamabe problem, Bull.AMS 17(1987), 37-81.

[L] Lelong-Ferrand, J.; Transformations conformes et quasi-conformes des varieties riemanniennes compactes applicationa la demonstration d'une conjecture de A. Lichnerowicz, C.R. Acad. Sci. Paris 269 (1969), 583-586.

[MS] McKean, H., and Singer, I.; Curvature and the eigenvalues of the Laplacian, J.D.G. 1(1967), 43-69.

[Mi] Milnor,J.; Eigenvalues of the Laplace operator on certain manifolds, P.N.A.S.51 (1964), 541.

[Mo1] Moser, J.; On a non-linear problem in differential geometry, Dynamical Systems (M. Peixoto ed.), Acad. Press. New York, 1973.

[Mo2] _____; A sharp form of an inequality by N. Trudinger, Indiana Univ. Math. Jour. 20(1971), 1077-1091.

[On] Onofri, E; On the positivity of the effective action in a theory of random surfaces, Comm. Math. Physics, 86(1982), 321-326.

[OPS1] Osgood, G; Phillips, R; and Sarnack, P; Extremals of determinants of Laplacians, preprint.

[OPS2] ____ ; Compact isospectral sets of Riemann surfaces, preprint.

[P] Poincare, H.; Les fonctions Fuchsiennes et l'equation $\Delta u = e^u$, Oeuvres t. II, Gauthier-Villars Paris (1916), 512-591.

[Pol] Polyakov, A.; Quantum geometry of the bosonic strings, Phys. Letters 103B(1981), 207-210.

[S1] Schoen, R; Conformal deformations of a Riemannian metric to constant scalar curvature, Jour. Diff. Geom., 20(1984), 479-496.

[S2] ____ ; The existence of weak solutions with prescribed singular behavior for a conformally invariant scalar equation, Comm. Pure and Appl. Math.41(1988), 317-392.

[SY1] Schoen, R.; and Yau, S.-T.; On the proof of the positive mass conjecture in general relativity, Comm. Math. Phys. 65(1979), 45-76.

[SY2] _____; Proof of the positive mass theorem II, Comm. Math.Phys. 79(1981), 231-260.

[SY3] _____; Conformally flat manifolds, Kleinian groups and scalar curvature, Inventiones Math. 92(1988), 47-71.

[Su] Sunada, T.; Riemannian coverings and isospectral manifolds, Ann. of Math. 121(1985), 169-186.

[T] Trudinger, N.; Remarks concerning the conformal deformation of Riemannian structure on compact manifolds, Ann. Scuo. Norm. Sup. Pisa 3(1968), 265-274.

[V] Vigneras, M.F.; Varietesriemanniennes isospetrales et non isometriques, Ann. of Math. 121(1980), 21-32.

[W] Wolpert, S; Asymptotics of the spectrum and the Selberg zeta function on the space of Riemann surfaces, preprint.

[Y] Yamabe, H.; On a deformation of riemannian structures on compact manifolds, Osaka Math. Jour. 12(1960), 21-37.

DEPARTMENT OF MATHEMATICS, UNIVERSITY OF CALIFORNIA, LOS ANGELES, CALIFORNIA 90024

DEPARTMENT OF MATHEMATICS, UNIV. OF SOUTHERN CALIF., LOS ANGELES, CALIFORNIA 90089

Contemporary Mathematics
Volume **101**, 1989

Analytic Torsion for Group Actions

John Lott*

I. Introduction

I would like to discuss some joint work with M. Rothenberg (U. of Chicago) on analytic torsion for group actions. Before stating the problem, let me sketch the story of the usual Ray-Singer analytic torsion. First, let M be a closed oriented smooth manifold with $\pi_1(M)$ nontrivial. Let $\rho : \pi_1(M) \to O(N)$ be an orthogonal representation. If the twisted cohomology $H^*(M, E_\rho)$ vanishes then one can define the Reidemeister torsion $\tau_\rho \in \mathbb{R}$ which is a homeomorphism invariant of M. Ray and Singer asked whether, as for many other topological quantities, one can compute τ_ρ by analytic methods. Given a Riemannian metric g, they defined the analytic torsion $T_\rho \in \mathbb{R}$ as a certain combination of the eigenvalues of the Laplacian acting on twisted differential forms. They showed that under the above acyclicity condition, T_ρ is independent of the metric g and conjectured that the analytic expression T_ρ equals the combinatorial expression τ_ρ [1]. This conjecture was proven independently by Cheeger [2] and Müller [3].

One can look at the above situation in the following way. The group $\pi_1(M)$ acts freely on the universal cover $\widetilde{M}$ and so one has an invariant for free group actions. One can then ask whether the Reidemeister torsion can be extended to an invariant for more general group actions. For a finite group acting (not necessarily freely) on a closed oriented PL manifold X, such a torsion τ was defined combinatorially by Rothenberg [4] and Illman [5]. One can then ask whether there is a corresponding analytic torsion when X is smooth, and whether the analytic torsion equals the combinatorial torsion. We will show that there is such an analytic torsion T, that $T = \tau$ if dim X is odd and that $T \neq \tau$ if dim X is even. We will also explain the reason for the nonequality in the latter case.

One can also ask if the analytic torsion can be defined for group actions in which there is no combinatorial torsion. We will show that such is indeed the case for certain discrete group actions and provide examples.

1980 *Mathematics Subject Classification* (1985 Revision). 57Q10, 57S15.

II. The Torsion of a Complex

For a survey article on Whitehead torsion, see [6]. Let us recall the definitions. Let V_{even} and V_{odd} be real finite-dimensional vector spaces and let

$$d_{even} : V_{even} \to V_{odd} \quad \text{and} \quad d_{odd} : V_{odd} \to V_{even}$$

be linear maps such that $d_{even} \circ d_{odd} = d_{odd} \circ d_{even} = 0$.

Put

$$B_{even} = \operatorname{Im} d_{odd}, \quad B_{odd} = \operatorname{Im} d_{even},$$

$$H_{even} = \operatorname{Ker} d_{even} / \operatorname{Im} d_{odd}, \quad H_{odd} = \operatorname{Ker} d_{odd} / \operatorname{Im} d_{even}.$$

Suppose that we are given volume forms

$$\omega_{\substack{even\\odd}} \text{ on } V_{\substack{even\\odd}} \quad \text{and} \quad \mu_{\substack{even\\odd}} \text{ on } H_{\substack{even\\odd}}.$$

Let us chose volume forms

$$\rho_{\substack{even\\odd}} \text{ on } B_{\substack{even\\odd}}.$$

From the Hodge decomposition of $V_{\substack{even\\odd}}$, we can find nonzero real numbers $m_{\substack{even\\odd}}$ such that

$$\omega_{even} = m_{even}\, \rho_{even} \wedge d^*_{even}\, \rho_{odd} \wedge \mu_{even} \quad \text{and}$$
$$\omega_{odd} = m_{odd}\, \rho_{odd} \wedge d^*_{odd}\, \rho_{even} \wedge \mu_{odd}.$$

Defn. The Reidemeister torsion of the complex is $\tau(\omega,u) = \ln(\,|\, m\ even / m\ odd\,|\,)$.
(Note that τ does not depend on the choice of volume forms $\rho_{\substack{even\\odd}}$).

Ex. Let M be an oriented smooth closed manifold with a triangulation K, whose universal cover is $\tilde{K}$, and let $\rho: \pi_1(M) \to O(N)$ be an orthogonal representation. Let $V_{\substack{\text{even}\\ \text{odd}}}$ be the twisted simplicial cochain groups:

$$V_{\substack{\text{even}\\ \text{odd}}} = C^*_{\substack{\text{even}\\ \text{odd}}}(K,E_\rho) = C^*_{\substack{\text{even}\\ \text{odd}}}(\tilde{K}) \otimes_\rho \mathbb{R}^N.$$

Let $d_{\substack{\text{even}\\ \text{odd}}}$ be the coboundary operators. In order to define the Reidemeister torsion τ_ρ, we need volume forms on $C^*_{\substack{\text{even}\\ \text{odd}}}(K,E_\rho)$ and $H^*_{\substack{\text{even}\\ \text{odd}}}(K,E_\rho)$.

For the cochain groups $C^*(K,E_\rho)$, let us define a preferred basis by $\{[\tilde{\sigma}\otimes e_i]\}$ where for each simplex σ of K, $\tilde{\sigma}$ is a fixed lifting of σ to $\tilde{K}$, and $\{e_i: 1 \le i \le N\}$ is a standard basis of $\mathbb{R}^N$. Using this preferred basis, we can define volume forms $\omega_{\substack{\text{even}\\ \text{odd}}}$ on $C^*_{\substack{\text{even}\\ \text{odd}}}(K,E_\rho)$.

For the cohomology groups $H^*_{\substack{\text{even}\\ \text{odd}}}(K,E_\rho)$, let us put a Riemannian metric g on M. Let $\Lambda^*(M,E_\rho)$ denote the twisted differential forms:

$$\Lambda^*(M,E_\rho) = \Lambda^*(\tilde{M}) \otimes_\rho \mathbb{R}^N.$$

The de Rham map

$$A: \Lambda^*(M,E_\rho) \to C^*(K,E_\rho)$$

induces an isomorphism between the twisted harmonic forms Ker Δ and the cohomology groups $H^*(K,E_\rho)$. Ker Δ has an inner product induced from the Hilbert space $\Lambda^*(M,E_\rho)$ and so we obtain an inner product on $H^*(K,E_\rho)$. Let $\mu_{\substack{\text{even}\\ \text{odd}}}$ be the volume forms on $H^*_{\substack{\text{even}\\ \text{odd}}}(K,E_\rho)$ derived from this inner product.

Prop. 1 (Whitehead [7], Ray-Singer [1]) The torsion τ_ρ is invariant under subdivision of the triangulation.

Note that if $H^*(K,E_\rho)$ vanishes then τ_ρ is metric-independent.

Prop. 2 (Chapman [8]) If $H^*(K,E_\rho)$ vanishes then τ_ρ is a homeomorphism invariant.

The original interest of the Reidemeister torsion was that it is not a homotopy invariant, and so can distinguish spaces which are homotopy-equivalent but not homeomorphic, such as certain lens spaces. More generally, consider the Clifford-Klein spaces S^n/Γ where Γ is a finite subgroup of $SO(n+1)$ which acts freely on S^n.

Prop. 3 (de Rham) The Clifford-Klein spaces of a given dimension are distinguished up to isometry by π_1 along with the Reidemeister torsions τ_ρ.

III. The Analytic Torsion

In order to motivate the Ray-Singer definition of the analytic torsion, let us first consider the situation of the example of section II when the complex is acyclic. We have an inner product on $C^*(K,E_\rho)$ coming from the preferred basis. Let $d^c = d^c{}_{even} \oplus d^c{}_{odd}$ be the coboundary operator and let d^{c*} denote the adjoint of d^c which is defined using the inner product. Define $\Delta_i^c = d^c d^{c*} + d^{c*} d^c$, the combinatorial Laplacian on $C^i(K,E_\rho)$.

One can show

$$\tau_\rho = -\frac{1}{2}\sum_{i=1}^{\dim M} (-)^i\, i \ln \det \Delta_i^c$$

$$= \frac{1}{2}\frac{d}{ds}\bigg|_{s=0} \frac{1}{\Gamma(s)}\int_0^\infty T^{s-1}\sum_i (-)^i\, i \operatorname{Tr} e^{-T\Delta_i^c}\, dT.$$

This motivates the definition of Ray-Singer : We no longer assume that $H^*(M,E_\rho)$ vanishes. Let g be a Riemannian metric on M. Let d be the exterior derivative on $\Lambda^*(M,E_\rho)$ and let d^* be its adjoint w.r.t. the L^2 inner product on $\Lambda^*(M,E_\rho)$. Put

$\Delta_i = d^*d + dd^*$ acting on $\Lambda^i(M,E_\rho)$ and

$\Delta = d^*d + dd^*$ acting on $\Lambda^*(M,E_\rho)$.

Let P be the projection onto Ker Δ and put $\Delta' = \Delta + P$. Let F be the operator on $\Lambda^*(M,E_\rho)$ which is multiplication by i on $\Lambda^i(M,E_\rho)$.

Defn. The analytic torsion of M is given by

$$T_\rho = \frac{1}{2}\frac{d}{dS}\Big|_{S=0} \frac{1}{\Gamma(S)}\int_0^\infty T^{S-1}\sum_{i=1}^{\dim M}(-)^i\, i\, \mathrm{Tr}\, e^{-T\Delta_i'}dT \tag{1}$$

$$= \frac{1}{2}\frac{d}{dS}\Big|_{S=0} \frac{1}{\Gamma(S)}\int_0^\infty T^{S-1}\,\mathrm{Tr}\,(-1)^F F\, e^{-T\Delta'}\,dT.$$

The trace of (1) is on trace-class operators on $\Lambda^*(M,E_\rho)$. The derivative at $S = 0$ is that of a meromorphic function (in S) which is defined by analytic continuation from the region Re S $> \frac{1}{2}\dim M$, and is analytic around $S = 0$.

Prop. 4. (Cheeger [2], Müller [3]) $T_\rho = \tau_\rho$

In order to motivate a more general definition of the analytic torsion, let us consider the case when $\pi_1(M)$ is finite. Let Π be the projection from $\Lambda^*(\widetilde{M}) \otimes \mathbb{R}^N$ to $\Lambda^*(M,E_\rho)$, namely

$$\Pi = \frac{1}{|\pi_1(M)|}\sum_{g\in\pi_1(M)} \rho(g)\, a(g^{-1})^*,$$

where a: G $\rightarrow$ Diff($\widetilde{M}$) are the deck transformations. Then we can write

$$T_\rho = \frac{1}{2}\frac{d}{dS}\Big|_{S=0} \frac{1}{\Gamma(S)}\int_0^\infty T^{S-1}\,\mathrm{Tr}\,(-1)^F F\,\Pi\, e^{-T\Delta'}\,dT,$$

where the trace is on trace-class operators on $\Lambda^*(\widetilde{M}) \otimes \mathbb{R}^N$.

This motivates the following generalization. Let G be a finite group and let $a : G \to \mathrm{Diff}\, X$ be an orientation preserving smooth action of G on an oriented smooth closed manifold X. Let $\rho: G \to O(N)$ be a representation of G. Choose a G-invariant metric h on X. Put

$$\Pi = \frac{1}{|G|} \sum_{g \in G} \rho(g)\, a(g^{-1})^{*} \text{, an operator on } \Lambda^{*}(X) \otimes \mathbb{R}^{N} .$$

Defn. The analytic torsion of the group action is given by

$$T_\rho = \frac{1}{2} \frac{d}{dS}\Big|_{S=0} \frac{1}{\Gamma(S)} \int_0^\infty T^{S-1}\, \mathrm{Tr}\,(-1)^F F\, \Pi\, e^{-T\Delta'}\, dT ,$$

where the trace is on trace-class operators on $\Lambda^*(X) \otimes \mathbb{R}^N$.

In order to know whether T_ρ is an invariant of the smooth group action, we want to know whether it depends on the metric h used in its definition. As the space of G-invariant metrics is path-connected, it is enough to look at the derivative of T_ρ along a curve of G-invariant metrics.

Prop. 5. If the ρ-equivariant cohomology $H^*(X,\rho)$ vanishes then T_ρ is independent of the G-invariant metric h. If $H^*(X,\rho)$ is nonzero, let $h(\varepsilon)$ be a 1-parameter family of G-invariant metrics. Then

$$\frac{d}{d\varepsilon} T_\rho = -\frac{1}{2} \mathrm{Tr}\big|_{\mathrm{Ker}\,\Delta}\, \Pi\, (-1)^F \left(\frac{d}{d\varepsilon} *\right) *^{-1} ,$$

where * is the Hodge dual operator.

The proof of this proposition is similar to the proof of the analogous proposition for the Ray-Singer torsion in that it uses the heat kernel asymptotics, but differs in that it is necessary to use off-diagonal heat kernel asymptotics.

IV. Relation of T_ρ with τ_ρ

One can ask if there is a combinatorial equivalent of our analytic torsion for finite group actions. A combinatorial torsion was defined in [4] as an element of a Whitehead group. We will need the speciaiization of this invariant to the real numbers, and will give a more rudimentarry construction. Let us take a G-triangularization [9] K of X and put

$$C^*(K,\rho) = C^*(K) \otimes_\rho \mathbb{R}^N .$$

We have a complex formed by

$$V_{\substack{\text{even}\\ \text{odd}}} = C^*_{\substack{\text{even}\\ \text{odd}}}(K,\rho)$$

and in order to define the torsion of the complex we need volume forms on $C^*_{\substack{\text{even}\\ \text{odd}}}(K,\rho)$ and $H^*_{\substack{\text{even}\\ \text{odd}}}(K,\rho)$. An important constraint on the volume forms is that the resulting torsion should be invariant under a G-subdivision of K.

For the volume forms $\mu_{\substack{\text{even}\\ \text{odd}}}$ on $H^*_{\substack{\text{even}\\ \text{odd}}}(K,\rho)$ we can use the de Rham isomorphism

$$A : \operatorname{Ker} \Delta \to H^*(K, \rho),$$

where the Laplacian acts on ρ-equivariant differential forms.

To define the volume forms $\omega_{\substack{\text{even}\\ \text{odd}}}$ on $C^*_{\substack{\text{even}\\ \text{odd}}}(K,\rho)$

let us choose a function

$$c : \{\text{subgroups of } G\} \to \mathbb{R}^+$$

which satisfies $c(g\,H\,g^{-1}) = c(H)$ for all $g \in G$, $H \subset G$. Let us define an inner product on $C^*(K, \rho)$ by requiring that

1. Cochains of different degree are orthogonal.

2. Two cochains of the same degree k but with support on disjoint orbits of k-simplices are orthogonal.

3. If $\eta_1, \eta_2 \in C^*(K, \rho)$ have support on the same orbit of k-simplices then $<\eta_1, \eta_2> \equiv c(H_\sigma) < \eta_1(\sigma), \eta_2(\sigma)>_{\mathbb{R}^N}$ where σ is any k-simplex in the orbit, $\eta(\sigma) \in \mathbb{R}^N$ is the evaluation of the cochain $\eta \in C^*(K, \rho)$ on the simplex σ and H_σ is the isotropy group of the simplex σ.

(Note that because ρ is an orthogonal representation, $<\eta_1, \eta_2>$ is independent of the choice of σ.)

Using this inner product, we can define the volume forms $\omega_{\substack{\text{even}\\ \text{odd}}}$ on $C^*_{\substack{\text{even}\\ \text{odd}}}(K,\rho)$ and the torsion τ_ρ of the complex.

Prop. 6. For each choice of the function c, the resulting torsion τ_ρ is subdivision invariant.

The question now arises as to what the function c should be in order to make τ_ρ equal T_ρ.

Prop 7. If dim X is odd then with the choice c(H) = 1 for all $H \subset G$, we have $T_\rho = \tau_\rho$.

The proof of the proposition is based on a surgery argument, as in [2,3]. Because both T_ρ and τ_ρ are invariant under a change of orientation, we have $(T - \tau)\ \rho(X) = \frac{1}{2}\ (T - \tau)\ \rho\ (X \cup \overline{X})$. One can equivariantly surger $X \cup \overline{X}$ to a disjoint union of spheres with orthogonal G-actions. As one surgers, one picks up a contribution to $T - \tau$ of 1/2 of (the $T - \tau$ of the double of the surgery pieces). Because dim X is odd, one can show by explicit calculation that $T - \tau$ vanishes on each of the doubles of the surgered pieces, as well as on the final disjoint union of spheres.

The case when dim X is even is rather different. Let us again take c(H) = 1 for all $H \subset G$.

Prop 8. If dim X is even then $T_\rho = 0$. If the G action is free then $\tau_\rho = 0$. However for general G actions, $\tau_\rho \neq 0$.

The vanishing of T_ρ comes from an argument using the Hodge duality operator on differential forms. The analogous argument for τ_ρ would relate the torsion of a triangulation K to that of the dual cell-complex K^*. If the action is free than both K and K^* can be G-complexes [9]. However, if the action is not free then it is impossible for both K and K^* to be G-complexes. This is why $\tau_\rho \neq 0$ in general.

One can see that when dim X is even, $T_\rho - \tau_\rho = -\tau_\rho$ is nonzero even for orthogonal actions on spheres. However, from the argument used to prove Prop. 7 we do derive an explicit equation for τ_ρ in terms of an equivariant surgery decompostion of X.

V. Analytic Torsion for Discrete Group Actions

Let G be a discrete group and let $a : G \to \mathrm{Diff}\ X$ be a smooth orientation-preserving action of G on an oriented smooth closed manifold X. Suppose that there is an invariant metric h for the G action i.e. the image of a has compact closure in Diff X. Let $f : G \to \mathbb{R}$ be a function on G with finite support such that $\sum_{g \in G} f(g)\, a\left(g^{-1}\right)^*$ vanishes on $H^*(X, \mathbb{R})$.

Define $I_f = \sum_{g \in G} f(g)\, T(g)$, where

$$T(g) = \frac{1}{2} \frac{d}{dS}\bigg|_{S=0} \frac{1}{\Gamma(S)} \int_0^\infty T^{S-1}\, \mathrm{Tr}\, (-1)^F F\, a\left(g^{-1}\right)^* e^{-T\Delta'} dT .$$

Prop. 9. I_f is independent of the G-invariant metric h.

Thus we have an invariant for smooth G-actions.

Ex. Consider the action of $\mathbb{Z}$ on S^1 given by

$$a(n)\left(e^{i\theta}\right) = e^{i(\theta + 2\pi n\alpha))}, \; 0 \le \alpha \le 1 .$$

Let us take $f = \delta_n - \delta_0$.

Prop 10.

$$I_f = \begin{cases} 0 \text{ if } n\alpha \in \mathbb{Z} \\ \gamma + \frac{1}{2} \psi\left(n\alpha - [n\alpha]\right) + \frac{1}{2} \psi\left(1 - n\alpha + [n\alpha]\right) \text{ if } n\alpha \notin \mathbb{Z} \end{cases} , \tag{2}$$

where γ is Euler's constant (.577...) and ψ is the logarithmic derivative of the gamma function, $\psi = \Gamma' / \Gamma$.

We know that any orientation-preserving action A of $\mathbb{Z}$ on S^1 which preserves a metric is smoothly conjugate to an action of the above type for some α_A. Because the expression I_f is invariant under a smooth conjugation of the action, its value for the action A is given by equation (2), using the corresponding value of α_A. It can be seen that the values of I_f as n ranges over $\mathbb{Z}$ determine α up to the symmetry $\alpha \to -\alpha$. Thus the invariant I determines orientation-preserving metric-preserving actions of $\mathbb{Z}$ on S^1 up to smooth conjugacy. More generally, we have

Prop 11. I determines special orthogonal actions of $\mathbb{Z}$ on S^{2n+1} up to smooth conjugacy.

It is hard to imagine what a combinatorial version of I could be.

VI. Discussion

We have shown that there is a natural definition of the analytic torsion for a discrete group action on a closed manifold which preserves a Riemannian metric. For a finite group action, we showed the relationship with the combinatorial torsion.

One case that we have not considered is that of a discrete group action on a noncompact manifold with the property that the quotient is compact. It should be possible to define an analytic torsion for such a situation by taking a trace not on trace-class operators, but on a type II_1 von Neumann algebra, as in the Atiyah-Singer L^2 index theorem for coverings [10]. For a free action, one should recover the Reidemeister torsion of the quotient space.

Another interesting case is that of a smooth action of a compact Lie group. There is a topological definition of the combinatorial torsion for such an action. However, if one naively defines the analytic torsion by using a projection onto ρ-equivariant differential forms, one finds that the result depends on the G-invariant Riemannian metric used in the definition. It is not clear what the correct definition of the analytic torsion should be for compact group actions.

I would like to thank M. Berger and the IHES for their support while part of this reseach was done.

References

[1] D.B. Ray and I.M. Singer, "R-Torsion and the Laplacian on Riemannian Manifolds", Adv. in Math. 7, p 145 (1971)

[2] J. Cheeger, "Analytic Torsion and the Heat Equation", Ann. Math. 109, p. 259 (1979)

[3] W. Müller, "Analytic Torsion and R-Torsion of Riemannian Manifolds", Adv. in Math. 28. p. 233 (1978)

[4] M. Rothenberg, "Torsion Invariants and Finite Transformation Groups", Proc. of Symp. in Pure Math. 32, AMS (1978)

[5] S. Illman, "Whitehead Torsion and Group Actions", Ann. Acad. Sci. Fenn. 588 (1974)

[6] J. Milnor, "Whitehead Torsion", BAMS 72, p. 358 (1966)

[7] J. Whitehead, "Simple Homotopy Types", Am. J. Math. 72, p.1 (1950)

[8] T. Chapman, "Topological Invariance of Whitehead Torsion", Am. J. Math. 96, p. 488 (1974)

[9] G. Bredon, "Introduction to Compact Transformation Groups", Academic Press, New York, London, 1972

[10] M. Atiyah, "Elliptic Operators, Discrete Groups and von Neumann Algebras", Astérisque 32, p. 43 (1976)

Note: The topological torsion for compact group actions has been studied by A. Stucker, Univ. of Chicago thesis, 1986, unpublished. The analytic torsion for group actions was previously studied by J. Cheeger in the preprinted version of Ref. 2, but this did not appear in the published version.

Department of Mathematics
University of Michigan
Ann Arbor, Michigan 48109

Contemporary Mathematics
Volume **101**, 1989

REGULARITY OF ENTROPY FOR GEODESIC FLOWS

G. Knieper[1] and H. Weiss[2]

In this paper, we announce partial solutions to the following question:

Let (M,g) be a compact n-dimensional Riemannian manifold and let ϕ_λ^t be the geodesic flow on SM. Let $\{g_\lambda\}$, $-\epsilon \leq \lambda \leq \epsilon$ be a C^k pertubation of $g=g_0$. How regular is the topological and measure theoretic entropy of ϕ_λ^t as a function of λ?

Let $h_{top}(\lambda) \equiv h_{top}(\phi_\lambda^1)$ and $h_m(\lambda) \equiv h_{\mu_\lambda}(\phi_\lambda^1)$, where μ_λ is the normalized Liouville measure with respect to g_λ.

Theorem 1 [KW1]

Let (M,g) be a compact n-dimensional Riemannian manifold with no conjugate points, and let $\{g_\lambda\}$, $-\epsilon \leq \lambda \leq \epsilon$ be a C^1 pertubation of $g=g_0$ through metrics having no conjugate points. Then $h_{top}(\lambda)$ is a Lipschitz function.

Theorem 2 [KW2]

Let (M^2,g) be a compact negatively curved surface, and let $\{g_\lambda\}$, $-\epsilon \leq \lambda \leq \epsilon$ be a C^3 pertubation of $g=g_0$ through metrics having negative curvature. Then $h_m(\lambda)$ is C^1.

Theorem 3 [KW3]

(a) Let (M^2,g) be a compact negatively curved surface, and let $\{g_\lambda\}$, $-\epsilon \leq \lambda \leq \epsilon$ be a C^3 pertubation of $g=g_0$ through metrics having negative curvature. Then $h_m(\lambda)$ is $C^{1,x\log x}$.

(b) Let (M,g) be a compact n-dimensional manifold with negative sectional curvature $-4<K\leq -1$, and let $\{g_\lambda\}$, $-\epsilon \leq \lambda \leq \epsilon$ be a C^3 pertubation of $g=g_0$ through metrics having negative sectional curvature $-4<K_\lambda \leq -1$. Then $h_m(\lambda)$ is $C^{1,\alpha}$, where $\alpha=\alpha(g,g_\lambda)$.

1985 Mathematics Subject Classification 58F17 58F30
[1]Partially supported by NSF Grant 8610730(2)
[2]Weizmann Fellow

Sketch Of Proofs

The proof of theorem 1 employ's the formula of Manning/Mañé & Freire [M],[FM] which relates the topological entropy to the exponential growth rate of volumes of balls in the universal covering. We find uniform bounds for the change of volume when the metric is smoothly perturbed. The proof is very soft and follows from the following lemma:

LEMMA

Let (M,g) be a compact n-dimensional Riemannian manifold, and let $\{g_\lambda\}$, $-\epsilon \leq \lambda \leq \epsilon$ be a C^1 pertubation of $g=g_0$. Then there exists Lipschitz functions $M(\lambda)$ and $m(\lambda)$, $M(0)=m(0)=1$ such that:

(a) For all $p,q \in M$, $$m(\lambda) \leq \frac{d_\lambda(p,q)}{d_0(p,q)} \leq M(\lambda)$$

(b) For all measurable sets $S \subset \widetilde{M}$, $$m(\lambda)^n \leq \frac{\widetilde{Vol}_\lambda(S)}{\widetilde{Vol}_0(S)} \leq M(\lambda)^n$$

PROOF

Exercise.

PROOF OF THEOREM 1

Fix $z_0 \in \widetilde{M}$. Part (a) of the lemma immediately implies:

$$\widetilde{B}_0(z_0, \tfrac{1}{M(\lambda)}) \subset \widetilde{B}_\lambda(z_0,r) \subset \widetilde{B}_0(z_0, \tfrac{1}{m(\lambda)}).$$

Hence,

$$\widetilde{Vol}_0(\widetilde{B}_0(z_0, \tfrac{1}{M(\lambda)})) \leq \widetilde{Vol}_0(\widetilde{B}_\lambda(z_0,r)) \leq \widetilde{Vol}_0(\widetilde{B}_0(z_0, \tfrac{1}{m(\lambda)})),$$

where $\widetilde{Vol}_0((\widetilde{B}_\lambda(z_0,r)))$ denotes the volume in the (lifted) g_0 metric of the ball of radius r in the g_λ metric about z_0.

Manning/Mañé & Freire [M],[FM] have shown that $h_{top}(\lambda) = \underset{r\to\infty}{\text{limit}} \frac{1}{r} \log(\widetilde{Vol}_\lambda(\widetilde{B}_\lambda(z_0,r)))$. Hence:

$$\frac{1}{M(\lambda)} h_{top}(0) \leq \underset{r\to\infty}{\text{limit}} \frac{1}{r} \log (\widetilde{Vol}_0(\widetilde{B}_\lambda(z_0,r))) \leq \frac{1}{m(\lambda)} h_{top}(0).$$

Part (b) of the lemma immediately implies that:

$$\underset{r\to\infty}{\text{limit}} \frac{1}{r} \log (\widetilde{Vol}_0(\widetilde{B}_\lambda(z_0,r))) = \underset{r\to\infty}{\text{limit}} \frac{1}{r} \log (\widetilde{Vol}_\lambda(\widetilde{B}_\lambda(z_0,r))) = h_{top}(\lambda).$$

We obtain:

$$\frac{1}{M(\lambda)}\, h_{top}(0) \le h_{top}(\lambda) \le \frac{1}{m(\lambda)}\, h_{top}(0)$$

or:

$$\left(\frac{1}{M(\lambda)} - 1\right) h_{top}(0) \le h_{top}(\lambda) - h_{top}(0) \le \left(\frac{1}{m(\lambda)} - 1\right) h_{top}(0).$$

The proof of theorem 2 is very geometric. It begins with Pesin's formula for the metric entropy as the average normal curvature of unstable horocycles [P]:

$$h_m(\lambda) \equiv h_{\mu_\lambda}(\phi_\lambda^1) = \int_{SM_\lambda} H^+(v,\lambda)\, d\mu_\lambda = -\int_{SM_\lambda} H^-(v,\lambda)\, d\mu_\lambda,$$

where $H^+(v,\lambda)$ $(H^-(v,\lambda))$ is the normal curvature of the unstable (stable) horocycle orthogonal to v in the g_λ metric and SM_λ is the unit tangent bundle with respect to g_λ. To show that the measure theoretic entropy is C^1, it suffices to show that the integrand is a C^1 function. Hopf showed that $H^\pm(v,\lambda)$ has a continuous partial derivative with respect to v [H]. We show that $H^\pm(v,\lambda)$ has a continuous partial derivative with respect to λ.

It is well known that $H^-(v,\lambda) = \lim_{r\to\infty} H_r^-(v,\lambda)$, where $H_r^-(v,\lambda)$ is the normal curvature of the unstable circle of radius r in g_λ, orthogonal to v, with normal derivative pointing in the unstable direction.

PROPOSITION 1

$$\left.\frac{\partial}{\partial\lambda}\right|_{\lambda=\lambda_0} H_r^-(v,\lambda) = \int_0^r \left.\frac{\partial}{\partial\lambda}\right|_{\lambda=\lambda_0} K(\lambda, \pi\circ\phi_\lambda^t(v)) \cdot S_r(t,\lambda,v)^2\, dt$$

where $K(\lambda,p)$ is the Gaussian curvature at the point $p\in M^2$ in the g_λ metric, ϕ_λ^t is the geodesic flow in the g_λ metric and $S_r(t,\lambda,v)$ is a stable Jacobi field in the g_λ metric along $\pi\circ\phi_\lambda^t(v)$ with $S_r(0,\lambda,v)=1$ and $S_r(r,\lambda,v)=0$.

PROOF

Calculation involving Jacobi equation and symplectic structure of the geodesic flow.

PROPOSITION 2

$$\frac{\partial}{\partial\lambda} H_r^-(v,\lambda) \text{ converges uniformly in } \lambda \text{ and v as } r\uparrow\infty.$$

PROOF

A standard comparison argument shows that the second term in the integrand is smaller than the square of a stable Jacobi field along $\pi\circ\phi_\lambda^t$. We prove that the first term in the integrand can be bounded by the growth of an unstable Jacobi field along $\pi \circ \phi_\lambda^t$. We then use the symplectic structure of the geodesic flow to show that the product of a stable and unstable Jacobi field is uniformly bounded. Another easy comparison argument allows us to conclude that the integrand is uniformly bounded by a exponentially decaying function.

To estimate the first term, we begin by applying the chain rule and obvious estimates to obtain:

$$\left|\frac{\partial}{\partial\lambda}\Big|_{\lambda=\lambda_0} K(\lambda,\pi\circ\phi_\lambda^t(v))\right| \leq \alpha + \beta \left\|\frac{\partial}{\partial\lambda}\Big|_{\lambda=\lambda_0} \phi_\lambda^t(v)\right\|_{TSM_{\lambda_0}} .$$

We need to estimate the rate at which corresponding geodesics in different metrics diverge.

PROPOSITION 3

$$\frac{\partial}{\partial\lambda}\Big|_{\lambda=\lambda_0} \phi_\lambda^t(v) = \int_0^t D\phi_{\lambda_0}^{t-s}(\phi_{\lambda_0}^s(v)) \cdot \frac{\partial}{\partial\lambda}\Big|_{\lambda=\lambda_0} \Phi(\lambda,\phi_{\lambda_0}^t(v))ds \quad ,$$

where $\Phi(\lambda,\cdot)$ is the infinitesimal generator of ϕ_λ^t and $D\phi_{\lambda_0}^t(p)$ denotes the differential of ϕ_λ^t at p.

PROOF

Linear ODE's.

Proposition 3 immediately implies:

$$\left\|\frac{\partial}{\partial\lambda}\Big|_{\lambda=\lambda_0} \phi_\lambda^t(v)\right\|_{TSM_{\lambda_0}} \leq \gamma \cdot \int_0^t \left\| D\phi_{\lambda_0}^{t-s}(\phi_{\lambda_0}^s(v))\right\|_{TSM_{\lambda_0}} ds .$$

The problem now reduces to estimating the differential of the geodesic flow (with respect to a fixed metric λ along ϕ_λ^t) by the growth of an unstable Jacobi field along $\pi \circ \phi_\lambda^t$. This is effected with several estimates of solutions of the Jacobi equation, some standard facts from linear ODE's, use of the symplectic structure of the geodesic flow, and some linear algebra.

To prove theorem 3(a), we need to show that $H^\pm(\lambda,v)$, the normal curvature of the stable and unstable horocycles orthogonal to v in the g_λ metric is $C^{1,x\log x}$, and to prove theorem 3(b), we need to show that $H^\pm(\lambda,v)$, the mean curvature of the stable and unstable horospheres orthogonal to v in the g_λ metric is $C^{1,\alpha}$. It suffices to show that $W^\pm(\lambda,v)$ (the stable and unstable horospherical foliations) is $C^{1,x\log x}$ for 3(a) and $C^{1,\alpha}$ for 3(b). Katok and Hurder [HK] have shown that $W^{\pm}(\lambda,v)$ is $C^{1,x\log x}$ in v for negatively curved surfaces and Hirsch and Pugh [HP] have shown that $W^{\pm}(\lambda,v)$ is $C^{1,\alpha}$ in v for 1/4 pinched negatively curved manifolds.

Both sets of authors prove their regularity theorems by first realizing the stable and unstable foliations as fixed points of a contraction map on a suitable space. K&H use a local contraction argument and H&P use a global contraction argument (the graph transform). Naively, we want to say that if the smooth contraction maps depend nicely on parameters then their fixed points will too. Unfortunately, life is much more difficult – the graph transform is a contraction only in the C^0 topology and Hurder and Katok's contraction is only in the C^α topology!

We prove a version of the Hirsch Pugh Shub C^r-C^α section theorem [HPS] with parameters and show that $W^{\pm}(\lambda,v)$ has at least as much regularity in λ as in v. We then quote the above mentioned theorems. As in [HP] , the theorem is proved inductively on the degree of smoothness, using auxiliary graph transforms at each step. We can use the implicit function theorem until the last step. Then we must analyze the regularity of the fixed points of a uniformly contracting family of operators which have less than differentiable dependence on an auxiliary parameter. This is easy but not well known.

References

[FM] A. Freire and R. Mañé, On the Entropy of the Geodesic Flow in Manifolds without Conjugate Points, Inventiones Mathematicae 69, 375–392 (1982).

[H] E. Hopf, Statistik der Losungen geodatischer Probleme vom unstabilen Typus II, Math Ann., 117 (1940),590-608.

[HP] M. Hirsch and C. Pugh, Smoothness of horocycle foliations, J. Differential Geometry 10(1975), 225–238.

[HPS] M.Hirsch, C.Pugh and M.Shub, Invariant Manifolds, Lect. Notes in Math. vol 583 (1977), Springer-Verlag, Berlin

[KW1] G. Knieper and H. Weiss, Regularity of Topological Entropy for Geodesic and Anosov Flows, in preparation.

[KW2] G. Knieper and H. Weiss, Regularity of Measure Theoretic Entropy for Geodesic Flows on Negatively Curved Surfaces, to appear.

[KW3] G. Knieper and H. Weiss, Regularity of Measure Theoretic Entropy for Geodesic and Anosov Flows, in preparation.

[M] A. Manning, Curvature Bounds for the Entropy of the Geodesic Flows on s Surface, J. of the London Math. Soc. 24, 351–358 (1981).

[P] Y. Pesin, Formulas for the Entropy of a Geodesic Flow on a Compact Riemannian Manifold without Conjugate Points, Math. Notes 24, 796–805 (1978).

GERHARD KNIEPER
INSTITUTE FOR ADVANCED STUDIES
PRINCETON, N.J. 08540

HOWARD WEISS
DEPARTMENT OF MATHEMATICS
CALIFORNIA INSTITUTE OF TECHNOLOGY
PASADENA C.A. 91125

Contemporary Mathematics
Volume **101**, 1989

TWISTOR AND GAUSS LIFTS OF SURFACES IN FOUR–MANIFOLDS

Gary R. Jensen & Marco Rigoli*

§1 Introduction. Let M be a Riemann surface, (N,h) a Riemannian 4–manifold and let $f:M \to N$ be a conformal immersion with induced metric (i.e., first fundamental form) $g = f^*h$. The area functional has a critical point at f (i.e., f is minimal) if and only if the mean curvature vector H of f vanishes. As a classical reference point, recall that if N is Euclidean 3–space, then f is minimal if and only if its Gauss map $\gamma_f:M \to S^2 \subset \mathbb{R}^3$ is anti–holomorphic. If N is Euclidean n–space, then Chern [Ch] generalized this result to the Gauss map $\gamma_f:M \to \tilde{G}_2(\mathbb{R}^n)$ into the Grassmannian of oriented 2–dimensional subspaces of $\mathbb{R}^n$. This latter space can be identified with the complex hyperquadric $Q_{n-2} \subset \mathbb{C}P^{n-1}$, by a biholomorphic isometry.

In the special case when N is Euclidean 4–space, the hyperquadric Q_2 splits biholomorphically and isometrically into a product of 2–spheres,

$$Q_2 = S^2 \times S^2, \tag{1.1}$$

and projection on each factor splits the Gauss map into factors, $\gamma_f = (\gamma_f^+, \gamma_f^-)$. Blaschke [Bl] and Hoffman–Osserman [HO1] proved that

$$K = \mathcal{J}(\gamma_f^+) + \mathcal{J}(\gamma_f^-) \ , \ K^\perp = \mathcal{J}(\gamma_f^+) - \mathcal{J}(\gamma_f^-),$$

where $\mathcal{J}(.)$ denotes the Jacobian of the map, and K and $K^\perp$ are the Gaussian and normal curvatures, respectively, of f. Integrating these equations, assuming M compact, and using the Chern–Gauss–Bonnet Theorem, they [HO1] generalized a result of

---

1980 Mathematics Subject Classification (1985 Revision). 53C42, 58E20.

*Expresses gratitude to the International Atomic Energy Agency and UNESCO for hospitality at the International Centre for Theoretical Physics, Trieste.

Chern–Spanier [CS]

$$\chi(M) = \deg(\gamma_f^+) + \deg(\gamma_f^-) \,, \quad \chi(TM^\perp) = \deg(\gamma_f^+) - \deg(\gamma_f^-),$$

where χ is the Euler characteristic and deg denotes the degree.

In this paper we allow N to be an arbitrary oriented Riemannian 4–manifold. The Gauss map is replaced by the Gauss lift into the Grassmann bundle $G_2(TN)$ of oriented tangent 2–planes of N. Although the splitting (1.1) holds in the fibers, this space does not in general split. However, the Penrose twistor spaces $Z_\pm$ provide fibrations $G_2(TN) \to Z_\pm$ and consequent factorizations, called the twistor lifts of f. For many generalizations the most interesting results occur when N is a ±self–dual Einstein space. In particular, a special case of our final result, Theorem 8.1, parametrizes a class of harmonic maps from compact Riemann surfaces into $\mathbb{C}P^3$ by compact oriented surfaces immersed in S^4 with parallel mean curvature.

This paper began as an attempt to understand the results of the paper by Eells and Salamon [ES]. Many of the results here were announced in [JR3] where this paper is referred to by the preliminary title "Surfaces in 4–manifolds". Throughout the paper we assume M and N are both connected. We use the Einstein summation convention (sum all repeated indices in a product), and the index conventions $1 \le a,b,c \le 4$; $1 \le i,j,k \le 2$; $3 \le \alpha,\beta,\gamma,\delta \le 4$; $1 \le p,q \le 6$. The paper is organized into eight sections:

§2 Isotropic surfaces in a Riemannian manifold. Let N be a connected n–dimensional Riemannian manifold. Let O(N) denote its principal O(n)–bundle of orthonormal frames. The $\mathbb{R}^n$–valued canonical form on O(N) is denoted $\theta = (\theta^a)$, and the $\mathfrak{o}(n)$–valued Levi–Civita connection and curvature forms on O(N) are denoted $\omega = (\omega^a_b)$ and $\Omega = (\Omega^a_b)$, respectively. Then

$$\Omega^a_b = \tfrac{1}{2} R_{abcd}\,\theta^c \wedge \theta^d \,,$$

where R_{abcd} are functions on $O(N)$ defining the Riemann curvature tensor of N. The structure equations of N are

$$d\theta^a = -\omega^a_b \wedge \theta^c, \quad d\omega^a_b = -\omega^a_c \wedge \omega^c_b + \Omega^a_b .$$

A local orthonormal frame field in N is a local section $e = (e_a)$ of $O(N)$. Its dual coframe field is $(e^*\theta^a)$, for which we will always omit the e^*. Similarly, the connection and curvature forms and components of the curvature tensor with respect to e will be denoted by ω^a_b, Ω^a_b and R_{abcd}, respectively, without explicit indication of e^*.

If N is non–orientable, then $O(N)$ is connected. Otherwise, $O(N)$ has two connected components, $O_{\pm}(N)$, and each of $O_{\pm}(N) \to N$ is a principal $SO(n)$–bundle.

Let M be an m–dimensional manifold, let $f:M \to N$ be an immersion, and let g denote the induced Riemannian metric on M. A local Darboux frame field along f is a local orthonormal frame field e in N such that $e_i \circ f$ is an oriented orthonormal frame field in M and $e_\alpha \circ f$ are normal to M; or, equivalently, $f^*\theta^i$ is an oriented orthonormal coframe in M and

$$f^*\theta^\alpha = 0 .$$

We will almost always suppress the writing of the f^*'s in this context. Exterior differentiation of this equation implies that on M

$$\omega^\alpha_i = h^\alpha_{ij}\theta^j ,$$

where h^α_{ij} are locally defined functions on M, symmetric in i and j. The second fundamental tensor of f is

$$II = h^\alpha_{ij}\theta^i\theta^j \otimes e_\alpha ,$$

a symmetric bilinear form on M with values in the normal bundle $TM^\perp$.

We let

$$H^\alpha = \tfrac{1}{2}(h^\alpha_{11} + h^\alpha_{22})$$

denote the components of the mean curvature vector, $H = H^\alpha e_\alpha$, of f. The Levi–Civita connection of N induces the Levi–Civita connection of g on M given by

$$\nabla e_i = \omega_i^j \otimes e_j \quad \text{and} \quad \nabla \theta^i = -\omega_j^i \otimes \theta^j ;$$

and a connection on $TM^\perp$ given by

$$\nabla e_\alpha = \omega_\alpha^\beta \otimes e_\beta ;$$

and thus in the standard way on $T^*M \otimes T^*M \otimes TM^\perp$. Then

$$\nabla II = Dh_{ij}^\alpha \otimes \theta^i \theta^j \otimes e_\alpha ,$$

where

$$Dh_{ij}^\alpha = dh_{ij}^\alpha - h_{kj}^\alpha \omega_i^k - h_{ik}^\alpha \omega_j^k + h_{ij}^\beta \omega_\beta^\alpha = h_{ijk}^\alpha \theta^k .$$

From the symmetry of II we have

$$h_{ijk}^\alpha = h_{jik}^\alpha ,$$

while by the Codazzi equations we have

$$h_{ijk}^\alpha - h_{ikj}^\alpha = -R_{\alpha ijk} . \tag{2.1}$$

It is easily verified that the covariant differential of H ,

$$\nabla H = (dH^\alpha + H^\beta \omega_\beta^\alpha) \otimes e_\alpha = H_j^\alpha \theta^j ,$$

is given by

$$H_j^\alpha = \tfrac{1}{2} h_{kkj}^\alpha . \tag{2.2}$$

We say that f is minimal if $H = 0$, and that f has parallel mean curvature vector if $\nabla H = 0$.

To construct global invariants from this local analysis, we must determine the transformation rules for changes of Darboux frame. For this purpose it is convenient to use the isomorphism

(2.3)
$$\rho: SO(2) \to U(1)$$
$$\begin{bmatrix} \cos t & -\sin t \\ \sin t & \cos t \end{bmatrix} \mapsto e^{it}$$

We define the Hopf transform from the space of real 2×2 symmetric matrices $h = (h_{ij})$ onto $\mathbb{C}$ by

(2.4)
$$L(h) = \tfrac{1}{2}(h_{11} - h_{22}) - ih_{12} .$$

The kernel of L consists of all scalar matrices, and L has the equivariance property

(2.5)
$$L({}^tAhA) = \rho(A)^2 L(h) ,$$

for any $A \in SO(2)$.

We restrict our attention now to the case $m = 2$, and we suppose that both M and N are oriented. A ± **oriented Darboux frame** along f will mean a Darboux frame $\{e_a\}$ such that $\{e_i\}$ is an oriented frame on M and $\{e_a\}$ is a ± oriented frame in $f^{-1}TM$. Thus $\{e_\alpha\}$ is a ± oriented frame of $TM^\perp$ which is oriented in the way compatible with the orientations of TM and TN, and the decompostion $f^{-1}TN = TM \oplus TM^\perp$.

An arbitrary change of oriented Darboux frame is given by

(2.6)
$$\tilde{e} = eG ,$$

where G is a locally defined function in M with values in

(2.7)
$$SO(2)\times SO(n-2) = \left\{\begin{bmatrix} A & 0 \\ 0 & B \end{bmatrix} : A \in SO(2), B \in SO(n-2)\right\} .$$

Under such a change the matrices $h^\alpha = (h^\alpha_{ij})$ of II transform by

(2.8)
$$\tilde{h}^\alpha = ({}^tB)^\alpha_\beta \, {}^tAh^\beta A ,$$

where tilded quantities are with respect to $\tilde{e}$. Writing L^α for $L(h^\alpha)$, we have

(2.9)
$$\tilde{L}^\alpha = \rho(A)^2 ({}^tB)^\alpha_\beta L^\beta ,$$

and

(2.10)
$$\tilde{H}^\alpha = ({}^tB)^\alpha_\beta H^\beta .$$

It is important to use the complex structure of M induced by g. If e is an oriented Darboux frame field along f, then its dual coframe (θ^a) defines a type (1,0) form

$$\varphi = \theta^1 + i\theta^2 , \tag{2.11}$$

which under a change (2.6) of oriented Darboux frame transforms to

$$\tilde{\varphi} = \rho(A)^{-1}\varphi . \tag{2.12}$$

Using the complex structure of M to decompose the second fundamental tensor by type,we have $\omega_1^\alpha + i\omega_2^\alpha = H^\alpha\varphi + L^\alpha\bar{\varphi}$, and thus

$$II = \tfrac{1}{2}\varphi\varphi\otimes L^\alpha e_\alpha + \varphi\bar{\varphi}\otimes H + \tfrac{1}{2}\overline{\varphi\varphi}\otimes\bar{L}^\alpha e_\alpha ,$$

where bars denote complex conjugation. The coefficients $L^\alpha e_\alpha$ and $\bar{L}^\alpha e_\alpha$ are local sections of the complexified normal bundle $TM^\perp_{\mathbb{C}} = TM^\perp \otimes \mathbb{C}$. The Riemannian metric on N induces a fibre metric on $TM^\perp$, which we extend to be complex linear and symmetric on $TM^\perp_{\mathbb{C}}$, and denote by $(.,.)$.

(2.13)Definition The isometric immersion f is **isotropic** at a point p of M if the complex normal vector $L^\alpha e_\alpha(p)$ is isotropic; that is, if

$$(L^\alpha e_\alpha, L^\beta e_\beta) = L^\alpha L^\alpha = 0$$

at p. We say that f is isotropic if it is isotropic at every point of M.

It is evident that the symmetric quartic form

$$\Lambda = L^\alpha L^\alpha \varphi^4 \tag{2.14}$$

is globally defined on M and vanishes at a point if and only if f is isotropic at that point. The function

$$u = \sum_\alpha |L^\alpha|^2 \tag{2.15}$$

is globally defined and C^∞ on M, and vanishes precisely at the umbilic points of f.

We specialize now to the case where $\dim N = 4$ when, as we shall see, the non–simplicity of SO(4) is reflected in the meaning of isotropicity. With respect to a local oriented Darboux frame e along f we define the complex valued functions

$$b = H^3 - iH^4, \quad S_+ = L^3 - iL^4 \ , \quad S_- = L^3 + iL^4 \ . \tag{2.16}$$

This strange convention is adopted to match that of the twistor lifts in §4. Under a change of oriented Darboux frame (2.6), these functions transform by

$$\tilde{b} = \rho({}^tB)b \ , \quad \tilde{S}_\pm = \rho(A^2B^{\pm 1})S_\pm \ . \tag{2.17}$$

The absolute values of these functions,

$$|b| = \|II\| \ , \quad s_\pm = |S_\pm|/\sqrt{2} \ , \tag{2.18}$$

are globally defined on M and their squares are of class C^∞ .

PROPOSITION 2.1 With respect to any oriented Darboux frame we have

$$\Lambda = S_+S_-\varphi^4 \ , \quad u = s_+^2 + s_-^2 \tag{2.19}$$

and

$$\begin{aligned} s_+^2 + s_-^2 &= \|H\|^2 - K + R_{1212} \\ s_+^2 - s_-^2 &= -K^\perp + R_{1234} \ , \end{aligned} \tag{2.20}$$

where $K^\perp$ is the curvature of the induced connection in the normal bundle; that is, $K^\perp$ is given by

$$d\theta_4^3 = K^\perp \theta^1 \wedge \theta^2 \ .$$

PROOF These equations follow from the Gauss equation

$$K = R_{1212} + \sum_\alpha \det h^\alpha \ , \tag{2.21}$$

the Ricci equation

$$K^\perp = R_{1234} + h^3_{k1}h^4_{k2} - h^3_{k2}h^4_{k1} \ , \tag{2.22}$$

and the easily derived formulas

(2.23)
$$\text{i)}\quad |L^{\alpha}|^2 = (H^{\alpha})^2 - \det(h^{\alpha})$$
$$\text{ii)}\quad L^3\bar{L}^4 - \bar{L}^3L^4 = i(h^3_{k1}h^4_{k2} - h^3_{k2}h^4_{k1})\,.$$
□

Definition At a point p in M the isometric immersion f is isotropic with **positive** (respectively, **negative**) spin if $s_+(p) = 0$ (respectively, $s_-(p) = 0$). It is isotropic with **positive** (**negative**) spin if it has the respective property at every point of M .

Remarks 1) This definition follows that of Bryant [B] for minimal surfaces in S^4 . In [Ca] Calabi observed that Λ is holomorphic when f is minimal and $N = S^4$. Thus when M is homeomorphic to S^2 , Λ must vanish identically for minimal f . He called isotropic minimal surfaces in S^4 pseudo–holomorphic curves. Our notion of isotropy corresponds to real isotropy of Eells–Wood [EW] and Chern [Ch2]. An isotropic f need not be minimal, even in S^4 . We discuss this further in §5 below.

2) If the orientation of N is reversed, thus reversing the orientation of $TM^{\perp}$, then s_+ and s_- are interchanged. Thus the notions of positive and negative spin are reversed by a reversal of orientation of N .

Isotropy can be defined geometrically in terms of the ellipse of curvature of the immersion (cf. [EGT] and [JR1]). Fix $p \in M$ and in T_pM consider the parametrized unit circle

$$X = X(t) = \cos t\, e_1 + \sin t\, e_2 \qquad 0 \le t \le 2\pi$$

where e is an oriented Darboux frame field along f . The ellipse of curvature at p is defined to be the curve in $TM_p^{\perp}$ given parametrically by

$$II(X,X) = H + \frac{L}{2}e^{2it} + \frac{\bar{L}}{2}e^{-2it},$$

where $L = L^{\alpha}e_{\alpha}$. This curve is a circle (with center H and radius $|L|^2/2$) if and only if L is isotropic. It degenerates to a line segment (possibly of zero length) if and only if $L \wedge \bar{L} = 0$ at p, which occurs if and only if $R_{1234} = K^{\perp}$ at p, by (2.22) and (2.23)ii).

THEOREM 2.1 Let $f:M \to N$ be an isometric immersion of a compact surface. Then

$$\int_M \|H\|^2 dA \geq 2\pi\chi(M) + \left|2\pi\chi(TM^\perp) - \int_M R_{1234} dA\right| - \int_M R_{1212} dA, \tag{2.24}$$

where $\chi(M)$ and $\chi(TM^\perp)$ are the Euler characteristics of M and its normal bundle. Equality holds if and only if f is isotropic with positive, or negative, spin.

PROOF Adding and subtracting the two equations in (2.20), we have

$$\|H\|^2 \geq K + |K^\perp - R_{1234}| - R_{1212}. \tag{2.25}$$

Integrating and using the Gauss–Bonnet theorem we obtain (2.24). Suppose f is isotropic with negative spin. Then from (2.20), $K^\perp - R_{1234} \leq 0$, and equality holds in (2.25), and hence also in (2.24). Similarly, equality holds in (2.24) if f is isotropic with positive spin. Conversely, suppose that

$$\int_M \|H\|^2 dA = 2\pi\chi(M) + 2\pi\chi(TM^\perp) - \int_M R_{1234} dA - \int_M R_{1212} dA.$$

(The same argument works if $2\pi\chi(TM^\perp) - \int_M R_{1234} < 0$.) From (2.20) we have

$$\|H\|^2 = 2s_+^2 + K + K^\perp - R_{1234} - R_{1212}.$$

Integrating and subtracting from the preceding equation we have $\int_M 2s_+^2 dA = 0$, that is, f is isotropic with positive spin. □

Remark Inequality (2.24) generalizes a result of Friedrich [F] (Theorem 1, p.272), and of Wintgen [W] obtained for $N = \mathbb{R}^4$. Indeed, in this case $R_{1212} = R_{1234} = 0$ and $\chi(TM^\perp) = 2q$, where q is the self–intersection number of the compact oriented surface $f(M)$ in $\mathbb{R}^4$. Thus, if g is the genus of M, then (2.24) reduces to Wintgen's inequality

$$\int_M \|H\|^2 dA \geq 4\pi(1 + |q| - g).$$

Equality in this case was first considered by Weiner [We].

PROPOSITION 2.2 Let $f:M \to N$ be an isometric immersion of a compact surface. If f is isotropic with positive (respectively, negative) spin, with $\chi(TM^\perp) = 0$ and $R_{1234} \geq 0$ (respectively, $R_{1234} \leq 0$), then f is totally umbilical.

PROOF Suppose $s_+ = 0$ (the case $s_- = 0$ is similar). From (2.20), the hypothesis $\chi(TM^\perp) = 0$, and the Chern–Gauss–Bonnet theorem we have

$$\int_M s_-^2 dA = -\int_M R_{1234} dA \leq 0,$$

and thus $s_- = 0$ also. Hence $u = 0$, and f is totally umbilical. □

Observe that in case f is totally umbilical and M is compact, then $2\pi\chi(TM^\perp) = \int_M R_{1234} dA$ by (2.20). In particular, if N is the constant curvature 4–sphere S^4, we have

PROPOSITION 2.3 Let $f:M \to S^4$ be a minimal surface where $M \approx S^2$. Then f is totally geodesic if and only if $\chi(TM^\perp) = 0$. If f is not totally geodesic, then $\chi(TM^\perp) = -4 - m$, where m is the total number of umbilical points counted with multiplicities (see Remark 1 below).

PROOF The first part follows from Proposition 2.2. Suppose f is not totally geodesic, or equivalently, that f is full in S^4. Then we apply Theorem 1 of [JR4] to obtain the desired estimates of $\chi(TM^\perp)$ (there f is isotropic with positive spin). □

Remarks 1. If a minimal immersion f is not totally umbilical then the umbilical points are isolated and have well defined multiplicities [JR4].

2. The above estimates of $\chi(TM^\perp)$ improve a result of Salamon [S2].

Now let $f:M \to N$ be a minimal immersion of a compact surface. Then from (2.20) and the Chern–Gauss–Bonnet theorem we have

$$\text{(2.26)}\quad \begin{aligned} \tfrac{1}{2}\{\chi(M) + \chi(TM^\perp)\} &= \frac{-1}{2\pi}\int_M s_+^2 dA + \frac{1}{4\pi}\int_M (R_{1212} + R_{1234}) dA \\ \tfrac{1}{2}\{\chi(M) - \chi(TM^\perp)\} &= \frac{-1}{2\pi}\int_M s_-^2 dA + \frac{1}{4\pi}\int_M (R_{1212} - R_{1234}) dA \end{aligned}$$

The left hand sides of (2.26) are the **twistor degrees** $d_{\pm}$ introduced by Eells–Salamon ([ES], §8), and thus (2.26) gives integral representations of the twistor degrees.

If N is Einstein and anti–self–dual (respectively, self–dual; see §3) with scalar curvature s, then (reading the +, respectively the −) $R_{1212} \pm R_{1234} = s/12$, and therefore

$$(2.27) \qquad d_{\pm} = \frac{-1}{2\pi}\int_M s_{\pm}^2 dA + \frac{s}{48\pi}A(M),$$

respectively, where A(M) is the area of M. Furthermore, f is then isotropic with positive (respectively, negative) spin if and only if

$$(2.28) \qquad d_{\pm} = \frac{s}{48\pi}A(M).$$

The necessity of this last statement for d_+ was first proved by Friedrich [F] and Poon [P] independently. (See also Salamon [S2].)

§3 Four–dimensional Riemannian geometry. The material of this section is well known (see, for example, [Be] or [S] for excellent expositons). We summarize here the essential points that we need and establish our notation and point of view. In this section we use the index conventions $1 \le i,j,k,l \le 3$, $1 \le a,b,c,d \le 4$.

The standard action of SO(4) on $\mathbb{R}^4$ (as column vectors) induces a representation of SO(4) on $\Lambda_2\mathbb{R}^4$ $(a(u\wedge v) = au\wedge av)$, which is reducible into irreducible factors $\Lambda_2\mathbb{R}^4 = \Lambda_+ \oplus \Lambda_-$, where the 3–dimensional subspaces $\Lambda_{\pm}$ are the ±1 eigenspaces of the Hodge *–operator on $\mathbb{R}^4$ with the orientation of the standard basis $\epsilon_1,\ldots,\epsilon_4$. Standard bases of $\Lambda_{\pm}$ are given by

$$(3.1) \qquad E^{\pm} = (E_1^{\pm}, E_2^{\pm}, E_3^{\pm})$$

where

$$E_1^{\pm} = (\epsilon_1\wedge\epsilon_2 \pm \epsilon_3\wedge\epsilon_4)/\sqrt{2},\ E_2^{\pm} = (\epsilon_1\wedge\epsilon_3 \pm \epsilon_4\wedge\epsilon_2)/\sqrt{2},\ E_3^{\pm} = (\epsilon_1\wedge\epsilon_4 \pm \epsilon_2\wedge\epsilon_3)/\sqrt{2}.$$

The standard metric on $\mathbb{R}^4$ induces an SO(4)–invariant inner product on $\Lambda_2\mathbb{R}^4$, and the restriction of the SO(4)–action to $\Lambda_{\pm}$ thus gives a 2:1 surjective homomorphism

$$\mu: SO(4) \to SO(3)\times SO(3)$$
$$a \to (a_+, a_-)$$

where for $a \in SO(4)$, $a_{\pm} = a|_{\Lambda_{\pm}}$ with respect to the bases (3.1) $(a_+E_i^+ = a_{+ij}E_j^+$ etc.)

There is an isomorphism $\mathfrak{o}(4) \cong \Lambda_2\mathbb{R}^4$ given by: the skew–symmetric matrix $X = (X_{ab}) \leftrightarrow \frac{1}{2}X_{ab}\epsilon_a\wedge\epsilon_b$. If $a \in SO(4)$, then the adjoint action of a on $\mathfrak{o}(4)$ corresponds to the above action of $SO(4)$ on $\Lambda_2\mathbb{R}^4$; namely, $Ad(a)X = aXa^{-1} \leftrightarrow a(\frac{1}{2}X_{ab}\epsilon_a\wedge\epsilon_b)$. The above decomposition of $\Lambda_2\mathbb{R}^4$ thus gives the Lie algebra isomorphism $\mathfrak{o}(4) \cong \mathfrak{o}(3)_+ \oplus \mathfrak{o}(3)_-$, where $\mathfrak{o}(3)_\pm \leftrightarrow \Lambda_\pm$.

Let N be a connected oriented Riemannian 4–manifold. Let $\theta = (\theta^a)$ and $\Omega = (\Omega^a_b)$ denote the canonical form and the curvature form of the Levi–Civita connection, respectively, on $O(N)$. For any $a \in O(4)$,

$$R_a^*\theta = a^{-1}\theta, \tag{3.2}$$

where R_a denotes right multiplication on $O(N)$ by a. If we define $\mathbb{R}^3$–valued 2–forms on $O(N)$ by

$$\alpha_\pm = \begin{bmatrix} \alpha^1_\pm \\ \alpha^2_\pm \\ \alpha^3_\pm \end{bmatrix} = \begin{bmatrix} \theta^1\wedge\theta^2 \pm \theta^3\wedge\theta^4 \\ \theta^1\wedge\theta^3 \pm \theta^4\wedge\theta^2 \\ \theta^1\wedge\theta^4 \pm \theta^2\wedge\theta^3 \end{bmatrix} /\sqrt{2}, \tag{3.3}$$

then

$$R_a^*\alpha_\pm = a_\pm^{-1}\alpha_\pm . \tag{3.4}$$

The curvature forms Ω^a_b are the components of the $\mathfrak{o}(4)$–valued 2–form Ω with respect to the standard basis of $\mathfrak{gl}(4;\mathbb{R})$, with the linear relations $\Omega^a_b = -\Omega^b_a$, because on $O(N)$, Ω takes values in $\mathfrak{o}(4)$. A fundamental property of the curvature form is that it is given by

$$\Omega^a_b = \tfrac{1}{2}R_{abcd}\theta^c\wedge\theta^d , \tag{3.5}$$

where the R_{abcd} are functions on $O(N)$ satisfying the symmetries of the Riemann curvature tensor: $R_{abcd} = -R_{abdc} = -R_{bacd} = R_{cdab}$.

If we express Ω in terms of the basis (3.1) of $\mathfrak{o}(4)$ and the 2–forms (3.3), then

$$\Omega = A_{ij}E^+_i\otimes\alpha^j_+ + B_{ij}E^-_i\otimes\alpha^j_+ + B_{ji}E^+_i\otimes\alpha^j_- + C_{ij}E^-_i\otimes\alpha^j_- , \tag{3.6}$$

where if $A = (A_{ij})$, $B = (B_{ij})$, and $C = (C_{ij})$, then

$${}^tA = A , \quad {}^tC = C , \quad \text{trace } A = \text{trace } C . \tag{3.7}$$

In matrix notation (3.6) becomes

(3.8) $$\Omega = E^{+} \otimes A \alpha_{+} + E^{-} \otimes B \alpha_{+} + E^{+} \otimes {}^{t}B \alpha_{-} + E^{-} \otimes C \alpha_{-} .$$

For any $a \in O(4)$, we have

(3.9) $$R_a^* \Omega = a^{-1} \Omega a ,$$

from which it follows that for any $a \in SO(4)$,

(3.10) $$R_a^* A = a_+^{-1} A a_+ , \quad R_a^* B = a_-^{-1} B a_+ , \quad R_a^* C = a_-^{-1} C a \; ,$$

(explicitly, for any $e \in O(N)$, $A(ea) = a_+^{-1}(e) A(e) a_+(e)$, etc.).

It will be handy to have explicit formulas relating R_{abcd} to A , B , and C . These are found by substituting (3.1) and (3.3) into (3.5). We have

(3.11)
$$\begin{aligned}
A_{11} &= \tfrac{1}{2}(R_{1212} + 2R_{1234} + R_{3434}), \quad A_{21} = \tfrac{1}{2}(R_{1312} + R_{1334} - R_{2412} - R_{2434}) \\
A_{31} &= \tfrac{1}{2}(R_{1412} + R_{1434} + R_{2312} + R_{2334}), \quad A_{22} = \tfrac{1}{2}(R_{1313} - 2R_{1324} + R_{2424}) \\
A_{32} &= \tfrac{1}{2}(R_{1413} - R_{1424} + R_{2313} - R_{2324}), \quad A_{33} = \tfrac{1}{2}(R_{1414} + 2R_{1423} + R_{2323})
\end{aligned}$$

(3.12)
$$\begin{aligned}
B_{11} &= \tfrac{1}{2}(R_{1212} - R_{3434}) = \tfrac{1}{4}(R_{11} + R_{22} - R_{33} - R_{44}) \\
B_{21} &= \tfrac{1}{2}(R_{1312} + R_{1334} + R_{2412} + R_{2434}) = \tfrac{1}{2}(R_{32} - R_{14}) \\
B_{31} &= \tfrac{1}{2}(R_{1412} + R_{1434} - R_{2312} - R_{2334}) = \tfrac{1}{2}(R_{42} + R_{13}) \\
B_{12} &= \tfrac{1}{2}(R_{1213} - R_{1224} - R_{3413} + R_{3424}) = \tfrac{1}{2}(R_{23} + R_{14}) \\
B_{22} &= \tfrac{1}{2}(R_{1313} - R_{2424}) = \tfrac{1}{4}(R_{11} - R_{22} + R_{33} - R_{44}) \\
B_{32} &= \tfrac{1}{2}(R_{1413} - R_{1424} - R_{2313} + R_{2324}) = \tfrac{1}{2}(R_{43} - R_{12}) \\
B_{13} &= \tfrac{1}{2}(R_{1214} + R_{1223} - R_{3414} - R_{3423}) = \tfrac{1}{2}(R_{24} - R_{13}) \\
B_{23} &= \tfrac{1}{2}(R_{1314} + R_{1323} + R_{2414} + R_{2423}) = \tfrac{1}{2}(R_{34} + R_{12}) \\
B_{33} &= \tfrac{1}{2}(R_{1414} - R_{2323}) = \tfrac{1}{4}(R_{11} - R_{22} - R_{33} + R_{44})
\end{aligned}$$

where $R_{ab} = R_{ba}$ are the components of the Ricci tensor: $R_{ab} = \Sigma R_{cacb}$.

(3.13)

$$C_{11} = \tfrac{1}{2}(R_{1212} - 2R_{1234} + R_{3434}),\ C_{21} = \tfrac{1}{2}(R_{1312} - R_{1334} + R_{2412} - R_{2434})$$
$$C_{31} = \tfrac{1}{2}(R_{1412} - R_{1434} - R_{2312} + R_{2334}),\ C_{22} = \tfrac{1}{2}(R_{1313} + 2R_{1324} + R_{2424})$$
$$C_{32} = \tfrac{1}{2}(R_{1413} + R_{1424} - R_{2313} - R_{2324}),\ C_{33} = \tfrac{1}{2}(R_{1414} - 2R_{1423} + R_{2323})$$

From (3.12) we see that

(3.14) N is Einstein if and only if $B = 0$.

The Weyl curvature form is the $\mathfrak{o}(4)$–valued 2–form $\Psi = (\Psi^a_b)$ on $O(N)$ defined by

$$\Psi^a_b = \Omega^a_b - \tfrac{1}{2}(R_{ac}\theta^c\wedge\theta^b + R_{bc}\theta^a\wedge\theta^c) + \tfrac{s}{6}\theta^a\wedge\theta^b ,$$

where $s = \Sigma R_{aa}$ is the scalar curvature. In terms of the bases (3.1) and (3.3) we have

$$\Psi = E^+\otimes(A - \tfrac{s}{12} I)\alpha_+ + E^-\otimes(C - \tfrac{s}{12} I)\alpha_- , \tag{3.15}$$

where I is the 3×3 identity matrix.

Let $e{:}U \subset N \to O_+(N)$ be a local oriented orthonormal frame field in N. For each point $p \in U$, $e(p) = (e_1,\dots,e_4)(p)$ is an oriented orthonormal frame of T_pN (which we interpret as an isomorphism $e(p){:}\mathbb{R}^4 \to T_pN$ given by $e(p)x = x^a e_a(p)$, where $x = (x^a)$) with dual coframe $e^*\theta(p)$. Thus $e(p)E^\pm$ are bases of $\Lambda_\pm T_pN$, the ±1 eigenspaces of the Hodge $*$ operator on TN ($e(p)E^+_1 = \frac{1}{\sqrt{2}}(e_1\wedge e_2 + e_3\wedge e_4)(p)$, etc.) with dual bases $e^*\alpha_\pm(p)$.

The curvature operator at p is $R(p) \in \mathfrak{S}^2(\Lambda_2T_pN) \subset \Lambda_2T_pN\otimes\Lambda_2T^*_pN$ given in terms of the basis $\{e_a\wedge e_b(p) : a < b\}$ by

$$R(p) = \tfrac{1}{4}R_{abcd}(e(p))e_a\wedge e_b(p)\otimes e^*(\theta^c\wedge\theta^d)(p) ,$$

and in terms of the basis $e(p)E^\pm$ by (e and e^* evaluated at p throughout)

$$R(p) = eE^+\otimes A(e)e^*\alpha_+ + eE^-\otimes B(e)e^*\alpha_+ + eE^+\otimes {}^tB(e)e^*\alpha_- + eE^-\otimes C(e)e^*\alpha_- . \tag{3.16}$$

The Weyl curvature operator at p preserves the ±1 eigenspaces of $*$, and thus $W(p) = W^+(p) + W^-(p)$, where $W^\pm(p){:}\Lambda_\pm T_pN \to \Lambda_\pm T_pN$ are given by

(3.17)
$$W^+(p) = eE^+\otimes(A(e) - \tfrac{s(p)}{12}I)e^*\alpha_+$$
$$W^-(p) = eE^-\otimes(C(e) - \tfrac{s(p)}{12}I)e^*\alpha_- .$$

If $e:U \to O_-(N)$ is a local negatively oriented orthonormal frame field in N, then for any point $p \in U$, $e(p)E^+$ is a basis of Λ_-T_pN and $e(p)E^-$ is a basis of Λ_+T_pN. Thus the expressions for $W^{\pm}(p)$ in (3.17) are reversed. In summary

(3.18) If e is positively oriented, then the matrix of W^+ (W^-) with respect to eE^+ (eE^-) is $A(e) - \frac{s}{12}I$ $(C(e) - \frac{s}{12}I)$; while if e is negatively oriented, then the matrix of W^+ (W^-) with respect to eE^- (eE^+) is $C(e) - \frac{s}{12}I$ $(A(e) - \frac{s}{12}I)$.

The oriented Riemannian manifold N is **self–dual** (respectively **anti–self–dual**) if at every point of N we have $W^- = 0$ (respectively $W^+ = 0$) [AHS], [Be]. By (3.18) the following are equivalent:

(3.19)
a) N is self–dual (anti–self–dual)
b) $C - \frac{s}{12}I = 0$ on $O_+(N)$ $(A - \frac{s}{12}I = 0$ on $O_+(N))$
c) $A - \frac{s}{12}I = 0$ on $O_-(N)$ $(C - \frac{s}{12}I = 0$ on $O_-(N))$

§4 Metric structure of the Twistor bundle. In this section $n = 2m$, and we return to the convention of index ranges given in §1. Let N be a connected $2m$–dimensional Riemannian manifold. The twistor space Z of N is defined to be the set of all pairs (p,J), where $p \in N$ and J is an orthogonal complex structure on T_pN; i.e., J is an orthogonal transformation of T_pN satisfying $J^2 = -$identity. The twistor projection

(4.1)
$$T:Z \to N$$

is defined by $T(p,J) = p$. As the set of orthogonal complex structures on T_pN depends only on the conformal class of the inner product on T_pN, it follows that Z depends only on the conformal structure of N.

The projection (4.1) is a fiber bundle over N with standard fiber $O(2m)/U(m)$. We associate Z to $O(N)$, the principal $O(2m)$–bundle of orthonormal frames on N. To do this we must first consider the representation of $U(m)$ in $O(2m)$.

Let

$$J_1 = \begin{bmatrix} 0 & -1 \\ 1 & 0 \end{bmatrix}, \text{ and } J_m = \begin{bmatrix} J_1 & & 0 \\ & \ddots & \\ 0 & & J_1 \end{bmatrix}.$$

Observe that $J_m \in SO(2m)$ and that $J_m^2 = -I_{2m}$. Then

$$U(m) \cong \{A \in SO(2m):\ {}^tAJ_mA = J_m\}. \tag{4.2}$$

At the Lie algebra level,

$$\mathfrak{u}(m) \cong \{A \in \mathfrak{o}(2m) : {}^tAJ_m + J_mA = 0\}\ . \tag{4.3}$$

It will be useful for us to see this explicitly when $m = 2$ as (see §3)

$$\mathfrak{u}(2) \cong \{\begin{bmatrix} 0 & a & b & c \\ -a & 0 & -c & b \\ -b & c & 0 & d \\ -c & -b & -d & 0 \end{bmatrix} : a,b,c,d \in \mathbb{R}\} = \operatorname{span}\{\mathfrak{o}(3)\_,\ E_1^+\}. \tag{4.4}$$

Let V be any oriented $2m$–dimensional inner product space. For any orthonormal frame $e = (e_1,\ldots,e_{2m})$ of V, define an orthogonal complex structure J_e on V by

$$J_e e_{2j-1} = e_{2j}\ ,\quad 1 \le j \le m\ ,\quad J_e^2 = -\text{identity}. \tag{4.5}$$

Thus, the matrix of J_e with respect to e is J_m. It is easily verified that any orthogonal complex structure on V is equal to J_e for some orthonormal frame e, and that $J_e = J_{\tilde{e}}$ if and only if $\tilde{e} = eA$ for some $A \in U(m) \subset SO(2m)$. The set of all orthogonal complex structures on V is $O(2m)/U(m)$ which has two connected components, corresponding to the two connected components of $O(2m)$. A component is selected by choosing an orientation on V, in which case $SO(2m)/U(m) = \{J_e : e \text{ is an oriented orthonormal basis of } V\}$.

From these pointwise considerations we see then that the twistor bundle is

$$Z = O(N)\times_{O(2m)}O(2m)/U(m) = O(N)/U(m)\ . \tag{4.6}$$

It is connected if N is non–orientable, while if N is oriented then Z has two connected components

$$Z_\pm = O_\pm(N)/U(m)\ , \tag{4.7}$$

where $O_\pm(N)$ are defined in §2. Let $\sigma{:}O(N) \to Z$ be the projection, and if N is oriented, let

$$\sigma_\pm{:}O_\pm(N) \to Z_\pm \tag{4.9}$$

be the separate projections. By (4.6) these are principal U(m)–bundles. In much of the literature (cf. Salamon [S]) Z_- is called the twistor space of N.

Up to constant positive factor, there is a unique O(2m)–invariant Riemannian metric on O(2m)/U(m). This metric, combined with the parallelism on O(N) defined by the canonical and Levi–Civita forms θ and ω, defines a natural 1–parameter family of metrics on Z which we now describe. This is a special case of a general construction defined in [JR2].

The unique, up to positive factor, Ad(O(2m))–invariant inner product on $\mathfrak{o}(2m)$ is

$$\langle X,Y\rangle = \operatorname{trace}\, {}^tXY , \tag{4.10}$$

for $X,Y \in \mathfrak{o}(2m)$. We let $\mathfrak{m}$ denote the orthogonal complement of $\mathfrak{u}(m)$ in $\mathfrak{o}(2m)$. Then $\mathfrak{o}(2m) = \mathfrak{u}(m) \oplus \mathfrak{m}$ decomposes ω into $\omega = \mu + \nu$, where

$$\mu = \tfrac{1}{2}(\omega - J_m \omega J_m) \text{ and } \nu = \tfrac{1}{2}(\omega + J_m \omega J_m) , \tag{4.11}$$

which in terms of components is

$$\begin{aligned} \mu^{2j-1}_{2k-1} &= \mu^{2j}_{2k} = (\omega^{2j-1}_{2k-1} + \omega^{2j}_{2k})/2 \\ \mu^{2j-1}_{2k} &= -\mu^{2j}_{2k-1} = (\omega^{2j-1}_{2k} - \omega^{2j}_{2k-1})/2 \\ \\ \nu^{2j-1}_{2k-1} &= -\nu^{2j}_{2k} = (\omega^{2j-1}_{2k-1} - \omega^{2j}_{2k})/2 \\ \nu^{2j-1}_{2k} &= \nu^{2j}_{2k-1} = (\omega^{2j-1}_{2k} + \omega^{2j}_{2k-1})/2 . \end{aligned} \tag{4.12}$$

When m = 2 these are the skew–symmetric matrices

$$\mu = \frac{1}{2}\begin{bmatrix} 0 & 2\omega^1_2 & \omega^1_3+\omega^2_4 & \omega^1_4-\omega^2_3 \\ & 0 & \omega^2_3-\omega^1_4 & \omega^2_4+\omega^1_3 \\ & & 0 & 2\omega^3_4 \\ & & & 0 \end{bmatrix}, \quad \nu = \frac{1}{2}\begin{bmatrix} 0 & 0 & \omega^1_3-\omega^2_4 & \omega^1_4+\omega^2_3 \\ & 0 & \omega^2_3+\omega^1_4 & \omega^2_4-\omega^1_3 \\ & & 0 & 0 \\ & & & 0 \end{bmatrix} \tag{4.13}$$

The fibers of σ are the integral submanifolds of the completely integrable system $\theta = 0$, $\nu = 0$.

We define a symmetric bilinear form Q_t on O(N), for any $t > 0$, by

$$Q_t = {}^t\theta\theta + t^2\langle\nu,\nu\rangle. \tag{4.14}$$

By (4.10) and (4.12) this is

$$Q_t = \Sigma\,(\theta^a)^2 + 4t^2 \sum_{j<k} [(\nu^{2j-1}_{2k-1})^2 + (\nu^{2j-1}_{2k})^2] \,. \tag{4.15}$$

It is easily checked that $R_a^* Q_t = Q_t$ for any $a \in U(m)$, where R_a denotes right multiplication by a on $O(N)$; and that Q_t is horizontal, meaning that it vanishes on any pair of vectors for which either of them is vertical with respect to σ. Thus there exists a unique Riemannian metric g_t on Z such that $\sigma^* g_t = Q_t$. With the Riemannian metric $Q_t + \langle\mu,\mu\rangle$ on $O(N)$ and g_t on Z, σ is a Riemannian submersion with totally geodesic fibers, as we shall see.

Let $U \subset Z$ be an open subset on which there is a local section $u : U \to O(N)$ of $\sigma : O(N) \to Z$. By (4.15) an orthonormal coframe for g_t on U is given by applying u^* to the 1–forms on $O(N)$

$$\theta^a\,,\; 2t\nu^{2j-1}_{2k-1}\,,\; 2t\nu^{2j-1}_{2k}\,,\; j < k\,. \tag{4.16}$$

For a uniform notation for this coframe we let

$$\begin{aligned} \theta^{jk-} &= -\theta^{kj-} = \nu^{2j-1}_{2k-1} \\ \theta^{jk+} &= -\theta^{kj+} = \nu^{2j-1}_{2k}\,, \end{aligned} \tag{4.17}$$

which, when $m = 2$, becomes (letting $12- = 5$ and $12+ = 6$)

$$\theta^5 = \tfrac{1}{2}(\omega^1_3 - \omega^2_4)\,,\; \theta^6 = \tfrac{1}{2}(\omega^1_4 + \omega^2_3)\,. \tag{4.18}$$

The Levi–Civita connection forms for g_t with respect to this orthonormal coframe are given by (where $j < k$ and $l < m$)

$$\begin{aligned}
\theta^a_b &= \omega^a_b - t^2(R_{2j,2k,ba} - R_{2j-1,2k-1,ba})\theta^{jk-} + t^2(R_{2j-1,2k,ba} + R_{2j,2k-1,ba})\theta^{jk+} \\
\theta^{jk-}_b &= \tfrac{t}{2}(R_{2j,2k,ab} - R_{2j-1,2k-1,ab})\theta^a \\
\theta^{jk+}_b &= -\tfrac{t}{2}(R_{2j-1,2k,ab} + R_{2j,2k-1,ab})\theta^a \\
\theta^{jk-}_{lm-} &= \delta^j_m \mu^{2l-1}_{2k-1} - \delta^j_l \mu^{2m-1}_{2k-1} + \delta^m_k \mu^{2j-1}_{2l-1} - \delta^l_k \mu^{2j-1}_{2m-1} \\
\theta^{jk-}_{lm+} &= \delta^j_m \mu^{2l}_{2k-1} - \delta^j_l \mu^{2m}_{2k-1} - \delta^k_m \mu^{2j}_{2l-1} + \delta^k_l \mu^{2j}_{2m-1} \\
\theta^{jk+}_{lm+} &= \theta^{jk-}_{lm-}
\end{aligned}$$

and of course $\theta^p_q = -\theta^q_p$ for $1 \le p,q \le m(m+1)$.

In the case $m = 2$ these are (letting $12- = 5$ and $12+ = 6$):

$$\begin{aligned}
&\theta^a_b = \omega^a_b + t^2(R_{13ba} - R_{24ba})\theta^5 + t^2(R_{14ba} + R_{23ba})\theta^6 \\
(4.19)\qquad &\theta^5_b = \tfrac{t}{2}(R_{24ab} - R_{13ab})\theta^a = -\theta^b_5, \quad \theta^6_b = -\tfrac{t}{2}(R_{14ab} + R_{23ab})\theta^a = -\theta^b_6 \\
&\theta^5_6 = \omega^1_2 + \omega^3_4 = -\theta^6_5
\end{aligned}$$

These last equations show that the fibers of σ are totally geodesic for the metric $Q_t + \langle\omega_o,\omega_o\rangle$ on $O(N)$.

With respect to the frame field (4.16) the components of the curvature tensor are

$$\begin{aligned}
T_{abcd} &= R_{abcd} - \tfrac{t^2}{2}\{\tfrac{1}{2}[(R_{24ca} - R_{13ca})(R_{24db} - R_{13db}) + \\
&(R_{14ca} + R_{23ca})(R_{14db} + R_{23db}) - (R_{24da} - R_{13da})(R_{24cb} - R_{13cb}) - \\
&(R_{14da} + R_{23da})(R_{14cb} + R_{23cb})] + (R_{24ba} - R_{13ba})(R_{24dc} - R_{13dc}) + \\
&(R_{14ba} + R_{23ba})(R_{14dc} + R_{23dc})\}
\end{aligned}$$

$$\begin{aligned}
&T_{abc5} = \tfrac{t}{2}(R_{13ba,c} - R_{24ba,c}), \quad T_{abc6} = \tfrac{t}{2}(R_{14ba,c} + R_{23ba,c}) \\
&T_{ab56} = \tfrac{t^2}{4}\{(R_{13ca} - R_{24ca})(R_{14bc} + R_{23bc}) - \\
&\qquad (R_{13bc} - R_{24bc})(R_{14ca} + R_{23ca})\} - R_{34ba} - R_{12ba} \\
(4.20)\qquad &T_{5bd5} = \tfrac{t^2}{2}(R_{13ba} - R_{24ba})(R_{24da} - R_{13da}) \\
&T_{5bd6} = \tfrac{1}{2}(R_{12db} + R_{34db}) + \tfrac{t^2}{4}(R_{14ba} + R_{23ba})(R_{24da} - R_{13da}) \\
&T_{6bc6} = -\tfrac{t^2}{4}(R_{14ba} + R_{23ba})(R_{14ca} + R_{23ca}) \\
&T_{5b56} = 0 = T_{6b56}, \quad T_{5656} = \frac{1}{t^2}
\end{aligned}$$

The remaining components are determined by the symmetries of the curvature tensor, and $R_{abcd,e}$ are the components of the covariant derivative of the curvature tensor of N . Contracting T_{pqrs} on the first and third index, we obtain the components T_{pq} of the Ricci tensor of g_t on Z :

$$\begin{aligned}
T_{bd} &= R_{bd} - \tfrac{t^2}{2}\{(R_{14ba} + R_{23ba})(R_{14da} + R_{23da}) + \\
&\qquad (R_{13ba} - R_{24ba})(R_{13da} - R_{24da})\} \\
T_{b5} &= \tfrac{t}{2}(R_{13ba,a} - R_{24ba,a}), \\
T_{b6} &= \tfrac{t}{2}(R_{14ba,a} + R_{23ba,a})
\end{aligned}$$

and

$$T_{55} = \frac{1}{t^2} + \frac{t^2}{4} \sum_{a,b} (R_{13ba} - R_{24ba})^2 ,$$
$$T_{56} = -\frac{t^2}{4} (R_{14ba} + R_{23ba})(R_{24ba} - R_{13ba})$$
$$T_{66} = \frac{1}{t^2} + \frac{t^2}{4} \sum_{a,b} (R_{14ba} + R_{23ba})^2 .$$

From the Ricci identitiy $R_{abcd;d} = R_{ca;b} - R_{cb;a}$, where $R_{ab;c}$ are the components of the covariant derivative of the Ricci tensor R_{ab} of N , we find that

(4.22)
$$T_{b5} = \frac{t}{2}(R_{b1,3} - R_{b3,1} - R_{b2,4} + R_{b4,2})$$
$$T_{b6} = \frac{t}{2}(R_{b1,4} - R_{b4,1} + R_{b2,3} - R_{b3,2}) .$$

These calculations are used to establish the following theorem which was first proved by Friedrich and Grunewald [FG] (the version Theorem 2.1 in [JR3] is incorrect).

THEOREM 4.1 Let N be a four dimensional oriented Riemannian manifold. Then the metric g_t on Z_- (respectively Z_+) is Einstein if and only if N is self–dual (respectively anti–self–dual) Einstein with positive scalar curvature $s = 6/t^2$ or $s = 12/t^2$.

§5 Almost complex structures on the twistor bundle We consider now two natural almost complex structures $J_\pm$ on the twistor space Z of a 2m–dimensional oriented Riemannian manifold N . We do this by defining locally the type (1,0) forms on Z . As usual, this we do by defining complex forms on O(N) and pulling them back to Z with local sections. We continue with the index conventions of §1.

It is convenient to begin with a more detailed description of the representation $U(m) \subset SO(2m)$ introduced in (4.2). If $\{\epsilon_i\}$ and $\{\epsilon_1, \epsilon_2\}$ denote the standard bases of $\mathbb{R}^m$ and $\mathbb{R}^2$, respectively, then we take as the standard basis of $\mathbb{R}^{2m} = \mathbb{R}^m \otimes \mathbb{R}^2$ the set of vectors $\{\epsilon_i \otimes \epsilon_1, \epsilon_i \otimes \epsilon_2\}$ ordered lexicographically. Thus $x \in \mathbb{R}^{2m}$ is given by $x = x^{2i-1} \epsilon_i \otimes \epsilon_1 + x^{2i} \epsilon_i \otimes \epsilon_2$. We define a real isomorphism $\alpha: \mathbb{R}^{2m} \to \mathbb{C}^m$ by

(5.1)
$$\alpha(x) = (x^{2j-1} + ix^{2j})\epsilon_j .$$

If we let $\rho: U(m) \to SO(2m)$ denote the faithful representation (4.2), then

(5.2)
$$\rho(A + iB) = A \otimes I_2 + B \otimes J_1 .$$

The induced representation on the Lie algebra we denote by ρ_*. It is easily verified that for any $a \in U(m)$ and any $x \in \mathbb{R}^{2m}$, we have

$$a\alpha(x) = \alpha(\rho(a)x) . \tag{5.3}$$

The same formula holds for $a \in \mathfrak{u}(m)$ and ρ_* in place of ρ.

Using (4.11) and (4.12), one can see that the orthogonal complement $\mathfrak{m}$ of $\mathfrak{u}(m)$ in $\mathfrak{o}(2m)$ has the simple description

$$\mathfrak{m} = \{X\otimes\begin{bmatrix} 1 & 0 \\ 0 & -1 \end{bmatrix} + Y\otimes\begin{bmatrix} 0 & 1 \\ 1 & 0 \end{bmatrix} : X,Y \in \mathfrak{o}(m)\} . \tag{5.4}$$

If we define the real isomorphism $\beta:\mathfrak{m} \to \mathfrak{o}(m;\mathbb{C})$ by

$$\beta(X\otimes\begin{bmatrix} 1 & 0 \\ 0 & -1 \end{bmatrix} + Y\otimes\begin{bmatrix} 0 & 1 \\ 1 & 0 \end{bmatrix}) = X + iY , \tag{5.5}$$

then for any $a \in U(m)$ and any $z \in \mathfrak{m}$,

$$\beta(\mathrm{Ad}(\rho(a))z) = a\beta(z)^t a , \tag{5.6}$$

and for any $u \in \mathfrak{u}(m)$

$$\beta(\mathrm{ad}(\rho_* u)z) = u\beta(z) + \beta(z)^t u . \tag{5.7}$$

On the bundle of orthonormal frames $O(N)$ we have the canonical form θ and the Levi–Civita connection form ω, which decomposes by (4.11) into $\omega = \mu + \nu$. We define a $\mathbb{C}^m$–valued 1–form φ and an $\mathfrak{o}(m;\mathbb{C})$–valued 1–form Φ on $O(N)$ by

$$\begin{aligned} &\varphi = \alpha(\theta) , \quad (\text{thus } \varphi^j = \theta^{2j-1} + i\theta^{2j}) \\ &\Phi = \beta(\nu) , \quad (\text{thus } \Phi^{jk} = \theta^{jk-} + i\theta^{jk+}) \qquad (\text{cf } (4.17). \end{aligned} \tag{5.8}$$

Then, letting R_b denote right multiplication by $b \in SO(2m)$,we have for any $a \in U(m)$

$$R^*_{\rho(a^{-1})}\varphi = a\varphi, \quad R^*_{\rho(a^{-1})}\Phi = a\Phi^t a . \tag{5.9}$$

It follows immediately that an almost complex structure J_+ is defined on Z by defining the type (1,0) vectors to be spanned by the pull back to $Z_\pm$ by local sections $u_\pm$ of $O_\pm(N) \to Z_\pm$, respectively, of the complex 1–forms

(5.10) $$\varphi^i, \quad \Phi^{jk}, \; j < k .$$

In fact, by (5.9) this span over $\mathbb{C}$ does not depend on the choice of $u_\pm$, and it is easily verified that

$$u_\pm^*(\bigwedge_i \varphi^i \wedge \bigwedge_{j<k} \Phi^{jk}) \neq 0$$

at every point. Another almost complex structure J_- is defined in the same way, but with Φ^{jk} replaced by their complex conjugates $\bar{\Phi}^{jk}$.

(5.11) **Remarks** 1. $O(2m)/U(m)$ is a Hermitian symmetric space whose $O(2m)$–invariant complex structure is defined by the left–invariant (1,0)–forms $\beta(\nu)$.

2. The almost complex structure J_+ was introduced by Atiyah, Hitchin and Singer [AHS] in their study of self–dual Yang–Mills equations in Euclidean 4–space, while J_- has been studied by Eells and Salamon [ES] for its relation to harmonic maps from Riemann surfaces into N.

When $m = 2$ we can use the notation of (4.18) to write

(5.12) $$\varphi^3 = \Phi^{12} = \theta^5 + i\theta^6.$$

If we let $\Omega_{\mathfrak{m}}$ denote the $\mathfrak{m}$–component of the curvature form Ω (cf. (4.11), and if we let $\hat{\mu}$ denote the $\mathfrak{u}(m)$–valued 1–form such that $\rho_*\hat{\mu} = \mu$ (thus $\hat{\mu}^j_k = \mu^{2j-1}_{2k-1} + i\mu^{2j-1}_{2k}$), then from the structure equations of N we obtain

(5.13) $$d\varphi = -\bar{\hat{\mu}}\wedge\varphi - \Phi\wedge\bar{\varphi}, \quad d\Phi = \beta(\Omega_{\mathfrak{m}}) - \hat{\mu}\wedge\Phi - \Phi\wedge\hat{\mu} .$$

By the Newlander–Nirenberg theorem, an almost complex structure is integrable if and only if the algebraic ideal generated by the (1,0)–forms is closed under exterior differentiation. Thus we conclude from the first equation in (5.13) a result proved in [ES]:

(5.14) J_- is never integrable.

From the second equation in (5.13) and the fact that the curvature form on $O(N)$ is horizontal, we conclude that

(5.15) J_+ is integrable if and only if $\beta(\Omega_{\mathfrak{m}}) \equiv 0$ modulo $\{\varphi\}$.

When $m = 2$ the 2×2 skew–symmetric matrices Φ and $\beta(\Omega_{\mathfrak{m}})$ are determined by their entries

(5.16)
$$\Phi^{12} = \varphi^3 = \tfrac{1}{2}(\omega_3^1 - \omega_4^2) + \tfrac{i}{2}(\omega_4^1 + \omega_3^2)$$

and

(5.17)
$$\beta(\Omega_{\mathfrak{m}})^{12} = \tfrac{1}{2}(\Omega_3^1 - \Omega_4^2) + \tfrac{i}{2}(\Omega_4^1 + \Omega_3^2) .$$

By (4.4), and using the notation of §3, we have $\mathfrak{m} = \mathrm{span}\{E_2^+, E_3^+\}$. Thus from (3.6) we have

(5.18)
$$\Omega_{\mathfrak{m}} = E_2^+\otimes \sum_1^3(A_{2j}\alpha_+^j + B_{j2}\alpha_-^j) + E_3^+\otimes\sum_1^3(A_{3j}\alpha_+^j + B_{j3}\alpha_-^j) .$$

Hence, as $\beta(E_2^+) = -\begin{bmatrix} 0 & 1 \\ -1 & 0 \end{bmatrix}/\sqrt{2}$ and $\beta(E_3^+) = i\begin{bmatrix} 0 & 1 \\ -1 & 0 \end{bmatrix}/\sqrt{2}$, we have

(5.19)
$$\beta(\Omega_{\mathfrak{m}})^{12} = \tfrac{1}{\sqrt{2}}[A_{2j}\alpha_+^j + B_{j2}\alpha_-^j + i(A_{3j}\alpha_+^j + B_{j3}\alpha_-^j)].$$

To express this in terms of φ and $\bar{\varphi}$, we use

(5.20)
$$\begin{aligned} \alpha_+^1 &= i(\varphi^1\wedge\bar{\varphi}^1 + \varphi^2\wedge\bar{\varphi}^2)/2\sqrt{2}\,, & \alpha_-^1 &= i(\varphi^1\wedge\bar{\varphi}^1 - \varphi^2\wedge\bar{\varphi}^2)/2\sqrt{2} \\ \alpha_+^2 &= (\varphi^1\wedge\varphi^2 + \bar{\varphi}^1\wedge\bar{\varphi}^2)/2\sqrt{2}\,, & \alpha_-^2 &= (\bar{\varphi}^1\wedge\varphi^2 + \varphi^1\wedge\bar{\varphi}^2)/2\sqrt{2} \\ \alpha_+^3 &= -i(\varphi^1\wedge\varphi^2 - \bar{\varphi}^1\wedge\bar{\varphi}^2)/2\sqrt{2}\,, & \alpha_-^3 &= -i(\bar{\varphi}^1\wedge\varphi^2 - \varphi^1\wedge\bar{\varphi}^2)/2\sqrt{2} \end{aligned}$$

which, when substituted into (5.19), gives

(5.21)
$$\beta(\Omega_{\mathfrak{m}})^{12} = P\varphi^1\wedge\varphi^2 + Q\bar{\varphi}^1\wedge\bar{\varphi}^2 + R\varphi^1\wedge\bar{\varphi}^1 + S\varphi^2\wedge\bar{\varphi}^2 + T\bar{\varphi}^1\wedge\varphi^2 + U\varphi^1\wedge\bar{\varphi}^2,$$

where

(5.22)
$$\begin{aligned} &P = (A_{22} + A_{33})/4\,, \quad Q = (A_{22} - A_{33} + 2iA_{23})/4 \\ &R = (-A_{31} - B_{13} + i(A_{21} + B_{12}))/4\,, \quad S = (-A_{31} + B_{13} + i(A_{21} - B_{12}))/4 \\ &T = (B_{22} + B_{33} + i(B_{23} - B_{32}))/4\,, \quad U = (B_{22} - B_{33} + i(B_{23} + B_{32}))/4\,. \end{aligned}$$

From these calculations we can conclude the result of [AHS]:

Theorem (5.23) Let N be an oriented Riemannian 4–manifold. Then J_+ on Z_- is integrable if and only if N is self–dual (and J_+ on Z_+ is integrable if and only if N is anti–self–dual).

Proof From (5.15) and (5.21) we have that J_+ on $Z_\pm$ is integrable if and only if

$Q = 0$ on $O_{\pm}(N)$, respectively; which, by (5.22), occurs if and only if

$$A_{23} = 0 \text{ and } A_{22} = A_{33} \tag{5.24}$$

on $O(N)_{\pm}$, respectively.

By (3.10) and the fact that the homomorphism $SO(4) \to SO(3)$ given by $a \mapsto a_{+}$ is surjective, it follows that A has the form (5.24) on $O_{\pm}(N)$ if and only if A is a scalar matrix. As trace $A = s/4$, it follows that (5.24) holds on $O_{\pm}(N)$ if and only if $A = \frac{s}{12} I$ on $O_{\pm}(N)$. Hence (5.23) follows from (3.18). □

We remarked above that the twistor space Z depends only on the conformal class of the metric h on N. To see how the almost complex structures $J_{\pm}$ on Z depend on the choice of metric, consider a conformally related metric $\tilde{h} = \lambda^2 h$, where λ is a nowhere zero C^{∞} function on N. Let $\tilde{\pi}:\tilde{O}(N) \to N$ denote the bundle of $\tilde{h}$–orthonormal frames on N, and let $\tilde{\theta}$, $\tilde{\omega}$ denote the canonical and Levi–Civita forms, respectively on $\tilde{O}(N)$. Then

$$\begin{aligned} F:\tilde{O}(N) &\to O(N) \\ (p,\tilde{e}) &\mapsto (p,\lambda(p)\tilde{e}) \end{aligned} \tag{5.25}$$

is a bundle isomorphism such that

$$\begin{aligned} F^{*}\theta &= \frac{1}{\lambda}\tilde{\theta} \\ F^{*}\omega^{a}_{b} &= \tilde{\omega}^{a}_{b} + \frac{\lambda_a}{\lambda}\tilde{\theta}^{b} - \frac{\lambda_b}{\lambda}\tilde{\theta}^{a} \end{aligned} \tag{5.26}$$

where we have written λ instead of $\lambda \circ \tilde{\pi}$, and where $d(\lambda\circ\tilde{\pi}) = \lambda_a \tilde{\theta}^{a}$, $\lambda_a \in C^{\infty}(\tilde{O}(N))$. Thus, using the notation of (4.18) with $t = 1/2$, we have

$$\begin{aligned} F^{*}\theta^{5} &= \frac{1}{2\lambda}(\lambda\tilde{\omega}^{1}_{3} + \lambda_1\tilde{\theta}^{3} - \lambda_3\tilde{\theta} - \lambda\tilde{\omega}^{2}_{4} - \lambda_2\tilde{\theta}^{4} + \lambda_4\tilde{\theta}^{2}) \\ F^{*}\theta^{6} &= \frac{1}{2\lambda}(\lambda\tilde{\omega}^{1}_{4} + \lambda_1\tilde{\theta}^{4} - \lambda_4\tilde{\theta}^{1} + \lambda\tilde{\omega}^{2}_{3} + \lambda_2\tilde{\theta}^{3} - \lambda_3\tilde{\theta}^{2}) \end{aligned} \tag{5.27}$$

and consequently, using the notation of (5.12), we have

$$\begin{aligned} F^{*}\varphi^{i} &= \frac{1}{\lambda(p)}\tilde{\varphi}^{i}, \quad i = 1,2 \\ F^{*}\varphi^{3} &= \tilde{\varphi}^{3} + \frac{1}{2\lambda}(\lambda_1 + i\lambda_2)\tilde{\varphi}^{2} - \frac{1}{2\lambda}(\lambda_3 + i\lambda_4)\tilde{\varphi}^{1}\,. \end{aligned} \tag{5.28}$$

We have U(2)–bundles $\sigma:O(N) \to Z$ and $\tilde{\sigma}:\tilde{O}(N) \to Z$ (see (4.8)) for which it is easily verified that $\sigma \circ F = \tilde{\sigma}$. Thus by definition of $J_{\pm}$ and $\tilde{J}_{\pm}$ on Z, defined by h and $\tilde{h}$, respectively, it follows from (5.28) that $J_+ = \tilde{J}_+$, for any conformal factor λ; while $J_- = \tilde{J}_-$ if and only if λ is constant. Thus J_+ is conformally invariant, while J_- is invariant only under change of scale.

§6 Hermitian structures on the twistor bundle. Consider the twistor space Z with metrics g_t of §4 and almost complex structures $J_{\pm}$. By (4.16) and (5.10), $(Z, J_{\pm}, g_t)$ is Hermitian and

$$\varphi^i,\ 2t\Phi^{jk},\ j<k \quad (\text{resp.},\ \varphi^i,\ 2t\bar{\Phi}^{jk}) \tag{6.1}$$

is a unitary coframe field for (Z, J_+, g_t) (resp., (Z, J_-, g_t)) when pulled back to Z by any section u of $\sigma:O(N) \to Z$. The associated (1,1)–form, i.e., Kaehler form, is then

$$\kappa_{\pm}(t) = \frac{i}{2}\Big[\sum_i \varphi^i \wedge \bar{\varphi}^i \pm 4t^2 \sum_{j<k} \Phi^{jk} \wedge \bar{\Phi}^{jk}\Big] \tag{6.2}$$

pulled back to Z by u*. Taking the exterior derivative of $\kappa_{\pm}(t)$, using the structure equations of N in the form (5.13), we find

$$d\kappa_{\pm}(t) = i \sum_{j<k} \{\Phi^{jk} \wedge \bar{\varphi}^j \wedge \bar{\varphi}^k - \varphi^j \wedge \varphi^k \wedge \bar{\Phi}^{jk} \pm 2t^2[\beta(\Omega_m)^{jk} \wedge \bar{\Phi}^{jk} - \Phi^{jk} \wedge \overline{\beta(\Omega_m)}^{jk}]\} \tag{6.3}$$

Suppose now that m = 2. Substituting (5.21) into (6.3) we find

$$\begin{aligned} d\kappa_{\pm}(t) = {} & -i\varphi^3 \wedge [(-1 \pm 2t^2 P)\bar{\varphi}^1 \wedge \bar{\varphi}^2 \pm 2t^2(\bar{Q}\varphi^1 \wedge \varphi^2 + \bar{R}\bar{\varphi}^1 \wedge \varphi^1 + \bar{S}\bar{\varphi}^2 \wedge \varphi^2 + \bar{T}\varphi^1 \wedge \bar{\varphi}^2 + \bar{U}\bar{\varphi}^1 \wedge \varphi^2)] \\ & + i[(-1 \pm 2t^2 P)\varphi^1 \wedge \varphi^2 \pm 2t^2(Q\bar{\varphi}^1 \wedge \bar{\varphi}^2 + R\varphi^1 \wedge \bar{\varphi}^1 + S\varphi^2 \wedge \bar{\varphi}^2 + T\bar{\varphi}^1 \wedge \varphi^2 + U\varphi^1 \wedge \bar{\varphi}^2)] \wedge \bar{\varphi}^3 \end{aligned} \tag{6.4}$$

(6.5) **Definition** Recall that an almost Hermitian manifold (Z, g, J) is **symplectic** if its associated (1,1) form κ is closed; it is **(1,2)–symplectic** if the (1,2) part of $d\kappa$ is zero; and it is **Kaehler** if it is symplectic and J is integrable.

THEOREM 6.1 Let N be an oriented Riemannian 4–manifold. The following are equivalent:

a) (Z_-, g_t, J_+) is (1,2)–symplectic;

b) N is self–dual Einstein with positive scalar curvature s and $t^2 = 12/s$;

c) (Z_-, g_t, J_+) is Kaehler Einstein.

PROOF Recall that the type (1,0) forms of J_+ on Z_- are given by the pull back of $\varphi^1, \varphi^2, \varphi^3$ to Z_- by any local frame u in Z_-. Furthermore, the twelve decomposable 3–forms giving $d\kappa_+(t)$ in (6.4) are linearly independent at each point when pulled back to Z_-. Thus, reading off the type (1,2) part from (6.4), we see that a) holds if and only if

$$P = 1/2t^2 \text{ and } R = S = T = U = 0 \tag{6.6}$$

on $O_-(N)$, which, by (5.22), holds if and only if

$$A_{22} + A_{33} = 2/t^2 \tag{6.7}$$

and

$$A = \begin{bmatrix} A_{11} & 0 & 0 \\ 0 & A_{22} & A_{23} \\ 0 & A_{32} & A_{33} \end{bmatrix}, \quad B = \begin{bmatrix} B_{11} & 0 & 0 \\ B_{21} & 0 & 0 \\ B_{31} & 0 & 0 \end{bmatrix} \tag{6.8}$$

on $O_-(N)$. By the transformation laws (3.10), and the fact that each of the homomorphisms $SO(4) \to SO(3)$, $a \mapsto a_\pm$, is surjective, it follows easily that A and B can have the form (6.8) at each point of $O_-(N)$ if and only if A is a scalar matrix and $B = 0$. If A is scalar, then $A = \frac{s}{12} I$, since trace $A = s/4$. Thus $s/6 = 2/t^2$ by (6.7), from which it follows that $s > 0$ and $t^2 = 12/s$. Hence a) is equivalent to b) by (3.14) and (5.23).

Suppose b) holds. Then J_+ on $O_-(N)$ is integrable by (5.23), and g_t is Einstein by (4.23). As remarked in the proof of (5.23), J_+ on $O_-(N)$ is integrable if and only if $Q = 0$ on $O_-(N)$. This, combined with (6.6) and (6.4), shows that (Z_-, g_t, J_+) is symplectic. Hence b) implies c).

A fortiori, c) implies a). □

Remarks (6.9) A similar result, with evident modifications, holds for (Z_+, g_t, J_+).

(6.10) This theorem and the next generalize Theorem 9.1 in [ES] where N was assumed to be S^4 with its canonical metric.

(6.11) By (4.23), if N is self–dual Einstein with positive scalar curvature s, then g_t on Z_- is Einstein when $t^2 = 6/s$. By our theorem, in this case (Z_-, g_t, J_+) is not even (1,2)–symplectic. (Cf. [FG]).

THEOREM 6.2 Let N be an oriented Riemannian 4–manifold. Then (Z_-, g_t, J_-) is (1,2)–symplectic if and only if N is self–dual Einstein (for any value of $t > 0$), while (Z_-, g_t, J_-) is symplectic if and only if N is self–dual Einstein with negative scalar curvature s and $t^2 = -12/s$.

PROOF The proof is similar to that of theorem 6.1 except that now the type (1,0) forms are the pull backs of φ^1, φ^2, $\bar{\varphi}^3$ to Z_- by local sections u. Thus, by (6.4), (Z_-, g_t, J_-) is (1,2)–symplectic if and only if $R = S = T = U = Q = 0$. As in the above proof, this is equivalent to N being self–dual Einstein. In this case $d\kappa_-(t) = i(1 + \frac{t^2 s}{12})(\varphi^3 \wedge \varphi^1 \wedge \varphi^2 - \varphi^1 \wedge \varphi^2 \wedge \bar{\varphi}^3)$, which can be zero if and only if $s < 0$ and $t^2 = -12/s$. □

Remarks (6.12) By results of Friedrich and Kurke [FK] and Hitchin [H], if N is a compact 4–dimensional self–dual Einstein space with positive scalar curvature, then it is isometric to either S^4 or $\mathbb{C}P^2$ with their canonical metrics. In these cases Z_- is $\mathbb{C}P^3$ or the flag manifold $\mathbb{F}(1,2)$, respectively. There is no known classification of 4–dimensional self–dual Einstein spaces with negative scalar curvature [V]. Examples are hyperbolic space and Hermitian hyperbolic space with their canonical metrics.

(6.13) The relevance to us of (1,2)–symplectic spaces comes from the result of Lichnerowicz [L]: If $f: M \to N$ is a holomorphic map from a Riemann surface to an almost Hermitian (1,2)–symplectic manifold, then f is harmonic.

§7 The Grassmann bundle. We briefly describe the geometry of the Grassmann bundle of oriented 2–planes tangent to N, $G : G_2(TN) \to N$. An element of $G_2(TN)$ is a pair (p,ξ) where $p \in N$ and ξ is a two–dimensional oriented subspace of T_pN. The Grassmann projection

$$G : G_2(TN) \to N \tag{7.1}$$

is defined by $G(p,\xi) = p$. The projection (7.1) presents $G_2(TN)$ as a fiber bundle over N with standard fiber the Grassmann manifold $\tilde{G}_2(4) = SO(4)/SO(2)\times SO(2)$. We define a map

$$\mu : O(N) \to G_2(TN) \tag{7.2}$$

by $\mu(e) = \{e_1,e_2\}$, where $e = (e_1,\ldots,e_4)$ is an orthonormal frame at a point $p \in N$ and $\{e_1,e_2\}$ is the oriented plane in T_pN spanned by e_1,e_2 with the orientation $e_1 \wedge e_2$. Thus

$$G_2(TN) \cong O(N)/SO(2)\times O(2).$$

Notice that μ restricted to $O_\pm(N)$ (which we denote $\mu_\pm$) is a principal $SO(2)\times SO(2)$ bundle. The fibers of μ are the integral submanifolds of the completely integrable system $\theta = 0$, $\zeta = 0$, where $\zeta = \begin{bmatrix} 0_\alpha & \omega^i_\alpha \\ \omega^\alpha_i & 0^\alpha \end{bmatrix}$.

For $t>0$ the Riemannian metric h_t on $G_2(TN)$ is characterized by

$$\mu^* h_t = P_t , \tag{7.3}$$

where P_t is the $O(2)\times O(2)$ invariant symmetric bilinear form on $O(N)$ given by

$$P_t = {}^t\theta\theta + \tfrac{1}{2}\langle t\zeta, t\zeta\rangle . \tag{7.4}$$

In terms of components, we have

$$P_t = \Sigma\, (\theta^a)^2 + t^2(\omega^\alpha_i)^2 . \tag{7.5}$$

An orthonormal frame for P_t is given by

$$\theta^a ,\ \theta^{\alpha i} = t\omega^\alpha_i . \tag{7.6}$$

If $u : U \to O(N)$ is a local section of (7.2), then an orthonormal coframe for h_t on U is given by

$$u^*\theta^a ,\ u^*\theta^{\alpha i} . \tag{7.7}$$

From the structure equations of $O(N)$ we find that the pull–back by u^* of the forms

$$\begin{aligned} \theta^a_b &= \omega^a_b + \frac{t^2}{2} R^\alpha_{i\,ab}\omega^i_\alpha \\ \theta^{\alpha i}_b &= -\frac{t}{2} R^\alpha_{i\,ab}\,\theta^a = -\theta^b_{\alpha i} \\ \theta^{\alpha i}_{\beta j} &= \delta_{\alpha\beta}\omega^i_j + \delta_{ij}\omega^\alpha_\beta \end{aligned} \tag{7.8}$$

gives the Levi–Civita connection forms of h_t with respect to the orthonormal frame (7.7).

From (7.8) one can compute the Riemann curvature tensor of h_t and its Ricci tensor. In contrast to the twistor bundle case of §4, there is no value of t for which h_t is Einstein.

We consider now two natural almost complex structures $J^G_\pm$ on the Grassmann bundle $G_2(TN)$. Using the coframe field (7.6) on $O_+(N)$ we let

$$\varphi^1 = \theta^1 + i\theta^2,\ \varphi^2 = \theta^3 + i\theta^4,\ \varphi^\alpha_G = \theta^{\alpha 1} + i\theta^{\alpha 2} = \omega^\alpha_1 + i\omega^\alpha_2 , \tag{7.9}$$

complex valued 1–forms on $O_+(N)$. Then J^G_+ (respectively, J^G_-) is defined by the condition that its type (1,0) forms are locally spanned by the pull–back of φ^j, φ^α_G (respectively, φ^j, $\overline{\varphi}^\alpha_G$) by any local section u of μ_+ of (7.2).

Using the structure equations of O(N) we find

$$(7.10)\qquad \begin{aligned} d\varphi^1 &= \tfrac{1}{2}(\varphi^3_G - i\varphi^4_G)\wedge\varphi^2 + \tfrac{1}{2}(\varphi^3_G + i\varphi^4_G)\wedge\overline{\varphi}^2 \\ d\varphi^2 &= -\tfrac{1}{2}(\overline{\varphi}^3_G + i\overline{\varphi}^4_G)\wedge\varphi^1 - \tfrac{1}{2}(\varphi^3_G + i\varphi^4_G)\wedge\overline{\varphi}^1 \end{aligned}$$

$$(7.11)\qquad d\varphi^\alpha_G = -i\varphi^\alpha_G\wedge\omega^1_2 - \omega^\alpha_\beta\wedge\varphi^\beta_G + \Omega^\alpha_1 + i\Omega^\alpha_2 .$$

From (7.10) and (7.11) if follows that J^G_- is never integrable, while J^G_+ is integrable if and only if

$$(7.12)\qquad \Omega^\alpha_1 + i\Omega^\alpha_2 = 0 \mod (\varphi^1, \varphi^2) .$$

Proposition 7.1 Let N be an oriented Riemannian 4–manifold. Then J^G_+ on $G_2(TN)$ is integrable if and only if N is anti–self–dual Einstein.

Proof We need to see that (7.12) holds if and only if N is anti–self–dual Einstein. By (3.6) and (5.20)

$$(7.13)\qquad \begin{aligned} \Omega^1_3 + i\Omega^2_3 &\equiv \tfrac{1}{4}(A_{22} - A_{33} + B_{22} + B_{33} + i(2A_{23} + B_{23} - B_{32}))\overline{\varphi}^1\wedge\overline{\varphi}^2 \\ \Omega^1_4 + i\Omega^2_4 &\equiv \tfrac{1}{4}(2A_{23} + B_{32} - B_{23} + i(A_{33} - A_{22} + B_{22} + B_{33}))\overline{\varphi}^1\wedge\overline{\varphi}^2 \\ &\qquad \mod (\varphi^1, \varphi^2). \end{aligned}$$

Hence (7.12) holds if and only if

$$\begin{aligned} A_{22} - A_{33} + B_{22} + B_{33} = 0 = 2A_{23} + B_{23} - B_{32} \\ 2A_{23} + B_{32} - B_{23} = 0 = A_{33} - A_{22} + B_{22} + B_{33} \end{aligned}$$

which holds if and only if $A_{23} = 0$, $A_{22} = A_{33}$, $B_{23} = B_{32}$, and $B_{22} + B_{33} = 0$.

By (3.10) this holds if and only if $A = \frac{S}{12} I$, and $B = 0$ on $O_+(N)$, which by (3.19) and (3.14) holds if and only if N is anti–self–dual Einstein. □

Remark Two other almost complex structures, $\tilde{J}^G_\pm$, can be defined on $G_2(TN)$ by pulling back with any positively oriented frame field the forms $\varphi^1, \overline{\varphi}^2, \varphi^\alpha_G$ or $\varphi^1, \overline{\varphi}^2, \overline{\varphi}^\alpha_G$ respectively. It is easily seen that this is equivalent to the structures obtained by pulling back $\varphi^1, \varphi^2, \varphi^\alpha_G$ (respectively, $\varphi^1, \varphi^2, \overline{\varphi}^\alpha_G$) by negatively oriented frame fields $e{:}U \subset G_2(TN) \to O_-(N)$. In fact, if $K = \mathrm{diag}(1,1,1,-1)$, then $R_K{:}O_-(N) \to O_+(N)$, $R_K \circ e{:}U \to O_+(N)$, and $R_K^*\varphi^1 = \varphi^1$, $R_K^*\varphi^2 = \overline{\varphi}^2$, $R_K^*\varphi^3_G = \varphi^3_G$, $R_K^*\varphi^4_G = -\varphi^4_G$. It follows that $\tilde{J}^G_-$ is never integrable, while $\tilde{J}^G_+$ is integrable if and only if N is self–dual Einstein.

§8 Twistor and Gauss lifts. Let $f{:}M \to N$ be an isometric immersion of an oriented surface into an oriented Riemannian 4–manifold. We define projections

$$\pi_\pm{:}G_2(TN) \to Z_\pm \tag{8.1}$$

of the Grassmann bundle of oriented tangent 2–planes of N onto the respective twistor spaces as follows. If $\zeta \subset T_pN$ is an oriented 2–dimensional subspace, then $\pi_+(p,\zeta)$ is the almost complex structure on T_pN given by the positive twist (i.e., rotation through $+\pi/2$) in each of ζ and its orthogonal complement $\zeta^\perp$ (with induced orientation from ζ and T_pN); while $\pi_-(p,\zeta)$ is the positive twist in ζ but the negative twist in $\zeta^\perp$. Observe that μ of (7.2) and $\sigma_\pm$ of (4.9) are related to $\pi_\pm$ by $\sigma_\pm = \pi_\pm \circ \mu_\pm$.

The twistor lifts of f,

$$\varphi_\pm{:}M \to Z_\pm, \tag{8.2}$$

are defined by: $\varphi_+(p)$ is the positive twist in f_*T_pM and in $f_*T_pM^\perp$, while $\varphi_-(p)$ is the positive twist in f_*T_pM but the negative twist in $f_*T_pM^\perp$. Thus $\varphi_\pm = \pi_\pm \circ \gamma_f$, where

$$\gamma_f{:}M \to G_2(TN)$$

is the Gauss lift: $\gamma_f(p) = f_*T_pM$ (with its orientation from M). These maps are illustrated by the commutative diagram

$$\text{(8.3)}\qquad \begin{array}{ccccccc} O_-(N) & \searrow^{\mu_-} & & {}^{\mu_+}\swarrow & O_+(N) \\ \sigma_- \downarrow & \swarrow^{\pi_-} & G_2(TN) & {}^{\pi_+}\searrow & \downarrow \sigma_+ \\ & & \uparrow \gamma_f & & \\ Z_- & \xleftarrow{\varphi_-} & M & \xrightarrow{\varphi_+} & Z_+ \\ & T \searrow & f \downarrow & \swarrow T & \\ & & N & & \end{array}$$

Recall the almost complex structures $J_\pm$ defined on Z in §5. The following result was first proved in [ES], Theorem 5.3.

Proposition 8.1 a) $\varphi_\pm$ is J_+ holomorphic if and only if f is isotropic with $\pm$ spin, respectively. b) $\varphi_\pm$ is J_- holomorphic if and only if f is minimal.

Proof Let $e = (e_1,\dots,e_4){:}U\subset M \to O_+(N)$ be a local oriented Darboux frame along f (see §2), and let $e_- = (e_1,e_2,e_3,-e_4) = R_K \circ e$, where $K = \mathrm{diag}(1,1,1,-1)$. We may assume the existence of local sections $u_\pm{:}Z_\pm \to O_\pm(N)$ such that $e_\pm = u_\pm \circ \varphi_\pm$, respectively. Thus (see (2.17) and (5.16))

$$\text{(8.4)}\qquad \begin{aligned} &\varphi_+^* u_+^* \varphi^1 = e^*\varphi^1 = \varphi, \qquad \varphi_+^* u_+^* \varphi^2 = e^*\varphi^2 = 0,\\ &\varphi_+^* u_+^* \varphi^3 = e^*\varphi^3 = -\tfrac{1}{2}\bar{b}\varphi - \tfrac{1}{2}\bar{S}_+\bar{\varphi}; \end{aligned}$$

while

$$\text{(8.5)}\qquad \begin{aligned} &\varphi_-^* u_-^* \varphi^1 = e_-^*\varphi^1 = e^* R_K^* \varphi^1 = \varphi, \qquad \varphi_-^* u_-^* \varphi^2 = 0,\\ &\varphi_-^* u_-^* \varphi^3 = e^* R_K^* \varphi^3 = -\tfrac{1}{2} b\varphi - \tfrac{1}{2}\bar{S}_-\bar{\varphi} \end{aligned}$$

since $R_K^*\varphi^1 = \varphi^1$, $R_K^*\varphi^2 = \varphi^2$, and $R_K^*\varphi^3 = \frac{1}{2}(\omega_3^1 + \omega_4^2) + \frac{i}{2}(\omega_3^2 - \omega_4^1)$. Thus a) follows from (8.4) and (8.5), respectively, while b) follows from (8.4) and (8.5) with φ^3 replaced by $\bar{\varphi}^3$. □

Recall from §6 that a unitary coframe for $(Z, J_\pm, g_t)$ is given (in $O(N)$) by $\varphi^1, \varphi^2, 2t\varphi^3$. By (8.4), using (2.17) and (2.18), we have

$$\text{(8.6)}\qquad \begin{aligned} &\varphi_+^* g_t = (1 + t^2\|H\|^2 + 2t^2 s_+^2)\varphi\bar{\varphi} + t^2\bar{b}S_+\varphi\varphi + t^2 b\bar{S}_+\bar{\varphi}\bar{\varphi}\\ &\varphi_-^* g_t = (1 + t^2\|H\|^2 + 2t^2 s_-^2)\varphi\bar{\varphi} + t^2 bS_-\varphi\varphi + t^2\bar{b}\bar{S}_-\bar{\varphi}\bar{\varphi}. \end{aligned}$$

These calculations prove the following.

PROPOSITION 8.2 Let $f:M \to N$ be an isometric immersion of an oriented surface into an oriented Riemannian 4–manifold. Let $\varphi_\pm:M \to Z_\pm$ be its twistor lifts. Let g_t be the Hermitian metric on $Z_\pm$ of §6. Then

(i) $\varphi_\pm$ is conformal if and only if either f is minimal or f is isotropic with ± spin, respectively;

(ii) $\varphi_\pm$ is an isometry if and only if f is minimal and isotropic with ± spin, respectively.

Let $\kappa_\pm$ be the Kaehler forms (6.2) of $(Z, J_\pm, g_t)$, respectively. From (8.4) and (8.5) we have

$$(8.7)\quad \begin{aligned} \varphi_\pm^*\kappa_+ &= (1 + t^2(\|H\|^2 - 2s_\pm^2))dA \\ \varphi_\pm^*\kappa_- &= (1 - t^2(\|H\|^2 - 2s_\pm^2))dA . \end{aligned}$$

Consequently

$$(8.8)\quad \begin{aligned} \varphi_\pm^*(\kappa_+ + \kappa_-) &= 2dA \\ \varphi_\pm^*(\kappa_+ - \kappa_-) &= 2t^2(\|H\|^2 - 2s_\pm^2)dA . \end{aligned}$$

Combining this with (2.20), we have

$$(K - R_{1212})dA = \frac{1}{2t^2}(\varphi_+^*\kappa_+ + \varphi_-^*\kappa_+) - t^{-2}dA = \frac{-1}{2t^2}(\varphi_+^*\kappa_- + \varphi_-^*\kappa_-) + t^{-2}dA$$

$$(K^\perp - R_{1234})dA = \frac{1}{2t^2}(\varphi_+^*\kappa_+ - \varphi_-^*\kappa_+) = \frac{-1}{2t^2}(\varphi_+^*\kappa_- - \varphi_-^*\kappa_-) .$$

The basic contact invariants $\|H\|$ and $s_\pm$ introduced in (2.18) can be interpreted in terms of the (1.0) and (0,1) energy densities of $\varphi_\pm$. If Z has the (J_+, g_t) structure (respectively the (J_-, g_t) structure), then from (8.4) and (8.5), we have

$$\begin{aligned} e_+'(\varphi_\pm) &= 1 + t^2\|H\|^2 \\ e_+''(\varphi_\pm) &= 2t^2 s_\pm^2 , \end{aligned} \quad \text{, respectively} \quad \begin{aligned} e_-'(\varphi_\pm) &= 1 + 2t^2 s_\pm^2 \\ e_-''(\varphi_\pm) &= t^2\|H\|^2 . \end{aligned}$$

We conclude this section by computing the tension field of φ_-. An analogous result holds for φ_+. As already remarked in §6, $(Z, J_\pm)$ in general is not (1,2)–symplectic (cf. Theorems 6.1 and 6.2). Nevertheless, we know examples (cf. [G] and [ES]) in which a suitable holomorphic map of a Riemann surface into a Hermitian, not (1,2)–symplectic, manifold is still harmonic. It is therefore instructive to face the problem directly and compute the tension of φ_-. One benefit of this is that it does allow us to discover a new feature for φ_-, even in a case for which (Z, J_+, g_t) is (1,2)–symplectic.

Let $u:U\subset Z_- \to O_-(N)$ be a local section of $\sigma_-:O_-(N) \to Z_-$. Recall from §4 that the Riemannian metric g_t on Z_- has a local orthonormal coframe field given by applying u^* to

$$(8.12)\qquad \theta^1,\dots,\theta^4,\ t(\omega_3^1 - \omega_4^2),\ t(\omega_4^1 + \omega_3^2)$$

on $O_-(N)$. For our present calculation it is convenient to modify (4.18) slightly and define

$$(8.13)\qquad \theta^5 = t(\omega_3^1 - \omega_4^2)\ ,\ \theta^6 = t(\omega_4^1 + \omega_3^2)\ .$$

We may choose u and a negatively oriented Darboux frame e along f such that $e = u\circ\varphi_-$, which means that $\varphi_-^*u^*\theta^p = e^*\theta^p$, $p = 1,\dots,6$. Let $\theta^j = e^*\theta^j$, an oriented orthonormal coframe in M, and let $e^*\theta^p = B_j^p\theta^j$, where the B_j^p are locally defined functions in M. If $\{E_p\}$ is the orthonormal frame field in Z_- dual to $\{u^*\theta^p\}$, then

$$d\varphi_- = B_j^p\theta^j\otimes E_p.$$

Thus $\nabla d\varphi_- = B_{jk}^p\theta^j\theta^k\otimes E_p$, where

$$(8.14)\qquad dB_j^p - B_j^p\omega_k^j + B_j^q\theta_q^p = B_{jk}^p\theta^k.$$

Here θ_q^p are the Levi–Civita connection forms of g_t listed in (4.19), appropriately modified as required by replacing (4.18) with (8.13). The tension field of φ is then given by

$$\tau(\varphi_-) = B_{jj}^p E_p.$$

We proceed to make these calculations.

From §2 we have $e^*\theta^j = \theta^j$, $e^*\theta^\alpha = 0$, $e^*\theta^5 = t(h_{2j}^4 - h_{1j}^3)\theta^j$, and $e^*\theta^6 = -t(h_{1j}^4 + h_{2j}^3)\theta^j$, from which it follows that

$$(8.15)\qquad B_k^j = \delta_k^j\ ,\ B_k^\alpha = 0\ ,\ B_k^5 = t(h_{2k}^4 - h_{1k}^3)\ ,\ B_k^6 = -t(h_{1k}^4 + h_{2k}^3).$$

Carrying out the calculations of (8.14) and using the notation of §3, we find (summing on k)

$$
\begin{aligned}
B^1_{kk} &= t^2\{-(A_{21}+B_{12})(h^4_{22}-h^3_{12})+(A_{31}+B_{13})(h^4_{12}+h^3_{22})\}\\
B^2_{kk} &= t^2\{(A_{21}+B_{12})(h^4_{21}-h^3_{11})-(A_{31}+B_{13})(h^4_{11}+h^3_{21})\}\\
B^3_{kk} &= h^3_{kk}+t^2\{(A_{22}+B_{22})(h^4_{21}-h^3_{11})+(A_{32}-B_{32})(h^4_{22}-h^3_{12})\\
&\qquad -(A_{32}+B_{23})(h^4_{11}+h^3_{21})-(A_{33}-B_{33})(h^4_{12}+h^3_{22})\}\\
B^4_{kk} &= h^4_{kk}+t^2\{(A_{32}+B_{32})(h^4_{21}-h^3_{11})+(B_{22}-A_{22})(h^4_{22}-h^3_{12})\\
&\qquad -(A_{33}+B_{33})(h^4_{11}+h^3_{21})-(B_{23}-A_{32})(h^4_{12}+h^3_{22})\}\\
B^5_{kk} &= t\{2(H^4_2-H^3_1)-A_{31}-B_{13}\}\\
B^6_{kk} &= -t\{2(H^4_1+H^3_2)-A_{21}-B_{12}\}.
\end{aligned}
$$

Recall (3.14), that N is Einstein if and only if $B = 0$ on $O(N)$, and (3.19), that N is self–dual if and only if $A = \frac{s}{12} I$ on $O\_(N)$, where s is the scalar curvature of N. Thus if N is self–dual Einstein, then

$$
\begin{aligned}
B^j_{kk} &= 0 & B^\alpha_{kk} &= 2H^\alpha(1-t^2 s/12)\\
B^5_{kk} &= 2t(H^4_2-H^3_1) & B^6_{kk} &= -2t(H^4_1+H^3_2).
\end{aligned}
$$

These calculations and their analogs for φ_+ yield the following results.

Theorem 8.1 Let $f:M \to N$ be an isometric immersion of a Riemann surface into a 4–dimensional self–dual (respectively, anti–self–dual) Einstein manifold with scalar curvature s, twistor space (Z, g_t) and twistor lifts $\varphi_\pm:M \to Z_\pm$.

a) If $st^2 \neq 12$, then f is minimal if and only if φ_- (respectively, φ_+) is harmonic;

b) If $st^2 = 12$, then φ_- (respectively, φ_+) is harmonic if and only if $H^4_1 = -H^3_2$ and $H^4_2 = H^3_1$; i.e., $\nabla H_{(p)}:T_pM \to T_pM^\perp$ is complex.

Remarks

(8.16) Suppose N is self–dual Einstein. Then by Theorem 6.2, (Z_-, g_t, J_-) is (1,2)–symplectic for any $t > 0$. Thus, by (6.13) and Proposition 8.1, if f is minimal, then φ_- is J_-–holomorphic, thus harmonic.

(8.17) If N is compact self–dual Einstein with $s > 0$, then it must be S^4 or $\mathbb{C}P^2$ with their canonical metrics (cf. Remark (6.12)). The twistor space Z_- of S^4 is $\mathbb{C}P^3$, and its metric g_t with $t^2 = 12/s$ is the Fubini–Study metric on $\mathbb{C}P^3$. Thus part b) of our theorem parametrizes a class of harmonic maps $\varphi\_:M \to \mathbb{C}P^3$ by immersed surfaces $f:M \to S^4$ whose mean curvature vector H satisfies $H_1^4 = -H_2^3$ and $H_2^4 = H_1^3$. As a consequence, there is a large class of harmonic maps $\varphi\_:M \to \mathbb{C}P^3$ which are not $J\_$–holomorphic.

BIBLIOGRAPHY

[AHS] M.F. Atiyah, H.J. Hitchin & I.M. Singer, "Self–duality in four dimensional Riemannian geometry," Proc. Royal Soc. Lond, Ser. A, 362 (1978), 425–461.

[Be] A.L. Besse, "Einstein Manifolds," Springer–Verlag, Berlin Heidelberg 1987.

[Bl] W. Blaschke, "Sulla geomctria differenziale delle superfici S_2 nello spazio euclideo S_4," Ann. Mat. Pura Appl. 28 (1949), 205–209.

[B] R.L. Bryant, "Conformal and minimal immersions of compact surfaces into the 4–sphere," J. Diff. Geom. 17 (1982), 455–473.

[Ca] E. Calabi, "Minimal immersions of surfaces in Euclidean spheres," J. Diff. Geom. 1 (1967), 111–125.

[Ch] S.S. Chern, "Minimal surfaces in an Euclidean space of N dimensions," Differential and Combinatorial Topology, Princeton Univ. Press (1965), 187–198.

[Ch2] ———, "On the minimal immersions of the two sphere in a spacc of constant curvature," Problems in Analysis, Princeton, N.J. (1970), 27–40.

[CS] S.S. Chern & E.H. Spanier, "A theorem on orientable surfaces in four–dimensional space," Comm. Math. Helv. 25 (1951), 205–209.

[EGT] J–H. Eschenburg, I.V. Guadalupe & R.A. Tribuzy, "The fundamental equations of minimal surfaces in $\mathbb{C}P^2$," Math. Ann. 270 (1985), 571–598.

[ES] J. Eells & S. Salamon, "Twistorial constructions of harmonic maps of surfaces into four–manifolds," Ann. Scuola Norm. Sup. Pisa 12 (1985), 589–640.

[EW] J. Eells & J.C. Wood, "Harmonic maps from surfaces to complex projective spaces," Advances in Math. 49 (1983), 217–263.

[F] Th. Friedrich, "On surfaces in four–spaces," Ann. Glob. Analysis and Geom. 2 (1984), 257–287.

[FG] Th. Friedrich & R. Grunewald, "On Einstein metrics on the twistor space of a four–dimensional Riemannian manifold," Math. Nachr. 123 (1985), 55–60.

[FK] Th. Friedrich & H. Kurke, "Compact four–dimensional self–dual Einstein manifolds with positive scalar curvature," Math. Nachr. 106 (1982), 271–299.

[G] A. Gray, "Minimal varieties and almost Hermitian submanifolds," Michigan Math. J. 12 (1965), 273–287.

[H] N. Hitchin, "Kaehlerian twistor spaces," Proc. Lond. Math. Soc., III. Ser., 43 (1981), 133–150.

[HO1] D.A. Hoffman & R. Osserman, "The geometry of the generalized Gauss map," Memoires Amer. Math. Soc. 28, Number 236, 1980.

[HO2] ———, "The Gauss map of surfaces in $\mathbb{R}^3$ and $\mathbb{R}^4$," Proc. Lond. Math. Soc. 50 (1985), 27–56.

[JR1] G.R. Jensen & M. Rigoli, "Minimal surfaces in spheres by the method of moving frames," Revue de l'Institute E. Cartan, 12 (1988), Nancy, France.

[JR2] ——— "Harmonic Gauss maps," Pacific J. of Math., to appear.

[JR3] ——— "Harmonic maps, twistor spaces and twistor lifts," Fibre Bundles: Their Use in Physics, ed. J. Ezin & A. Verjovski, World Scientific Press, Singapore, 1988.

[JR4] ——— "Minimal surfaces in spheres," Rend. Sem. Mat. Univers. Politec. Torino Special Volume (1983), 75–98.

[L] A. Lichnerowicz, "Applications harmoniques et varietes Kaehleriennes," Symp. Math III Bologno (1970), 341–402.

[P] Y.S. Poon, preprint.

[RV] E.A. Ruh & J. Vilms, "The tension field of the Gauss map," Trans. Amer. Math. Soc. 149 (1970), 569–573.

[S] S. Salamon, "Topics in four dimensional Riemannian geometry," in Geometry Seminar "Luigi Bianchi", ed. E. Vesentini, Lecture Notes in Math. 1022 Springer Verlag 1983, 33–124.

[S2] ———, "Degrees of minimal surfaces in 4–manifolds," preprint (1986).

[V] A. Vitter, "Self–dual Einstein metrics," Contemporary Math. 51 (1986), 113–120.

[We] J.L. Weiner, "On an inequality of P. Wintgen for the integral of the square of the mean curvature," preprint (1986).

[W] P. Wintgen, "On the total curvature of surfaces in E^4," Coll. Math. vol. xxxix, Fasc. 2 (1978), 289–296.

DEPARTMENT of MATHEMATICS
CAMPUS BOX 1146
WASHINGTON UNIVERSITY
ST. LOUIS, MO 63130

INTERNATIONAL CENTER for
THEORETICAL PHYSICS
STRADA COSTIERA 11, MIRAMARE
34100 TRIESTE, ITALY

Current Address:
Dipartimento di Matematica
Citta Universitaria
Viale A. Doria 6
95125 Catania, Italy

Contemporary Mathematics
Volume **101**, 1989

Affine Differential Geometry of Complex Hypersurfaces*

Weiqi Gao

Abstract. Affine invariants of complex hypersurfaces are defined. These invariants characterize the complex hypersurface up to a complex affine motion.

0. Introduction

The study of affine differential geometry has traditionally focused on hypersurfaces in the real affine space. Although K. Weise [40, 41] had already defined the affine normal space of a n-dimensional manifold immersed into the $(n + k)$-dimensional affine space for $k \leq n(n + 1)/2$ in the 1930's, the study of affine geometry of submanifolds of codimension $k \geq 2$ has been limited, and publications in this direction are few [1, 11, 22, 23, 40, 41, 42]. In recent years, there is a revived interest in affine differential geometry. In two independant studies K. Abe [1] and F. Dillen, L. Vracken and L. Verstaelen [11] introduced theories of complex hypersurfaces in complex affine spaces. They generalized the theory of affine differential geometry of real hypersurfaces to the complex case, and defined the complex affine normal line. Their theories treated a special class of codimension two submanifolds of the affine space. The general codimension two submanifolds were discussed in S. Wilkinson [42], where along K. Weise's line of arguement [40, 41], three affine invarint tensors were constructed, which under some suitable nondegeneracy conditions characterize the submanifold up to a general affine motion.

In this article we use Wilkinson's method to study complex hypersurfaces in the complex affine space. We showed that the affine invariants constructed by Abe [1] and Dillen-Vracken-Verstaelen [11] can actually be used to describe a complex hypersurface up to a complex affine motion.

---

* This paper is in final form and no version of it will be submitted for publication elsewhere.
1980 Mathematical Subject Classification (1985 Revision) 53A13.

The first section outlines the general theory of submanifolds in affine spaces and defines the Wilkinson invariant tensors. Wilkinson's theorems on codimension one and two submanifolds are stated. The second section links the general theory to the complex hypersurface case and proves our main theorem.

1. Submanifolds in the affine space

The $(n+k)$-dimensional real affine space $\mathbf{A}^{n+k}$ is the $(n+k)$-dimensional Euclidean space $\mathbf{R}^{n+k}$ equipt with the standard flat connection. The special affine group $SA(n+k, \mathbf{R})$ consists of all the unimodular affine transformations

$$\underline{A}: \mathbf{R}^{n+k} \to \mathbf{R}^{n+k}$$
$$\underline{A}(x) = Ax + b$$

where $A \in SL(n+k, \mathbf{R})$, $b \in \mathbf{R}^{n+k}$.

Identify the tangent bundle $T(\mathbf{A}^{n+k})$ with $\mathbf{A}^{n+k} \times \mathbf{R}^{n+k}$, then the subbundle

$$LA(n+k, \mathbf{R}) = \{(x, e_1, \dots, e_{n+k}) \mid x \in \mathbf{A}^{n+k}, e_\alpha \in T_x(\mathbf{A}^{n+k}) \cong \mathbf{R}^{n+k}, \alpha = 1, \dots, n+k, \det(e_1, \dots, e_{n+k}) = 1\}$$

is the bundle of affine frames. It is a principal bundle with structure group $SL(n+k, \mathbf{R})$.

The differentials of x and e_α can be written as

$$dx = \omega^\alpha e_\alpha, \qquad de_\alpha = \omega_\alpha^\beta e_\beta, \tag{1}$$

where ω^α and ω_α^β are 1-forms. The summation convention is used here and throughout this article.

The range of the Greek letters is $\{1, \dots, n+k\}$.

The structure equations are obtained by differentiating (1) as well as the equation $\det(e_1, \dots, e_{n+k}) = 1$:

$$d\omega^\alpha = \omega^\beta \wedge \omega_\beta^\alpha, \qquad d\omega_\alpha^\beta = \omega_\alpha^\gamma \wedge \omega_\gamma^\beta, \tag{2}$$

$$\omega_\alpha^\alpha = 0.$$

Let $x\colon M^n \to \mathbf{A}^{n+k}$ be a smooth immersion of a n-dimensional manifold into $\mathbf{A}^{n+k}$. Since we shall study only local properties, we may assume that x is an embedding. An affine frame field defined along x(M) is adapted to M if its first n legs span the tangent space of M. Using an adapted affine frame field on M, we have $\omega^A = 0$, $A = n+1, \dots, n+k$, since dx is in the space spanned by $e_1, \dots, e_n$. Differentiating these equations and using the structure equations (2) we get

$$\omega^\beta \wedge \omega_\beta^A = \omega^i \wedge \omega_i^A = 0,$$

where i runs in the set $\{1, \dots, n\}$.
By Cartan's lemma there are functions h_{ij}^A, symmetric in i and j, such that

$$\omega_i^A = h_{ij}^A\, \omega^j. \tag{3}$$

Differentiating (3), we have

$$\omega_i^\gamma \wedge \omega_\gamma^A = (d\, h_{ij}^A) \wedge \omega^j + h_{ij}^A\, \omega^\gamma \wedge \omega_\gamma^j$$

i. e.,

$$(h_{jk}^A\, \omega_i^j + h_{ij}^A\, \omega_k^j - d\, h_{ik}^A - h_{ik}^B\, \omega_B^A) \wedge \omega^k = 0.$$

Thus by Cartan's lemma, there are functions h_{ikl}^A symmetric in k and l such that

$$h_{jk}^A\, \omega_i^j + h_{ij}^A\, \omega_k^j - d\, h_{ik}^A - h_{ik}^B\, \omega_B^A = h_{ikl}^A \omega^l. \tag{4}$$

Clearly h_{ikl}^A is symmetric in all its subscripts.
Differentiating (4) and using Cartan's lemma again, we find functions h_{iklm}^A, symmetric in its subscripts, such that

$$h_{ikj}^A\, \omega_l^j + h_{ijl}^A\, \omega_k^j + h_{jkl}^A\, \omega_i^j - d\, h_{ikl}^A - h_{ikl}^B\, \omega_B^A - (h_{ij}^A\, h_{kl}^B + h_{kj}^A\, h_{il}^B + h_{lj}^A\, h_{ik}^B)\, \omega_B^j \tag{5}$$
$$= h_{iklm}^A\, \omega^m.$$

Now the various nondegeneracy conditions can be formulated. First we consider the second order invariants. In [41], K. Weise defined the quantity T through the formulas

$$\Delta^{A_1 \dots A_n} = \det \begin{pmatrix} h_{11}^{A_1} & \dots & h_{1n}^{A_n} \\ & \dots & \\ h_{n1}^{A_1} & \dots & h_{nn}^{A_n} \end{pmatrix},$$

$$S_{\underline{i \dots n}\ \underline{i \dots n}}^{A_1 \dots A_n} = \frac{1}{n!} \sum_{\sigma \in S_n} \sigma(\Delta^{A_1 \dots A_n}),$$

$$T_{\underline{i \dots n}\ \underline{i \dots n}\ \underline{i \dots n}\ \underline{i \dots n}}^{A_1 \dots A_{2n}} = \frac{1}{(2n)!} \sum_{\sigma \in S_{2n}} \sigma(S_{\underline{i \dots n}\ \underline{i \dots n}}^{A_1 \dots A_n} \times S_{\underline{i \dots n}\ \underline{i \dots n}}^{A_{n+1} \dots A_{2n}}),$$

$$T = \sum_{\sigma's \in S_k} \text{sgn}(\sigma_1) \dots \text{sgn}(\sigma_{2n-1})\ T_{\underline{1 \dots n}\ \underline{1 \dots n}\ \underline{1 \dots n}\ \underline{1 \dots n}}^{n+1\ \sigma_1(n+1) \dots \sigma_{2n-1}(n+1)} \dots T_{\underline{1 \dots n}\ \underline{1 \dots n}\ \underline{1 \dots n}\ \underline{1 \dots n}}^{n+k\ \sigma_1(n+k) \dots \sigma_{2n-1}(n+k)}.$$

where S_n, S_{2n} and S_k are the permutation groups of the set $\{1, \dots, n\}$, $\{1, \dots, 2n\}$, and $\{n + 1, \dots, n + k\}$ respectively. Weise showed that the non-vanishing of T is invariant by the affine frame change $(e_1, \dots, e_{n+k}) \to (e_1', \dots, e_{n+k}')$ given by $e_\alpha' = a_\alpha^\beta e_\beta$ with $\det(a_\alpha^\beta) = 1$, $\det(a_i^j) = 1$, and $a_i^\beta = 0$. Such an affine frame change is called adapted to M. Moreover, if $T \ne 0$, then h_{ij}^A has a complement $\check{h}_A^{ij}$ in the sense that

$$(6) \qquad \check{h}_A^{ij}\ h_{jk}^A = k\ \delta_k^i$$

$$\check{h}_A^{ij}\ h_{ij}^B = n\ \delta_A^B.$$

Assume $T \ne 0$. Assume further that the tensor

$$E_{mB}^{Al} = \delta_B^A\ \delta_m^l + \frac{2}{nk}\ \check{h}_B^{ij}\ \check{h}_C^{kl}\ h_{im}^C\ h_{jk}^A,$$

has a complement $\check{E}_{mB}^{Al}$ in the sense that

$$(7) \qquad \check{E}_{lC}^{Bk}\ E_{mB}^{Al} = \delta_C^A\ \delta_m^k.$$

This condition is satisfied if and only if the quantity $E = \det(E_{m+n(B-1)}^{l+n(A-1)}) \ne 0$. Then after a suitable affine frame change adapted to M, we have the Weise's apolarity condition

(8) $$\check{h}^{ij}_A \check{h}^{kl}_B h^B_{ijk} = 0.$$

Under this condition, the space spanned by the vectors $e_{n+1}, \dots, e_{n+k}$ at each point of M is well-defined. This space is called the affine normal space. Weise showed that a necessary condition for $T \neq 0$ is $k \leq n(n+1)/2$ and conjectured that the condition $E \neq 0$ is not an additional assumption at all. In the hypersurface case E is actually a constant depending only on n.

The best way to describe the nondegeneracy condition for the third order invariants is to look at the cubic form

(9) $$P = h^A_{ijk}\, \omega^i \otimes \omega^j \otimes \omega^k \otimes e_A,$$

called the Fubini-Pick form. This form can be viewed as a linear transformation in several ways:

(10) $$P: S^3(TM) \to N,$$

(11) $$P: S^2(TM) \to T^*M \otimes N,$$

(12) $$P: TM \to S^2(T^*M) \otimes N,$$

where S^3 and S^2 indicates symmetric tensor products and N is the affine normal space. Hence we have several different nondegeneracy conditions by requiring any one of the above linear transformations to have full rank.

In [42] Wilkinson defined three invariant tensors on M:

(13) $$A^{ij}_{kl} = \check{h}^{ij}_A h^A_{kl},$$

(14) $$B^{ij}_{klm} = \check{h}^{ij}_A h^A_{klm},$$

(15) $$C^{ij}_{klmn} = \check{h}^{ij}_A h^a_{klmn}$$

and proved that when $k = 1$ or 2 these invariants, under suitable nondegeneracy conditions, characterize the immersion up to a general affine motion. Specifically, when $k = 2$ and $n \geq 3$, Wilkinson also assumed that the matrices (h^{n+1}_{ij}) and (h^{2n+2}_{ij}) are simultaneously diagonolized so that

(16) $$h^A_{ij} = \delta_{ij} h^A_i.$$

Then the condition

(17) $$h_i^{n+2} h_j^{n+1} \neq h_j^{n+2} h_i^{n+1}, \quad i \neq j,$$

implies both conditions (6) and (7). And the following theorem is true

Theorem. If $x, \tilde{x}: M^n \to \mathbf{A}^{n+2}$ are two immersions satisfying $n \geq 3$ and the nondegeneracy conditions (16) and (17), and if $A = \tilde{A}$, $B = \tilde{B}$, then there exist $D \in GL(n+2, \mathbf{R})$ and $b \in \mathbf{R}^{n+2}$ such that locally $\tilde{x} = Dx + b$.

When $k = 2 = n$ or $k = 1$, a nondegeneracy condition for the third order invariants as well as the fourth order information is needed.

Theorem. If $x, \tilde{x}: M^2 \to \mathbf{A}^4$ are two immersions for which (16) and (17) hold and either (10) or (12) has full rank, and if $A = \tilde{A}$, $B = \tilde{B}$, and $C = \tilde{C}$, then there exist $D \in GL(4, \mathbf{R})$ and $b \in \mathbf{R}^4$ such that locally $\tilde{x} = Dx + b$.

For hypersurfaces, conditions (6) and (7) is equivalent to $\det(h_{ij}^{n+1}) \neq 0$.

Theorem. If $x, \tilde{x}: M^n \to \mathbf{A}^{n+1}$ are two immersions for which $\det(h_{ij}^{n+1}) \neq 0$ and (11) has full rank, and if $A = \tilde{A}$, $B = \tilde{B}$, and $C = \tilde{C}$, then there exist $D \in GL(n+1, \mathbf{R})$ and $b \in \mathbf{R}^{n+1}$ such that locally $\tilde{x} = Dx + b$.

Theorem. If $x, \tilde{x}: M^n \to \mathbf{A}^{n+1}$ are two strictly convex immersions for which (10) has full rank, and if $A = \tilde{A}$, $B = \tilde{B}$, and $C = \tilde{C}$, then there exist $D \in GL(n+1, \mathbf{R})$ and $b \in \mathbf{R}^{n+1}$ such that locally $\tilde{x} = Dx + b$.

2. Complex hypersurfaces

In the canonical basis $e_1, \ldots, e_{2n+2}$ of $\mathbf{R}^{2n+2}$, the standard complex structure $J: \mathbf{R}^{2n+2} \to \mathbf{R}^{2n+2}$ is given by $J e_\alpha = e_{n+1+\alpha}$, $J e_{n+1+\alpha} = -e_\alpha$. We identify $\mathbf{C}^{n+1}$ with the $+i$ eigenspace of J spanned by the basis $v_\alpha = \frac{1}{2}(e_\alpha - i J e_\alpha)$, $\alpha = 1, \ldots, n+1$, in $\mathbf{R}^{2n+2} \otimes_{\mathbf{R}} \mathbf{C}$, then the map $\iota: \mathbf{R}^{2n+2} \to \mathbf{C}^{n+1}$, $\iota(x) = \frac{1}{2}(x - i J x)$ sends $(x_1, \ldots, x_{n+1}, y_1, \ldots, y_{n+1}) \in \mathbf{R}^{2n+2}$ to $(x_1+iy_1, \ldots, x_{n+1}+iy_{n+1}) \in \mathbf{C}^{n+1}$. Then the linear transformations in $SL(2n+2, \mathbf{R})$ that preserves the complex structure form a subgroup $SA_0(n+1, \mathbf{C})$ of the complex general linear group $GL(n+1, \mathbf{C})$, consisting of linear transformations A that satisfies $|\det A| = 1$. The complex affine transformation group $SA(n+1, \mathbf{C})$ is then defined as the semi-direct product of $SA_0(n+1, \mathbf{C})$ and $\mathbf{C}^{n+1}$. Thus any $\underline{A} \in SA(n+1, \mathbf{C})$ is of the form

$$\underline{A}: \mathbf{C}^{n+1} \to \mathbf{C}^{n+1}$$

$$\underline{A}(x) = Ax + b$$

where $A \in SA_0(n+1, \mathbf{C})$, and $b \in \mathbf{C}^{n+1}$.

Now we consider a holomorphic immersion $\chi: M \to A^{n+1}$, where M is an n-dimensional complex manifold and A^{n+1} is the $(n+1)$-dimensional complex affine space. Since χ is holomorphic, the complex tangent space at a point $p \in M$ is mapped into a complex subspace of the complex tangent space at $\chi(p) \in A^{n+1}$. Thus an (real) affine frame field $e_1, \ldots, e_n, e_{n+2}, \ldots, e_{2n+1}, e_{n+1}, e_{2n+2}$ can be found along M such that

$$J e_\alpha = e_{n+1+\alpha}, \quad \alpha = 1, \ldots, n+1 \tag{18}$$

and that the field is adapted to M. Then $v_\alpha = \frac{1}{2}(e_\alpha - i\, e_{n+1+\alpha})$, $\alpha = 1, \ldots, n+1$ satisfies $|\det(v_1, \ldots, v_{n+1})| = 1$ and $v_1, \ldots, v_n$ are tangential to M. We call such a field complex affine frame field adapted to M. Note that any such fields arises in the above manner.

Let $\omega^1, \ldots, \omega^n, \omega^{n+2}, \ldots, \omega^{2n+1}, \omega^{n+1}, \omega^{2n+2}$ be the dual basis of the above frame $e_1, \ldots, e_n, e_{n+2}, \ldots, e_{2n+1}, e_{n+1}, e_{2n+2}$, then the dual of $v_1, \ldots, v_{n+1}$ is given by $\psi^\alpha = \omega^\alpha + i\, \omega^{n+1+\alpha}$, $\alpha = 1, \ldots, n+1$. If $x: M \to \mathbf{R}^{2n+2}$ is the underlying smooth immersion, than all the discussion of section one can be applied. The condition (18) implies the following relations among the various quantities:

$$\omega_\beta^{n+1+\alpha} = -\,\omega_{n+1+\beta}^{\alpha}, \tag{19}$$

$$\omega_{n+1+\beta}^{n+1+\alpha} = \omega_\beta^\alpha,$$

$$h_{ij}^{n+1} = h_{n+1+i\, j}^{2n+2}, \tag{20}$$

$$h_{i\, n+1+j}^{n+1} = h_{n+1+i\, n+1+j}^{2n+2},$$

$$h_{n+1+i\, j}^{n+1} = -\,h_{ij}^{2n+2},$$

$$h^{n+1}_{n+1+i\, n+1+j} = - h^{2n+2}_{i\, n+1+j},$$

(21) $$h^{n+1}_{abk} = h^{2n+2}_{ab\, n+1+k},$$

$$h^{n+1}_{ab\, n+1+k} = - h^{2n+2}_{abk},$$

(22) $$h^{n+1}_{abck} = h^{2n+2}_{abc\, n+1+k},$$

$$ha^{n+1}_{abc\, n+1+k} = - h^{2n+2}_{abck},$$

where a, b, c runs in the set $\{1, ..., n, n + 2, ..., 2n + 1\}$.

To put all these information into complex form, we let

(23) $$v_\alpha = \frac{1}{2}(e_\alpha - i\; e_{n+1+\alpha}),$$

$$\psi^\alpha = \omega^\alpha + i\, \omega^{n+1+\alpha},$$

$$\psi^\beta_\alpha = \omega^\beta_\alpha + i\, \omega^{n+1+\beta}_\alpha,$$

$$h_{ij} = h^{n+1}_{ij} + i\, h^{2n+2}_{ij},$$

$$h_{ijk} = h^{n+1}_{ijk} + i\, h^{2n+2}_{ijk},$$

$$h_{ijkl} = h^{n+1}_{ijkl} + i\, h^{2n+2}_{ijkl}.$$

Then equations (1) – (5) takes the simpler form

(24) $$d\chi = \psi^\alpha v_\alpha,$$

$$dv_\alpha = \psi^\beta_\alpha v_\beta,$$

$$d\psi^\alpha = \psi^\beta \wedge \psi\beta^\alpha_\beta,$$

$$d\psi^\beta_\alpha = \psi^\gamma_\alpha \wedge \psi^\beta_\gamma,$$

$$\psi^{n+1}_i = h_{ij}\, \psi^j,$$

$$h_{ij}\, \psi^j_k + h_{jk}\, \psi^j_i - d\, h_{ik} - h_{ik}\, \psi^{n+1}_{n+1} = h_{ikl}\, \psi^l,$$

$$h_{ikj}\, \psi^j_l + h_{ijl}\, \psi^j_k + h_{jkl}\, \psi^j_i - d\, h_{ikl} - h_{ikl}\, \psi^{n+1}_{n+1}$$

$$- (h_{ij}\, h_{kl} + h_{kj}\, h_{il} + h_{lj}\, h_{ik})\, \psi^j_{n+1} = h_{iklm}\, \psi^m.$$

This is exactly the generalization to the complex case the formulas for real hypersurfaces.

Denote by H^{n+1} and H^{2n+2} the matrices (h^{n+1}_{ab}) and (h^{2n+2}_{ab}) and $\check{H}_{n+1}$ and $\check{H}_{2n+2}$ the matrices $(\check{h}^{ab}_{n+1})$ and $(\check{h}^{ab}_{2n+2})$. Then the first Weise nondegeneracy condition (6) becomes

$$\text{(6')} \qquad \begin{aligned} &\check{H}_{n+1} H^{n+1} + \check{H}_{2n+2} H^{2n+2} = 2\, I_{2n}, \\ &\operatorname{tr} \check{H}_{n+1} H^{n+1} = \operatorname{tr} \check{H}_{2n+2} H^{2n+2} = 2n, \\ &\operatorname{tr} \check{H}_{n+1} H^{2n+2} = \operatorname{tr} \check{H}_{2n+2} H^{n+1} = 0, \end{aligned}$$

where I_{2n} denotes the $(2n) \times (2n)$ identity matrix. By (20) we have $H^{2n+2} = - J H^{n+1}$ where J is the matrix

$$\begin{pmatrix} 0 & I_n \\ -I_n & 0 \end{pmatrix}.$$

Hence

$$\frac{1}{2} (\check{H}_{n+1} - \check{H}_{2n+2} J) H^{n+1} = I_{2n}.$$

Similarly $H^{n+1} = J H^{n+2}$ and

$$\frac{1}{2} (\check{H}_{2n+2} + \check{H}_{n+1} J) H^{2n+2} = I_{2n}.$$

Therefore H^{n+1} and H^{2n+2} are both invertible.

Let H be the matrix (h_{ij}), then H is invertible if and only if

$$\text{(25)} \qquad \det \begin{pmatrix} h^{n+1}_{ij} & h^{2n+2}_{ij} \\ -h^{2n+2}_{ij} & h^{n+1}_{ij} \end{pmatrix} \neq 0.$$

This is equivalent to

$$\det \begin{pmatrix} h^{n+1}_{ij} & h^{n+1}_{i\, n+1+j} \\ h^{n+1}_{i\, n+1+j} & -h^{n+1}_{ij} \end{pmatrix} \neq 0,$$

i.e., H^{n+1} (hence $H^{2n+2} = - J H^{n+1}$) is invertible. Moreover, if H_{n+1} and H_{2n+2} are the inverses of H^{n+1} and H^{2n+2} respectively, then

$$H_{n+1} H^{n+1} + H_{2n+2} H^{2n+2} = 2\, I_{2n},$$

$$\operatorname{tr} H_{n+1} H^{n+1} = \operatorname{tr} H_{2n+2} H^{2n+2} = 2n,$$

$$\operatorname{tr} H_{n+1} H^{2n+2} = \operatorname{tr} J = 0 = \operatorname{tr} H_{2n+2} H^{n+1},$$

i.e., the first Weise nondegeneracy condition (6) is satisfied with $\breve{H}_A = H_A$, $A = n+1, n+2$. Thus the tensors A^{ij}_{kl}, B^{ij}_{klm}, and C^{ij}_{klmn} are the real parts of the holomorphic tensors $\breve{h}^{ij} h_{kl}$, $\breve{h}^{ij} h_{klm}$, and $\breve{h}^{ij} h_{klmn}$ respectively.Assume (25), the tensor E^{Ad}_{cB} has the form

$$(E^{Ad}_{cB}) = \begin{cases} (1 + \frac{1}{n}) I_{2n}, & \text{if } A = B, \\ -\frac{1}{n} J, & \text{if } A = n+1,\ B = 2n+2, \\ \frac{1}{n} J, & \text{if } A = 2n+2,\ B = n+1. \end{cases}$$

The tensor $\breve{E}^{Ad}_{cB}$ given by

$$(\breve{E}^{Ad}_{cB}) = \begin{cases} \frac{n+1}{n+2} I_{2n}, & \text{if } A = B, \\ \frac{1}{n+2} J, & \text{if } A = n+1,\ B = 2n+2, \\ -\frac{1}{n+2} J, & \text{if } A = 2n+2,\ B = n+1. \end{cases}$$

is an inverse to (E^{Ad}_{cB}) in the sense of (7).

This shows that the Weise nondegeneracy condition (6) is equivalent to the nondegeneracy condition (25), and condition (7) follows either of the condiotns (6) or (25) Assume the holomorphic immersion χ satisfies condition (25). Then an invariant affine normal space is defined by the Weise apolarity condition (8), which can now be written

(8') $$h^{ij} h_{ijk} = 0,$$

where h^{ij} is the element of the inverse matrix H^{-1} of H.

If e_{n+1} and e_{2n+2} span the invariant affine normal space, then replace e_{n+1} and e_{2n+2} by $J e_{n+1}$ and $J e_{2n+2}$ in the adapted complex affine frame, (8') still holds. Hence the invariant affine normal space is a complex subspace of the tangent space. This space is called the complex affine normal line, and we can choose a basis e_{n+1} and e_{2n+2} for it such that $e_{2n+2} = J e_{n+1}$.

Mow that we have chosen the normal vectors canonically, we would like to choose an adapted complex frame field on M with a suitable tangential part so that the matrices H^{n+1} and

H^{2n+2} have a simpler form. Notice that these are symmetric matrices satisfying $J^{-1} H^{n+1} J = - H^{n+1}$ and $H^{n+1} H^{2n+2} = - H^{2n+2} H^{n+1}$. A linear algebraic arguement will show that we can choose an complex affine frame adapted to M so that

(26) $$H^{n+1} = I_{n,n} = \begin{pmatrix} I_n & 0 \\ 0 & -I_n \end{pmatrix},$$

$$H^{2n+2} = J_{n,n} = \begin{pmatrix} 0 & I_n \\ I_n & 0 \end{pmatrix}.$$

Using such a frame, (4) becomes

(27) $$\omega_i^k + \omega_k^i = h_{ikl}^{n+1} \omega^l, \quad i \neq k,$$

$$-\omega_i^{n+1+k} + \omega_{n+1+k}^i = h_{i\, n+1+k\, l}^{n+1} \omega^l, \quad i \neq k,$$

$$2\omega_{\underline{i}}^{i} - \omega_{n+1}^{n+1} = h_{iil}^{n+1} \omega^l,$$

$$-\omega_{\underline{i}}^{n+1+i} + \omega_{n+1+\underline{i}}^{i} - \omega_{2n+2}^{n+1} = h_{i\, n+1+i\, l}^{n+1} \omega^l.$$

In the last two equations the index i is not summed. We indicate this by underlining the i's that is not summed.

From the third equation we have

(28) $$\omega_{n+1}^{n+1} = \frac{1}{n} \left(2 \sum_{i=1}^{n} \omega_i^i - \sum_{i=1}^{n} h_{iil}^{n+1} \omega^l\right).$$

The fourth equation is equivalent to

$$2\, \omega_{n+1+\underline{i}}^{i} - \omega_{2n+2}^{n+1} = h_{i\, n+1+i\, l}^{n+1} \omega^l.$$

Hence

(29) $$\omega_{2n+2}^{n+1} = \frac{1}{n} \left(2 \sum_{i=1}^{n} \omega_{n+1+i}^{i} - \sum_{i=1}^{n} h_{i\, n+1+i\, l}^{n+1} \omega^l\right).$$

This shows that the normal connection ω^{n+1}_{n+1} and ω^{n+1}_{2n+2} are determined by the tangential connection and the h^A_{ij}'s and the h^A_{ikl}'s.

The apolarity condition (8) implies that the 1-forms ω^a_A are determined by the h^A_{ij}'s, the h^A_{ikl}'s and the h^A_{iklm}'s (see Wilkinson [42]).

Combining (27), (28), (29) we get

$$\text{(30)} \qquad \omega^k_i + \omega^i_k = h^{n+1}_{ikl}\,\omega^l, \ i \neq k,$$

$$-\omega^{n+1+k}_i + \omega^i_{n+1+k} = h^{n+1}_{i\,n+1+k\,l}\,\omega^l, \ i \neq k,$$

$$2\,\omega^i_{\underline{i}} - \frac{1}{n}\,(2\sum_{i=1}^n \omega^i_i - \sum_{i=1}^n h^{n+1}_{iil}\,\omega^l) = h^{n+1}_{iil}\,\omega^l,$$

$$2\,\omega^i_{n+1+\underline{i}} = \frac{1}{n}\,(2\sum_{i=1}^n \omega^i_{n+1+i} - \sum_{i=1}^n h^{n+1}_{i\,n+1+i\,l}\,\omega^l) = h^{n+1}_{i\,n+1+i\,l}\,\omega^l.$$

Let ω^b_a and $\tilde{\omega}^b_a$ be two tangential connections that satisfy

$$d\omega^a = \omega^b \wedge \omega^a_b$$

and the equationa (4) and (19) (thus also (27) and (30). Then $(\omega^a_b - \tilde{\omega}^a_b) \wedge \omega^b = 0$. By Cartan's lemma there exist functions u^a_{bc}, symmetric in b, c, such that $\omega^a_b - \tilde{\omega}^a_b = u^a_{bc}\,\omega^c$. Substract from (30) the same equations with $\tilde{\omega}^a_b$ we get

$$\text{(31)} \qquad u^k_{ic} + u^i_{kc} = 0, \ i \neq k,$$

$$-u^{n+1+k}_{ic} + u^i_{n+1+k\,c} = 0, \ i \neq k,$$

$$2\,u^i_{\underline{i}c} - \frac{2}{n}\sum_{i=1}^n u^i_{ic} = 0,$$

$$2\,u^i_{n+1+\underline{i}\,c} - \frac{2}{n}\sum_{i=1}^n u^i_{n+1+i\,c} = 0,$$

while equations (19) give us

$$u^{i}_{j\,n+1+k} = -\,u^{n+1+i}_{jk} = u^{i}_{n+1+j\,k} = u^{n+1+i}_{n+1+j\,n+1+k},$$

$$u^{i}_{n+1+j\,n+1+k} = -\,u^{n+1+i}_{j\,n+1+k} = -\,u^{i}_{jk} = -\,u^{n+1+i}_{n+1+j\,k}.$$

From these equations we have, for i, j, k distinct,

$$u^{i}_{jk} = 0,\quad u^{i}_{j\,n+1+k} = 0,$$

also

$$u^{i}_{ic} = u^{j}_{jc},\quad u^{i}_{n+1+i\,c} = u^{j}_{n+1+j\,c},\quad u^{k}_{ii} = -\,u^{i}_{ik},\quad u^{n+1+k}_{ii} = u^{i}_{i\,n+1+k}.$$

Thus u^{i}_{ic} is independant of i, therefore we can simply denote u^{i}_{ic} by u_c. Then $u^{i}_{n+1+i\,j} = u_{n+1+j}$, $u^{i}_{n+1+i\,n+1+j} = -\,u_{n+1+j}$, $u^{k}_{ii} = -\,u_k$, $u^{n+1+k}_{ii} = u_{n+1+k}$.

We would like to have all the u_c's zero. In order to get this, we consider equation (5). Substruct from the equation (5) the same equation with $\tilde{\omega}^{b}_{a}$, and recall that ω^{b}_{B} is determined by the h^{A}_{ij}'s, the h^{A}_{ikl}'s and the h^{A}_{iklm}'s, we get

$$h^{n+1}_{abd}\,u^{d}_{ce} + h^{n+1}_{adc}\,u^{d}_{bc} + h^{n+1}_{dbc}\,u^{d}_{ae} - \frac{2}{n}\,h^{n+1}_{abc}\sum_{i=1}^{n} u^{i}_{ie} - \frac{2}{n}h^{2n+2}_{abc}\sum_{i=1}^{n} u^{i}_{n+1+i\,e} = 0.$$

In particular

$$h^{n+1}_{ijl}\,u^{l}_{ke} + h^{n+1}_{ij\,n+1+l}\,u^{n+1+l}_{ke} + h^{n+1}_{ilk}\,u^{l}_{je} + h^{n+1}_{i\,n+1+l\,k}\,u^{n+1+l}_{je} + h^{n+1}_{ljk}\,u^{l}_{ic}$$
$$+ h^{n+1}_{n+1+l\,jk}\,u^{n+1+l}_{ie} - \frac{2}{n}\,h^{n+1}_{ijk}\sum_{i=1}^{n} u^{i}_{ie} - \frac{2}{n}h^{2n+2}_{ijk}\sum_{i=1}^{n} u^{i}_{n+1+i\,e} = 0.$$

For $e = m \neq i, j, k$, we get

$$h^{n+1}_{ijk}\,u_m + h^{n+1}_{ijm}\,u_k + h^{n+1}_{imk}\,u_j + h^{n+1}_{mjk}\,u_i + h^{2n+2}_{ijk}\,u_{n+1+m} \tag{32}$$
$$+ h^{2n+2}_{ijm}\,u_{n+1+k} + h^{2n+2}_{imk}\,u_{n+1+j} + h^{2n+2}_{mjk}\,u_{n+1+i} = 0.$$

For $e = n+1+m$, $m \neq i, j, k$, we get

(33) $$h^{n+1}_{ijk}\, u_{n+1+m} + h^{n+1}_{ijm}\, u_{n+1+k} + h^{n+1}_{imk}\, u_{n+1+j} + h^{n+1}_{mjk}\, u_{n+1+i}$$
$$- h^{2n+2}_{ijk}\, u_m - h^{2n+2}_{ijm}\, u_k - h^{2n+2}_{imk}\, u_j - h^{2n+2}_{mjk}\, u_i = 0.$$

If h^A_{abc} has an inverse in the sense that there exists a complement tensor $\check{h}^{abc}_A$ such that (20) type relation exists and

(34) $$\check{h}^{ijk}_A\, h^A_{ijl} = \delta^k_l,$$
$$\check{h}^{ijk}_A\, h^B_{ijl} = \check{h}^{ijk}_B\, h^A_{ijl},$$

then the system of equations (32) and (33) yield $u_c = 0$, i.e., $u^i_{ic} = 0$, for all c. Thus we get $u^a_{bc} = 0$ for all a, b, c, therefore $\omega^a_b = \tilde{\omega}^a_b$. Therefore all the 1-forms ω^α_β are uniquely determined by the h^A_{ij}'s, the h^A_{ikl}'s and the h^A_{iklm}'s. From these facts and the techniques of [42], we have the following

Theorem. If $\chi, \tilde{\chi} : M \to \mathbf{A}^{n+1}$ are two holomorphic immersions satisfying $n \geq 2$ and the nondegeneracy conditions (25) and (34), and if $A = \tilde{A}$, $B = \tilde{B}$, and $C = \tilde{C}$, then there exist $D \in SA_0(n+1, \mathbf{C})$ and $b \in \mathbf{C}^{n+1}$ such that locally $\tilde{\chi} = D\chi + b$.

For the proof we refer the reader to Wilkinson [42, theorem 2.4].

Bibliography

[1] Abe, K., Affine geometry of complex hypersurfaces I, preprint.

[2] Berwald, L., Die Grundgleichungen der Hyperflächen in Euklidischen $\mathbf{R}_{n+1}$ gegenüber den inhaltstreuen Affinitäten, Monatschefte fur Math. und Phys. **32** (1922), 89-106.

[3] Blaschke, W., Vorlesungen über Differentialgeometie II, Springer, Berlin, 1923.

[4] Calabi, E., Improper affine hyperspheres of convex type and a generalization of a theorem by K. Joergens, Mich. Math. J. **5** (1958), 105-126.

[5] Calabi, E., Hypersurfaces with maximal affinely invariant area, Amer. J. Math. **104** (1982), 91-126.

[6] Cheng, S.-Y. and Yau, S.-T., Complete affine hypersurfaces. Part I. The completeness of affine metrics, Comm. Pure & Appl. Math. **39** (1986), 839-886.

[7] Chern, S.-S., Affine minimal hypersurfaces, minimal submanifolds anf geodesics, Kaigai Publications, Tokyo, 1978.

[8] Deicke, A., Über die Finsler-Räume mit $A_i = 0$, Arch. Math. **4** (1953), 45-51.

[9] Deprez, J., Semi-parallel hypersurfaces, Rend. Semin. Mat. Torino. **44** (1986), 303-316.

[10] Dillen, F., The complex version of a theorem by Berwald, preprint.

[11] Dillen, F., Vracken, L. and Verstaelen, L., Complex affine differential geometry, preprint.

[12] Flanders, H., Local theory of affine hypersurfaces, J. d'Analyse Math. **15** (1965), 353-387.

[13] Gao, W. and Wong, P.-M., The unimodular Monge-Ampere equation I: the two dimensional cese, preprint.

[14] Gigana, S., On a conjecture by E. Calabi, Geom. Dedicata **11** (1981), 387-396.

[15] Gruber, P. und Höbinger, J., Kennzeichnungen von Ellipsoiden mit Anwendungen, Jahrbuch Uberblicke Math. 9-29, Mannheim 1976.

[16] Guggenheimer, H. W., Differential geometry, McGraw-Hill, New Nork, 1963.

[17] Gunning, R. C. and Rossi, H., Analytic functions of several complex variables, Prentice-Hall, Englewood Cliffs, NJ, 1965.

[18] Hano, J. and Nomizu, K., On isometric immersions of the hyperbolic plane into the Lorentz-Minkowski space and the Monge-Ampere equation of a certain type, Math. Ann. **262** (1983), 245-253.

[19] Harris, S. G. and Nomizu, K., On the convexity of spacelike hypersurfaces with nonpositive curvatures, Geom. Dedicata **13** (1983), 347-350

[20] Jensen, G., Higher order contact of submanifolds of homogeneous spaces, Lecture notes in math. 610, Springer-Verlag, New York, 1977.

[21] Jörgens, K., Über die Lösungen der Differentialgleichung $rt - s^2 = 1$, Math. Ann. **127** (1954), 130-134.

[22] Klingenberg, W., Zur affinen Differentialgeometrie. Teil I: Über p-dimensionale Minimalflächen und Sphären im n-dimensionalen Raum, Math. Z. **54** (1951), 65-80.

[23] Klingenberg, K., Zur affinen Differentialgeometrie. Teil II: Über zweidimensionale Flächen im vierdimensionalen Raum, Math. Z. **54** (1951), 184-216.

[24] Nakajima, S., Über die Isoperimetrie der Ellipsoide und Eiflächen mit konstanter mittlerer Affinkrümmung im (n+1)-dimensionalen Raum, Japan J. Math. **2** (1927), 193-196.

[25] Nomizu, K., What is affine differential geometry? Proc. Conf. on Differential Geometry, Munster, 1982, 42-43.

[26] Nomizu, K., On completeness in affine differential geometry, Geom. Dedicata **20** (1986), 43-49.

[27] Petty, C. M., Ellipsoids, convexity and its applications, Collect. Surv. (1983), 264-276.

[28] Pogorelov, A. V., On the improper affine hyperspheres, Geom. Dedicata **1** (1972), 33-46.

[29] Reilly, R. C., Affine geometry and the form of the equation of a hyperdurface, Rocky Mount. J. Math. **16** (1986), 553-565.

[30] Sasaki, T., On affine isoperimetric inequality for a strongly convex closed hypersurface in the unimodular affine space A_{n+1}, Komamoto J. Sci. (Math), **16** (1984), 23-38.

[31] Schirokov, P. A. and A. P., Affine Differentialgeometrie, Teubner, Leipzig, 1962.

[32] Schneider, R., Zur affinen Differentialgeometrie im Grossen I, Math. Z. **101** (1967), 375-406.

[33] Schneider, R., Zur affinen Differentialgeometrie im Grossen II: Über eine Abschatzung der Pickschen Invariante auf Affinsphären, Math. Z. **102** (1967), 1-8.

[34] Schwenk, A., Eigenwertprobleme des Laplace-Operators und Anwendungen auf Untermagnigfaltigkeiten, Dissertation TU Berlin 1984.

[35] Schwenk, A. and Simon, U., Hypersurfaces with constant equiaffine mean curvature, Arch. Math. **46** (1986), 85-90.

[36] Simon, U., Minkowskische Integralformeln und ihre Anwendungen in der Differentialgeometrie im Grossen, Math. Ann. **173** (1967), 307-321.

[37] Simon, U., Zur Relativegeometrie: Symmetrische Zusammenhänge auf Hyperflächen, Math. Z. **106** (1968), 36-46.

[38] Simon, U., Hypersurfaces in equiaffine differential geometry, Geom. Dedicata **17** (1984), 157-168.

[39] Spivak, M., A comprehensive intruduction to differential geometry. III, Publish or Perish, Berkeley, 1979.

[40] Weise, K., Der Berührungstensor zweier Flächen und die Affingeometrie der F_p im A_m. (Teil I), Math. Z. **43** (1938), 469-480.

[41] Weise, K., Der Berührungstensor zweier Flächen und die Affingeometrie der F_p im A_m. (Teil II), Math. Z. **44** (1939), 161-184.

[42] Wilkinson, W., General affine differential geometry for low codimension immersions, Math. Z. **197** (1988), 583-594.

[43] Yau, C.-M., The affine comornal of convex hypersurfaces, preprint.

DEPARTMENT OF MATHEMATICS
UNIVERSITY OF NOTRE DAME
NOTRE DAME, IN 46556

Contemporary Mathematics
Volume **101**, 1989

DOMAINS WITH NONCOMPACT AUTOMORPHISM GROUPS

Kang-Tae Kim

ABSTRACT: We introduce a version of so-called scaling technique which may be widely applied in the study of domains with locally convex boundries and with noncompact automorphism groups. A generalization of the theorem of R. Greene and S. Krantz is stated and proved. Moreover, we state more general theorems which extend the theorems of B. Wong ([10]), Rosay ([9]), R. Greene and S. Krantz ([5]), and sketch the ideas of the proof.

1. INTRODUCTION. By an automorphism of a domain Ω in $\mathbb{C}^n$, we mean a one-to-one holomorphic mapping from Ω onto itself. We denote by $\mathrm{Aut}(\Omega)$ the group of all the automorphisms of Ω. This group is known to be a Lie group with the topology of uniform convergence on compact subsets.

Our starting point is the following well-known theorem of B. Wong ([9]) and J.-P. Rosay [10]):

1980 Mathematics Subject Classification (1985 Revision). 32F15

THEOREM. Any domain in $\mathbb{C}^n$ with a C^2 smooth strongly pseudoconvex boundary point at which an orbit of an interior point under the automorphism group action accumulates is biholomorphic to a ball.

This beautiful theorem plays an important role in the study of the strongly pseudoconvex domains and their automorphism groups (cf. [4]). Our goal is to generalize this theorem to a broader collection $\mathcal{C}_{2k}$ of domains in $\mathbb{C}^n$ defined as follows:

$\Omega \in \mathcal{C}_{2k}$ if and only if it is a bounded domain satisfying the following four conditions:

(0) $0 \in \mathbb{C}^n$ is a boundary point of Ω.

(1) $\partial\Omega$ is real analytic near 0 and is of finite type $2k$ at 0. Here we use the notion of finite type due to J. D'Angelo ([1]).

(2) There exist an open neighborhood V of 0 in $\mathbb{C}^n$ and a one-to-one holomorphic mapping $F : V \to \mathbb{C}^n$ such that $F(V \cap \Omega)$ is convex.

(3) There exist a point $q \in \Omega$ and a sequence $\{f_j\}$ of automorphisms of Ω such that $\lim_{j\to\infty} f_j(q) = 0$ in $\mathbb{C}^n$.

In this category of domains, we have:

THEOREM 1. Let Ω_1 and Ω_2 be members of $\mathcal{C}_{2k}$. If there exist open neighborhoods U_1 and U_2 of 0 in $\mathbb{C}^n$ and a biholomorphic mapping $f : U_1 \to U_2$ with

$f(0) = 0$ and with $f(U_1 \cap \partial\Omega_1)$ coinciding with $U_2 \cap \partial\Omega_2$ up to and including $2k$-th order terms in the Taylor expansions of their defining functions at 0, then Ω_1 and Ω_2 are biholomorphic to each other.

This theorem is not a consequence of Wong-Rosay theorem because the collection $\mathfrak{C}_{2k}$ contains many weakly pseudoconvex domains. Of course, this theorem, as it is stated, is not a direct generalization of Wong-Rosay theorem because of the real analyticity assumption. However, in case when the domain Ω is strongly pseudoconvex at 0 the same technique we will discuss to prove this result in the later chapters will prove Wong-Rosay theorem without assuming any extra smoothness of $\partial\Omega$ at 0 beyond C^2 regularity. Also notice that the condition (0) in the definition of $\mathfrak{C}$ is for the sake of simplicity on notations only. On the other hand (2) is rather a strong assumption. It entered because of a technical reason one may see in the following chapter. However, we conjecture that Theorem 1 would be proved without convexity condition. Also, the real analyticity condition should be replaced by C^{2k} smoothness of $\partial\Omega$ at 0. However, we do not know how to prove these at this moment. We would like to mention, also, that R. Greene and S. Krantz proved a weaker form of this theorem much earlier which was the major inspiration in proving this result. Actually we will prove the following a bit improved version of their theorem:

THEOREM. (Greene-Krantz [5]). Let $D_{m_2 \cdots m_n}$ be the bounded domain in $\mathbb{C}^n$ defined by

$$D_{m_2 \cdots m_n} := \{(z_1, \ldots, z_n) \mid |z_1|^2 + |z_2|^{2m_2} + \ldots + |z_n|^{2m_n} < 1\}.$$

Let $\Omega \subset \mathbb{C}^n$ be a complete hyperbolic domain satisfying:

(1) $p = (1, 0, \ldots, 0) \in \partial\Omega$.

(2) there are neighborhoods U and V of p in $\mathbb{C}^n$ such that, up to a biholomorphism, $U \cap \partial\Omega$ and $V \cap \partial D_{m_2 \cdots m_n}$ coincide.

If there are a sequence of points q_j converging to $q \in \Omega$ and a sequence of automorphisms φ_j of Ω such that $\lim_{j \to \infty} \varphi_j(q_j) = p$, then Ω is biholomorphic to $D_{m_2 \cdots m_n}$.

Moreover, we would like to point out that one can actually get more as in the following theorem:

THEOREM 2. Let Ω be a bounded domain belonging to the collection $\mathcal{C}_{2\ell}$ for some ℓ . Then Ω is biholomorphic to the domain defined by the inequality

$$0 > \operatorname{Re} z_1 + P_{m_2}(z_2) + \ldots + P_{m_2}(z_n)$$

$$+ \sum_{i_2, \ldots, i_n \geq 0} Q_{i_2 \cdots i_n}(z_2, \ldots, z_n)$$

where:

(1) P_{m_k} $(k = 2, \ldots, m)$ are positive real valued homogeneous polynomials of degree $m_k \leq 2\ell$;

(2) $Q_{i_2 \ldots i_n}$ are either identically 0 or real homogeneous polynomials of degree Σi_ℓ with fixed degree i_ℓ in variables $z_\ell, \bar{z}_\ell$ for each ℓ ; and

(3) $(i_2, \ldots, i_n)$ varies over the set of $(n-1)$-tuples of non-negative integers, at least two of whose entries are non-zero, satisfying

$$\frac{i_2}{m_2} + \ldots + \frac{i_n}{m_n} = 1 .$$

In this paper, we introduce a new technique and give alternative proofs to Wong-Rosay theorem as well as the theorem of Greene-Krantz.

These proofs are simpler and more intuitive than the original ones. However, the proofs of Theorems 1 and 2 are rather complicated and long, and the detailed proof will appear somewhere else. In this paper, though, we try to give at least the ideas through the proofs of Wong-Rosay theorem and the theorem of Greene-Krantz.

2. LOCALIZATION AND SCALING. Now we introduce a technique to prove Theorems 1 and 2. As one sees from the statments, most conditions are local at the boundary point 0 . So we will need the localization theorem:

LEMMA 1. (cf. [9]) Let Ω be a bounded domain in $\mathbb{C}^n$. Suppose Ω has a C^0 local peaking function at $p \in \partial\Omega$, i.e., a continuous function $\varphi : U \to \mathbb{C}$ defined on a neighborhood U of p in $\mathbb{C}^n$ which is complex analytic on $U \cap \Omega$ satisfying $\varphi(p) = 1, |\varphi(q)| < 1 \ \forall q \in \overline{\Omega} \cap U \setminus \{p\}$. If there is a sequence $\{f_j\} \subset \mathrm{Aut}(\Omega)$ such that $f_j(q_0) \to p$ as $j \to \infty$ for some fixed $q_0 \in \Omega$, then, for any compact set $K \subset\subset \Omega$ and for any $\varepsilon > 0$, there is $N > 0$ such that $f_j(K) \subset B_\epsilon(0) \cap \Omega$ for any $j \geq N$.

This follows from a standard normal family argument and the maximum principle. For a detailed proof, see [9].

Before going any further we introduce another collection $\mathcal{C}$ of bounded domains in $\mathbb{C}^n$ as follows:

A bounded domain Ω belongs to $\mathcal{C}$ if and only if it satisfies (0), (2) and (3) in the definition of $\mathcal{C}_{2k}$ together with the condition

(*) Ω admits a C^0 local peaking function at $0 \in \partial\Omega$.

Notice that $\mathcal{C}$ is a strictly larger set than $\mathcal{C}_{2k}$ for any k. Now using the lemma above we modify the theorem of S. Frankel ([3]) as follows:

THEOREM 3. Let the bounded domain Ω in $\mathbb{C}^n$ belong to $\mathcal{C}$. Then the sequence

$$\{\varphi_j : \Omega \to \mathbb{C}^n \mid \varphi_j(z) = d(F \circ f_j)(q_0)^{-1}(F \circ f_j)(z))\}$$

is a normal family of holomorphic mappings every subsequential limit of which is a biholomorphism into. Here F is as in (2) in the definition of $\mathcal{C}_{2k}$.

In the Chapter 1 of [3], the same conclusion was proved under a stronger assumption that Ω covers a compact quotient holomorphically and their proof was partly based on the existence of an orbit non-tangential to the boundary $\partial\Omega$ of Ω at 0. Here, we do not assume any restriction on the shape of the orbit $\{f_j(q_0)\}$. This mild improvement turns out to be essential in proving the theorems we stated in the previous chapter. Now we begin to prove Theorem 3.

First we repeat the theorem of S. Frankel (Theorem 1, [3]):

(*) The sequence $\{\omega_j(z) := [df_j(q_0)]^{-1}(f_j(z) - f_j(q_0))\}$ is a normal family, assuming that Ω is convex near 0.

To prove this, we consider a compact set $K \subset\subset \Omega$ and the sequence $F_j : K \times K \to \Omega$ defined by

$$F_j(z, \zeta) = \omega_j^{-1}\left(\frac{\omega_j(z) + \omega_j(\zeta)}{2}\right).$$

This is well-defined for large j because of Lemma 1 and (3) in the definition of $\mathcal{C}$. Since $F_j(q_0, q_0) = 0$ for any j, and since Ω is a bounded domain, $\{F_j\}$ is a normal family (cf. page 74, [7]). Comparing the terms in

$\frac{\partial^2}{\partial z_k \partial \zeta_\ell}(\omega_j \circ F_j)$ and $\frac{\partial^2}{\partial z_k \partial z_\ell}(\omega_j \circ F_j)$ at $(z . z)$ one obtains the differential inequality

$$\|D^2 \omega_j\|_\infty \le C_K \|D\omega_j\|_\infty$$

where C_K is a constant depending only on K. Since $D\omega_j(0) = I$, one gets $\|D\omega_j\|_\infty \le \tilde{C}_K$ on K. Since K was chosen arbitrarily, $\{\omega_j\}$ is a normal family.

Also notice that subsequential limits are one-to-one holomorphic. Since $d\omega_j(0) = I$, Hurwitz' theorem on $\det(d\omega_j(z))$ guarantees that subsequential limits of $\{\omega_j\}$ are one-to-one locally. But since ω_j's are 1-1 and since the convergence is uniform on all compact subsets, they are actually 1-1.

Now to finish the proof of Theorem 3, also assuming that Ω is convex near 0 we prove

(∗∗) $\{[df_j(q_0)]^{-1}(f_j(q_0))\}$ <u>is a bounded sequence</u>.

Because of Lemma 1, $\|df_j(q_0)\| \to 0$ as $j \to \infty$. Therefore, in a fixed ball $B_R(0)$ with radius R centered at 0, $B_R(0) \cap \omega_j(\Omega)$ is convex for large j. Hence ω_j has a subsequence $\{\omega_{j_k}\}$ such that $\{\omega_{j_k}\}$ converges to a convex domain in the sense of local Hausdorff convergence of sets. By choosing a subsequence if necessary, we may assume that $\{\omega_{j_k}\}$ converges and write $\lim_{k\to\infty} \omega_{j_k} = \omega$. Then clearly $\omega(\Omega) = \lim \omega_{j_k}(\Omega)$. Since $\omega(q_0) = 0$ and since $\omega(\Omega)$ is convex, if $W_{j_k}(q_0)) \to \infty$ as $k \to \infty$, $\omega(\Omega) = \mathbb{C}^n$. But this will result that $\mathbb{C}^n$ is biholomorphic to a bounded domain, which is absurd.

So far, we pretended as if Ω is actually convex near $0 \in \partial\Omega$. However, it is easy to see that the same argument holds even if we replace f_j by $F \circ f_j$ where F is as in (2) of the definition of $\mathcal{C}_{2k}$.

All the arguments in the above can be summarized into

SCALING LEMMA. Let Ω be a bounded pseudoconvex domain in $\mathbb{C}^n$ belonging to the class $\mathcal{C}$. Then there is a sequence $\{A_j\} \subset GL_n(\mathbb{C})$ such that

(1) $\|A_j^{-1}\| \to 0$ as $j \to \infty$, and

(2) $\lim_{j\to\infty} A_j(\Omega) = \hat{\Omega}$ exists and biholomorphic to Ω, where the limit is taken in the sense of local Hausdorff distances in $\mathbb{C}^n$.

3. CHARACTERIZATION OF OVAL DOMAINS. Here we will give alternative proofs to Wong-Rosay theorem and the theorem of Greene-Krantz. Actually the theorem of Greene-Krantz stated in section 1 is an improvement from their original theorem in [5] since

we eliminated the extra assumption on $\partial\Omega$ its being C^3 smooth. After giving these proofs, we will present a very rough sketch of the proof of the Theorems 1 and 2. We start with the two-dimensional case:

PROOF OF GREENE-KRANTZ'S THEOREM. We prove for the domain

$$E_m = \{(z,\omega) \in \mathbb{C}^2 \,|\, |z+1|^2 + |\omega|^{2m} < 1\}\,.$$

It is enough to show that the scaling of Ω at $(0,0)$ is uniquely and canonically determined, up to biholomorphic equivalences, by the local shape of $\partial\Omega$ near $(0,0)$. From the scaling lemma, we have a sequence

$\{A_j\} \subset GL_2(\mathbb{C})$ with $\|A_j^{-1}\| \to 0$ as $j \to \infty$ such that

$\lim_{j\to\infty} A_j\Omega = \hat{\Omega}$ and $\hat{\Omega}$ is biholomorphic to Ω. For the sake of convenience, we introduce the following notations:

$$A_j^{-1} := (a^j_{\alpha\beta}) = \begin{pmatrix} a^j_{11} & a^j_{12} \\ a^j_{21} & a^j_{22} \end{pmatrix}.$$

$$\Omega_j = \text{"}A_j\Omega\text{"} = A_j(F(\Omega \cap V))\,.$$

(here, F and V are as in (2) of the definition of $\mathcal{C}$)

$$\hat{\Omega} = \lim_{j\to\infty} A_j\Omega\,.$$

Since $F(\Omega \cap V)$ is repreented by $|z+1|^2 + |\omega|^{2m} < 1$ in a small neighborhood of $(0,0)$, we try to scale this inequality by $\{A_j\}$. Rewrite the inequality as

$$2\operatorname{Re} z < -|z|^2 - |\omega|^2 .$$

Hence, Ω_j in a large ball $B_R(0)$ may be represented by

$$2\operatorname{Re}(a^j_{11} z + a^j_{12}\omega) < -|a^j_{11} z + a^j_{12}\omega|^2 - |a^j_{21} z + A^j_{22}\omega|^{2m} .$$

Without loss of generality we may assume that $\{a^j_{12}/a^j_{11}\}$ is bounded. By choosing subsequences if necessary we may also assume that $\frac{a^j_{11}}{|a^j_{11}|} \to \alpha$ as $j \to \infty$. Then we have

$$2\operatorname{Re}\left(\frac{a^j_{11}}{|a^j_{11}|} z + \frac{a^j_{12}}{|a^j_{11}|}\omega\right) < -\left|\frac{a^j_{11}}{\sqrt{|a^j_{11}|}} z + \frac{a^j_{12}}{\sqrt{|a^j_{11}|}}\omega\right|^2$$
$$+\left|\frac{a^j_{21}}{\sqrt[2m]{|a^j_{11}|}} z + \frac{a^j_{22}}{\sqrt[2m]{|a^j_{11}|}}\omega\right|^{2m} .$$

Note that the first term in the right-hand side converges to 0. Because of the scaling lemma, we must have a complete hyperbolic domain as a limit of this process. Thus choosing a subsequence from $\{A_j\}$ if necessary, we will have a limit: $2\operatorname{Re}(\alpha z + \beta\omega) < -|\gamma z + \delta\omega|^{2m}$.

Of course $\det\begin{pmatrix}\alpha & \beta \\ \gamma & \delta\end{pmatrix} \neq 0$, since otherwise this limit would not be hyperbolic in Kobayashi's sense. Hence Ω is biholomorphic to the domain in $\mathbb{C}^2$ defined by

$2\operatorname{Re}\zeta < -|\xi|^{2m}$, which is known to be biholomorphic to E_m. Now the same argument proves the theorem of Greene-Krantz in its full version as stated in Section 1.

Actually from this argument one obtains the following compactness result:

COROLLARY. Let Ω be a bounded domain in $\mathbb{C}^n$ with boundary point p such that there exist an open neighborhood U of p in $\mathbb{C}^n$ and a biholomorphic mapping $F : U \to \mathbb{C}^n$ with $F(p) = (1,0,\ldots,0)$ and with $F(U \cap \Omega)$ coinciding with

$$D_{m_1 \cdots m_n} = \{(z_1,\ldots,z_n) \mid |z_1|^{2m_1} + \ldots + |z_n|^{2m_n} < 1\}$$

near $(1,0,\ldots,0)$ where $m_j \geq 2$ for all j. Then either Ω is biholomorphic to some E_m or there do not exist a point $q \in \Omega$ and a sequence $\{\varphi_j\} \subset \operatorname{Aut}\Omega$ such that $\lim_{j\to\infty} \varphi_j(q) = (1,0,\ldots,0)$.

A slightly different result is recently given by A. Kodama in this direction. His theorem is that with set theoretic coincidence between Ω and $D_{m_1\cdots m_n}$ (with all $m_j \geq 2$) instead of the biholomorphic coincidence as above at $(1,0,\ldots,0)$ one derives that Ω does not admit a sequence

$$\{\varphi_j\} \subset \operatorname{Aut}(\Omega) \text{ such that } \lim_{j\to\infty} \varphi_j(q) = (1,0,\ldots,0)$$

for some $q \in \Omega$.

Now we will prove Wong-Rosay theorem: Let Ω be C^2 strongly pseudoconvex at $(0,0) \in \partial\Omega$. Then it is known that after some holomorphic change of coordinates at $(0,0)$, Ω may be strongly convex. Hence one can find constants C_1 and C_2 such that

$G_1 \subset \Omega \subset G_2$ near $(0,0)$ where $G_1 := \{|z_1+1|^2+C_1|z'|^2 < 1\}$ and

$G_2 := \{|z_1+1|^2+C_2|z'|^2 < 1\}$. Here $z' = (z_2,\ldots,z_n)$ and

$|z'|^2 = |z_2|^2+\ldots+|z_n|^2$. Then in a fixed large ball,

$A_jG_1 \subset \Omega_j \subset A_jG_2$ for j large. By scaling lemma, Ω_j will converge to a complete hyperbolic domain. Hence neither can $\lim\limits_{j\to\infty} A_jG_2$ be of lower dimensional set, nor $\lim\limits_{j\to\infty} A_jG_1$ gets too large to contain a complex line. So as before we will

$$\text{get } \frac{a^j_{1k}}{|a^j_{11}|} \to \alpha_k, \frac{a^j_{\ell k}}{\sqrt{|a^j_{11}|}} \to \gamma_{\ell k}, (\ell \geq 2) \text{ with } \det\begin{pmatrix}\alpha\\ \gamma\end{pmatrix} \neq 0 .$$

Then, as before, we scale Ω to get the result.

For the Theorems 1 and 2, we do basically the same thing. One can show that the scaled domain is uniquely determined, up to biholomorphic equivalences, only by the local shape of the boundary of the original domains. The crux of the matter is to show that there are fixed ratios between the rows of the matrices A_j^{-1} in the scaling. This is long and complicated. However, once one shows that, scaling the domains once the proofs will be complete. Detailed proofs [6] will appear somewhere else.

BIBLIOGRAPHY

1. J. D'Angelo, "Real Hypersurfaces, orders of contact, applications", Ann. Math., 115 (1982), 615-637.

2. E. Bedofrd and S. Pinčuk, "Domains in $\mathbb{C}^2$ with a noncompact automorphism group", Preprint.

3. S. Frankel, "Bounded convex domains with compact quotients are symmetric spaces in complex dimension two", Thesis, Stanford Univ., California, 1986.

4. R. Greene and S. Krantz, "Deformation of complex structures, estimates for the $\bar{\partial}$ equation, and stability of Bergman kernel", Advances in Math., 43 (1982), 1-86.

5. ———, "Characterization of certain weakly pseudoconvex domains with noncompact automorphism groups", Complex Analysis, Lecture notes in Math., v. 1268, Springer-Verlag, N.Y., 1987.

6. K. T. Kim, "Complete localization of domains with noncompact automorphism groups", Preprint.

7. S. Kobayashi, "Hyperbolic manifolds and holomorphic mappings", Marcel-Dekker, New York, 1972.

8. S. Pinčuk, "Holomorphic inequivalences of some classes of domains in $\mathbb{C}^n$", Math. USSR Sbornik, no. 1, 39 (1981).

9. J.-P. Rosay, "Sur une caracterisation de la boule parmi les domaines de $\mathbb{C}^n$ par son group d'automorphismes", Ann. Inst. Fourier (Grenoble), no. 4, 29 (1979), 91-97.

10. B. Wong, "Characterization of the unit ball in $\mathbb{C}^n$ by its automorphism group". Invent. Math. 41 (1977), 253-257.

DEPARTMENT OF MATHEMATICS
BROWN UNIVERSITY
PROVIDENCE, RHODE ISLAND 02912

Contemporary Mathematics
Volume **101**, 1989

AFFINE APPROACH TO COMPLEX GEOMETRY

Sidney Frankel

Abstract. We briefly discuss our results on convex domains with compact quotients in the more general context of applications of affine geometry to geometric function theory in several complex variables. The emphasis here is on connections between complex analysis and affine geometry. We survey the relation between our techniques and those used in earlier works on related problems.

1 Introduction

Our goal in this paper is to survey some emerging interaction of geometric function theory in several complex variables, with affine and projective geometry. More specifically, we will

1. Introduce the reader to the rescale blow up technique, [6,7]. We emphasize here that our technique provides a method for reducing problems involving bounded domains in several complex variables, to problems in affine geometry.

2. We sketch the application to convex domains with compact quotients from [7]. In particular, we show how to produce continuous families of automorphisms

3. We suggest that one can develop a coherent general theory of bounded domains based on affine geometry. A detailed account is begun in [8]. An important part of this theory is presented in terms of the questions of existence, uniqueness, comparison and 'structure' of the rescale blow up . We briefly indicate some aspects in this paper.

4. We discuss a body of results in geometric function theory of several complex variables, from the period 1974-1984, with specific reference to papers of B. Wong, Rosay, Greene

1980 Mathematics Subject Classification (1985 Revision). 32H20.

and Krantz, Pinchuk, Bedford and Fornaess and Sibony. These papers all involve the idea of *boundary localization* in some form or other. This idea underlies the rescale blow up technique. We focus on how the rescale blow up technique can be applied to each, sketching proofs of some of these results in terms of this technique. This motivates us to suggest some non-trivial generalizations of these results. The recent work of K.T. Kim is a good example of how rescale blow up leads to very general results. We hope that our presentation of the above results as applications of affine geometry illustrates some of the 'unifying effect' of our approach.

5. We briefly discuss a body of results in affine and projective geometry from the period 1952-1972. We are especially interested in the papers of Kuiper, Benzecri, Koszul, Vey and Katz-Vinberg. In particular the example of Katz-Vinberg gives a convex, hyperbolic cone domain, which is not homogeneous, but with a co-compact holomorphic automorphism group. These papers all use some aspect of rescale blow up . It seems that the complex analysis community was not aware of this work, despite its relevance, and that they independently rediscovered the boundary localization idea.

   W. Goldman has recently produced a very clear set of notes on affine and projective geometry covering among other things Benzecri's work, and including a useful bibliography. We thank him for bringing this literature to our attention.

6. Topics that come to mind, involving complex analysis and affine geometry, that we do not discuss here include the theory of the Monge-Ampere equation and the remarkable result of Lempert, [25], stating that for any convex domain Ω the Kobayashi and Caratheodory metrics are identical. Topics such as boundary values of bounded holomorphic functions and extension of holomorphic embeddings to the boundary are not discussed here, but are taken up in [8] .

   In his report to the International Congress, [26], Lempert explains how his result nicely parallels some classical results in affine geometry. The latter was all but forgotten by 'modern mathematics'. We refer the reader to Lempert's article which illustrates how over the last thirty years many ideas from classical affine geometry were independently rediscovered by complex analysts.

   The first mentioned paper of Lempert motivated this author to investigate automorphism groups of convex domains. We eventually did find a technique that leads to very general results, theorem 2.1, though it does not involve any application of Lempert's work, it does rely on 'convexity' in the usual sense. This is surprising because in complex analysis pseudoconvexity is widely regarded as the natural concept.

I would like to thank Jonathan Goodman for many useful conversations and for educating me on the subject of Poincare normal forms and Hartman-Grossman theory, K.T. Kim for working through my thesis and motivating me to simplify, and generalize the techniques there, Y. T. Siu for encouragement and guidance as well as many useful suggestions, and S. Paik for being a perceptive listener.

2 Main Results

The affine blow-up was developed to prove the main result of [7] ;

Theorem 2.1 *Let Ω be a convex bounded domain in $\mathbf{C}^n$ and suppose there is a subgroup $\Gamma \subset Aut(\Omega)$ such that*

(Γ1). Γ is discrete and fixed-point free

(Γ2). Γ is co-compact (in Ω)

then Ω is biholomorphic to a bounded symmetric domain .

This confirms a conjecture cited by Yau in [38], p.140.

The proof is in two parts. Much of this paper is concerned with the type of affine geometry which we use in the proof of theorem 2.2. We discuss the proof in §6 after covering some preparatory material.

Theorem 2.2 *Let Ω be a convex bounded domain in $\mathbf{C}^n$ and suppose there is a subgroup $\Gamma \subset Aut(\Omega)$ such that*

(Γ2). Γ is co-compact (in Ω)

then $\mathrm{Aut}_0(\Omega\ )$ *is non-trivial, in fact there is a convex holomorphic embedding $w : \Omega \to \mathbf{C}^n$* such that $w(\Omega)$ *is invariant under a 1-parameter group of translations .*

The second part of the proof does not involve the affine blow-up technique. It is devoted to proving a theorem in the complex differential geometry of compact manifolds , which we apply together with the affine blow-up technique to prove our main result. The main result of the second part is;

Theorem 2.3 *Let Ω be a convex hyperbolic domain in $\mathbf{C}^n$ and suppose there is a subgroup $\Gamma \subset Aut(\Omega)$ such that*

(Γ1). Γ is discrete and fixed-point free

(Γ2). Γ is co-compact (in Ω)

then if $\mathrm{Aut}_0(\Omega)$ *is non-trivial, Ω has a factor* Ω_1 such that

1. $\Omega = \Omega_1 \times \Omega_2$ *where Ω_1 is non-trivial and is biholomorphic to a bounded symmetric domain .*
2. *Γ has a finite index normal subgroup* Γ' such that $\Gamma' = \Gamma_1' \times \Gamma_2'$ *where* $\Gamma' \subset \mathrm{Aut}(\Omega_1) \times \mathrm{Aut}(\Omega_2)$ *and* $\Gamma_j' = \Gamma' \cap \mathrm{Aut}(\Omega_j)$.

The second part of the proof is another long story which we don't elaborate on in this paper. It uses alot of Lie group theory, together with intrinsic metrics. Classification theorems such as the Iwasawa decomposition and the Levi-Malcev decomposition, and structure of nilpotent groups are applied. This is natural to expect, as we are trying to find a canonical model for a domain Ω with a continuous automorphism group. In the future we hope to investigate the more direct links of this group theory to the affine geometry of domains. This should lead to simplifications of the proof.

Other sections of part two use the theory of discrete subgroups of Lie groups, and group cohomology. Siu recently suggested a way of eliminating the former by using the stability of bundles with Hermitian Einstein metrics (Kobayashi-Lubke).

We refer the reader to [7] for details of the proof of theorem 2.3. There will also be presentations by Siu appearing in [36].

The example of Katz and Vinberg enables us to construct a convex, Kobayashi hyperbolic cone domain Ω in $\mathbf{C}^3$, which is not homogeneous but does have a co-compact holomorphic automorphism group. Ω has a bounded embedding in $\mathbf{C}^3$. This shows, quite surprisingly, that the hypothesis that Γ is discrete is essential, even to show that Ω is homogeneous. It is the subtlety of exploiting this hypothesis that accounts for the difficulty in proving part two. W. Goldman has examined a moduli space of domains with co-compact groups, generalizing the constructions of Katz and Vinberg in a geometric fashion, [10].

In my thesis I did the complex two dimensional case of part two, and in [7] the general case. Part one of the proof was done in the thesis for any dimension.

3 The Basic Constructions

We begin our presentation with definitions of some constructs in affine geometry, (from our preprints);

The basic concepts can be adapted to affine or projective geometry over the real or complex numbers. For the sake of brevity, we restrict ourselves here to affine geometry over **C** . This suffices for applications to several complex variables.

Definition 3.1 *The affine groups;*

$\mathcal{A}(n, \mathbf{C}) = \{A : \mathbf{C}^n \to \mathbf{C}^n,\ Ax = Mx + b,\ M \in \mathrm{GL}(\mathrm{n}, \mathbf{C}), b \in \mathbf{C}^n\}$,

$\mathcal{A}(n, \mathbf{R}) = \{A : \mathbf{R}^n \to \mathbf{R}^n,\ Ax = Mx + b,\ M \in \mathrm{GL}(\mathrm{n}, \mathbf{R}), b \in \mathbf{R}^n\}$,

We say a domain $\Omega \subset \mathbf{C}^n$ is complex affine hyperbolic if any complex affine-linear embedding of the complex line into Ω,

$$z \mapsto z \cdot A + B$$

for $A, B \in \mathbf{C}^n$ is a constant map. Equivalently, given a complex (affine-linear) line L in $\mathbf{C}^n$, $L - L \cap \Omega$ contains at least two points.

There is an analogous notion of real affine hyperbolic, the term 'sharp' is used in affine geometry literature.

Proposition 3.2 *The following are equivalent;*

1. *Ω is convex and Kobayashi hyperbolic.*

2. *Ω is convex and complex affine hyperbolic.*

3. *Ω is convex, and there is a bounded holomorphic embedding $w : \Omega \to \mathbf{C}^n$.*

4. *For all L (as above), $L \cap \Omega$ is convex and $L - L \cap \Omega \neq \emptyset$.*

(See p.7 of [7] .)

Definition 3.3 *Given*

1. *domains Ω, $\hat{\Omega}$, both open, convex and affine-hyperbolic ,*

2. *$A_i \in \mathcal{A}(n; \mathbf{C})$ (the affine group)* such that $\Omega_i = A_i\Omega \to \hat{\Omega}$

3. *$p \in \partial\Omega$, $x_i \in \Omega$* such that $x_i \to p$, $A_i x_i = Z_0$,

   (or $A_i x_i \to Z_0$, this makes no essential difference).

We say that $\hat{\Omega}$ is an affine blow-up of Ω , by A_i, (or by x_i), at p. We also use the terms *rescale blow up* or *just blow up*.

We can also consider 'real affine blow-ups' where the group $A(n,\mathbf{R})$ replaces $A(n,\mathbf{C})$, and where Ω ,$\hat{\Omega}$ are 'real affine-hyperbolic '.

In §6 we give a holomorphic version of this construction along with motivation for and history of the concept. The main idea of the next definition is to single out sequences whose asymptotic properties are determined in a simple way by the germ of $\partial\Omega$ at p.

Definition 3.4 *Given a blow-up as above, the sequence A_i is affine admissible (for Ω) if the sequence A_ip is bounded.*

The sequence x_i is affine admissible if for any *sequence A_i that blows-up Ω at x_i, A_i is* affine admissible.

$U \subset \Omega$ is a domain of affine admissible approach (to p) if for any *sequence $x_i \in$ U* such that *$x_i \to p$, x_i is affine admissible. We say that $x_i \to p = 0$ is radial* if *there is a vector v and $t_i \in \mathbf{R}$* such that *$x_i = t_iv$ and A_i is radial if there is a radial sequence x_i* such that *A_i blows-up Ω at x_i.*

For our purposes, radial approach is no stronger a hypothesis than 'approaching within a cone C on p' (such that $C \cap B(p, \epsilon) \subset \Omega$), we use them interchangeably here. There are affine admissible A_i that aren't radial, in particular when Ω is a cone, but these are in some sense exceptional. For sufficiently regular boundaries there is an inward pointing normal that is intrinsic to the affine structure. Any admissible blow up is radial with respect to this direction.

Now we can pose the basic problems, note that these are problems in affine geometry. These issues are discussed in detail in [8] .

1. Given a sequence $x_i \in \Omega$, does there exist a blow-up A_i of Ω at x_i? The answer is yes if Ω is convex near p. The existence proof amounts to a simple geometric compactness theorem which we prove in [8] ;

   Theorem 3.5 *The space of pointed, affine-hyperbolic convex domains in $\mathbf{C}^n$, modulo affine equivalence, $(A(n,\mathbf{C}))$, is compact with respect to local Hausdorff distance.*

   The proof only uses elementary affine geometry. (Hausdorff distance is used, but no measure theory, or Hausdorff measure. Though no geometric measure theory is involved, we hope that some real geometric measure theory will eventually be applied to get better results/ weaker hypotheses here.) This theorem is a simple case of Gromov compactness, [13].

2. Given a sequence $x_i \in \Omega$, is the blow-up at x_i unique? *i.e.* given a pair $\hat{\Omega}_j$ of blow-ups of Ω at x_i, is there an affine A such that $\hat{\Omega}_1 = A\hat{\Omega}_2$? The answer is yes.

3. Fix p and let $x_i \to p$ be any admissible sequence , what are necessary and sufficient conditions on Ω , near p, for the blow-up to be independent of the sequence?

4. Let x_i be any admissible sequence , what are necessary and sufficient conditions on Ω for the blow-up to be independent of the sequence? Benzecri considered this in the context of real projective geometry, he showed that a sufficient condition is that the projective automorphiom group G has a compact fundamental domain $K \subset \Omega$. (We call such G a *co-compact group.*) Benzecri used a technique that we call 'the blow-up by automorphisms'. We will describe it below. This is covered in the notes by W. Goldman (from a slightly different point of view).

5. Given a pair of domains Ω_i such that $p \in \partial\Omega_i$, what are necessary and sufficient conditions on Ω_i near p for their blow-ups $\hat{\Omega}_i$ to be identical? This is what we call the 'comparison problem', the analysis is essentially the same as that involved in the uniqueness problems. This amounts to considering the significance of various notions of osculation.

6. What special properties do domains $\hat{\Omega}$, obtained by blowing-up, enjoy? Using Zorn's lemma one can produce an $\hat{\Omega}$ which enjoys the property that it can be blown up to itself. We suggest these should be called 'self reproducing domains'. The most natural question to investigate is; understand the automorphism group of $\hat{\Omega}$. There are examples, of affine blow-ups over $\mathbf{R}$ exhibiting all kinds of pathology (compared to the real analytic case). If a domain Ω is 'sufficiently regular' then $\hat{\Omega}$ admits automorphisms , this is closely related to uniqueness of blow-ups.

The most surprising fact is the existence of continuous groups of automorphisms for any affine blow up over $\mathbf{C}$, theorem 2.2.

4 Examples of Blow-ups

We describe some basic examples of blow-ups, these examples are important as motivation.

Example 4.1 Given $\Omega \subset \mathbf{C}^n$ and $A \in \mathcal{A}(n, \mathbf{C})$ such that $A\Omega = \Omega$ and $x \in \Omega$ such that $A^i x \to p \in \partial\Omega$, let $x_i = A^i x$ and $A_i = A^{-i}$, then $\hat{\Omega} = \Omega$. The simplest example is the upper-half-plane $\{z \in \mathbf{C}^n : \operatorname{Im} z > 0\}$ with $Az = z/2$. Siegel domains [30](1st,2nd,3rd type and 'generalized') give analogous examples in higher dimensions. Since $Ap = p$ these are admissible sequences .

Given a canonical embedding of a bounded symmetric domain , any affine blow up is one of its Siegel domain realizations. The blow up of a strictly pseudoconvex domain is always the Siegel domain realization of the ball.

Example 4.2 Given Ω convex affine-hyperbolic in $\mathbf{C}^1$, for any admissible blow-up of Ω at $p \in \partial\Omega$, *$\hat{\Omega}$ is the tangent cone of Ω at p.* Note that in $\mathbf{C}^1$, the A_i are conformal. One expects admissible blow-ups to be much better behaved than non-admissible blow-ups, but the complex one-dimensional case is an exception, for if the blow-up is not admissible then $\hat{\Omega}$ is simply the upper-half-plane , [8] .

Given Ω convex affine-hyperbolic in $\mathbf{C}^n$, *if the tangent cone of Ω at $p \in \partial\Omega$ is affine-hyperbolic* then for any admissible blow-up of Ω at $p \in \partial\Omega$, $\hat{\Omega}$ is the tangent cone of Ω at p.

By contrast, one can produce pathological examples of blow-ups where Ω has a unique supporting hyperplane at p, but $\hat{\Omega}$ is a non-trivial cone, [8] . These were found in conversations with Jonathan Goodman.

We shall use $diag(A, B)$ to denote a block diagonal matrix with blocks A, B.

Example 4.3 Suppose A_i^j blow up $\Omega^j, j = 1, 2$ then $diag(A_i^1, A_i^2)$ blows up $\Omega = \Omega^1 \times \Omega^2$ to $\hat{\Omega} = \hat{\Omega}^1 \times \hat{\Omega}^2$, also $diag(A_i^1, I)$ blows up $\Omega = \Omega^1 \times \Omega^2$ to $\hat{\Omega} = \hat{\Omega}^1 \times \Omega^2$.

Example 4.4 A *tube domain* in $\mathbf{C}^n$ is a domain of the form $\mathbf{R}^n + T$ with $T \subset J\mathbf{R}^n$, where J is the complex structure of $\mathbf{C}^n$. If T is defined by $y > |x|^\alpha, 1 \leq \alpha < \infty$ and $A = diag(1/2, (1/2)^\alpha)$, then example (4.1) applies.

For non-admissible blow-ups $\hat{\Omega}$ would still be a tube domain, with T defined by $y = |x|^2$.

Proof. Note that a convex tube domain is affine-hyperbolic iff the convex tube T is 'real affine-hyperbolic ' *i.e.* $\forall \ell \subset J\mathbf{R}^n$, a real affine line, $\ell - \ell \cap T \neq \emptyset$

Let $q_i \in \partial\Omega$ be colinear with $p = 0$ and x_i. Then non-admissible implies

$$\frac{|q_i - x_i|}{|p - x_i|} \to 0$$

and exploiting the automorphism group of T to send q_i to $q \neq 0$, $\hat{\Omega}$ is just the admissible blow-up of Ω at q. **qed**

For $\alpha = \infty$ the analogous example is $T = \{y > 0, 0 < x < 1\}$, $A = diag(1, (1/2))$, this is biholomorphic to the bidisc, every blow up of the bidisc is (biholomorphic to) the bidisc.

Suppose T is defined by $y > C|x|^\alpha + O(|x|^{\alpha+\epsilon})$, then $A_i = A^{-i}$ blows up Ω and $\hat{\Omega}$ is defined by $y > C|x|^\alpha$. Suppose T is defined by

(1) $$y > \begin{cases} C|x|^\alpha, & x > 0; \\ C'''|x|^{\alpha'}, & x < 0 \end{cases} \qquad (\alpha' > \alpha)$$

Then $\hat{\Omega}$ is defined by

(2) $$y > \begin{cases} C|x|^\alpha, & x > 0; \\ 0, & x < 0 \end{cases}$$

5 Affine-Structure-Functions

We now present a geometric approach to the normal family theorem proved in [6] . We also call it a 'distortion theorem' because it does not presume that target spaces are uniformly bounded. This will enable us to apply the affine blow up in the context of holomorphic geometry.

Let $f : \Omega \to \mathbf{C}^n$ where f is holomorphic , $\Omega \subset \mathbf{C}^n$ is a domain, (open, connected). Let $N_f \subset \Omega \times \Omega$ be $\{(x, y) : \frac{1}{2}(f(x) + f(y)) \in f(\Omega)\}$. If f is an embedding, then N_f is an open neighborhood of the diagonal, $\{(x, x) : x \in \Omega\}$. We define the affine-structure-function $\phi_f : N_f \to \Omega$ by

(3) $$\phi_f(x, y) = f^{-1}(\frac{1}{2}(f(x) + f(y)))$$

If f is an immersion we can still define ϕ_f on a neighborhood of the diagonal, but a maximal domain of definition is not uniquely determined.

We summarize some basic facts;

Lemma 5.1

1. ϕ_f *is holomorphic*

2. $N_f = \Omega \times \Omega$ *iff* f *is an embedding and* $f(\Omega)$ *is convex*

3. *(F1)* $\phi(x,x) = x$

4. *(F2)* $\phi(x,y) = \phi(y,x)$

5. *(F3)* $\phi(x,\phi(y,z)) = \phi(\phi(x,y),\phi(x,z))$

6. *For* $A \in \mathcal{A}(n,\mathbf{C})$, $\phi_{Af} = \phi_f$

7. *For* $\gamma \in \mathrm{Aut}(\ \Omega\ )$, $\phi_{f\gamma}(x,y) = \gamma^{-1}\phi_f(\gamma x,\gamma y)$

8. *If* Ω *is hyperbolic and* N *is a fixed neighborhood of the diagonal, then the class of* ϕ_f such that $N \subset N_f$ *is a normal family, (we restrict* ϕ_f *to* N*).*

Proposition 5.2

1. $\phi_f = \phi_g$ *iff there is an* $A \in \mathcal{A}(n,\mathbf{C})$ such that $\phi_{Af} = \phi_f$

2. *Given* $\phi : N \to \Omega$ *there is an immersion* f such that $\phi_f = \phi$ *iff* ϕ *satisfies the functional equations (Fn),* $n = 1,2,3$.

Proof.

For the first claim, fix $x \in \Omega$, and let

(4)
$$\phi_{f,x}(y) = \phi_f(x,y),\ \phi_{f,x}^{(i)}(y) = \phi_{f,x}(\phi_{f,x}^{(i-1)}(y))$$

be the iterates.

$d\phi_{f,x}(x) = \frac{1}{2}I$ so x is an attracting fixed-point , and the basin of attraction $N_{f,x}$ is a neighborhood of x. Letting $K \subset\subset N_{f,x}$ be a smaller neighborhood of x, $\forall\epsilon\exists i(K,\epsilon)$, such that $\forall i > i(x,\epsilon)$ $\phi_{f,x}^{(i)}(K) \subset B(x,\epsilon)$.

We can assume without loss of generality that $f(x) = 0$, (by choice of coordinates), so $\phi_{f,x}^{(i)}(y) = f^{-1}(\frac{f(y)}{2^i})$. It follows that

(5)
$$df^{-1}(x)f(y) = \lim 2^i\phi_{f,x}^{(i)}(y)$$

Hence $\phi_f = \phi_g$ implies that $f(y) = df(x)dg^{-1}(x)g(y)$ in a neighborhood of x. One easily sees that the same holds on the whole connected component.

For the second claim, we define $\phi_x^{(i)}(y)$ as above, noting that (F1) and (F2) imply $\phi_x(x) = x$ and

(6)
$$d\phi_x(x) = \frac{1}{2}I$$

so the basin of attraction $N_{\phi,x}$ is a neighborhood of x, and

$$\phi(y', z') = \frac{y' + z'}{2} + o(|y'| + |z'|)$$

Iterating (F3) we have

(7) $$(\mathrm{F3}^i) \quad \phi_x^{(i)}(\phi(y,z)) = \phi(\phi_x^{(i)}(y), \phi_x^{(i)}(z))$$

We want to define

(8) $$f(y) = \lim 2^i \phi_x^{(i)}(y)$$

for, if this converges then by F3^i and equation (6) $\phi_{f,x} = \phi$.

We apply Poincare's theory of normal forms, [1]; there are no resonances by equation (6), so ϕ_x is conjugate to its linear part, *i.e.* there is a holomorphic change of coordinates u such that $\phi_x(u(y)) = u(\frac{1}{2}y)$. Working in the coordinates z given by $z = u(y)$, the sequence in equation (8) trivially converges to ϕ_x, so letting $f = u\phi_x u^{-1}$, $\phi_f = \phi$ on N_{ϕ_x}. qed

Remark 5.3

1. The name 'affine-structure-function ' is justified by proposition 5.2 which show there is a one to one correspondence of embeddings modulo the affine group to the affine-structure-functions .
2. The claims of proposition 5.2 do not really depend on the holomorphicity of ϕ, f. They are probably true for C^1 maps. Intuitively, ϕ determines f in a very simple way; choose p_i to be the vertices of an n-simplex, $p_i \mapsto f(p_i) \Rightarrow \phi(\frac{1}{2}(p_i + p_j) = \phi(f(p_i), f(p_j))$ and iterating ϕ determines the map on a dense subset of the simplex.
3. One can begin to set up a dictionary, translating properties of f to properties of ϕ and vice versus. An example is $p \in \partial\Omega$, such that '$f(p) = \infty$', this translates to $x_i \to p \Rightarrow \forall y \in \Omega,\ \phi(x_i, y) \to \partial\Omega$. For example $f(z) = \frac{1}{1-z}$, taking the unit disc to the half-plane, and '$f(1) = \infty$'.

   Points at infinity are especially relevant to blow-ups since the translation groups we produce fix points at infinity. This is potentially useful in parametrizing the translation groups for some applications.

As a simple application of the normal form theory in proposition 5.2 we get a proof of the type of distortion theorem presented in [7] , p.13. The technique of proof offered there has technical advantages over the technique we proceed to present now, but the proof below is more geometrically intuitive.

Theorem 5.4 *Given* $f_i : \Omega \to \mathbf{C}^n$ *holomorphic and* such that $f(0) = 0$, $df(0) = I$, *if* $\bigcap_i N_{f_i} = N$ *is a neighborhood of the diagonal then* f_i *has a subsequence that converges,* $f_{i_j} \to f$, f *is holomorphic and* $N_f \supset N$*. In particular* f *is an (unbranched) immerson.*

Proof. Let $\phi_i = \phi_{f_i}$, by lemma 5.1 there is a convergent subsequence $\phi_{i_j} \to \phi : N \to \Omega$. ϕ is holomorphic and satisfies the functional equations (Fn) of lemma 5.1. By proposition 5.2, there is an f such that $\phi_f = \phi$, $f(0) = 0$, $df(0) = I$ f is holomorphic and $N_f \supset N$. In fact the construction of f from the normal form theory just involves solving linear equations, [1], and we can easily check that $\phi_{i_j} \to \phi \Rightarrow f_{i_j} \to f$ as power series. Uniform convergence on compacta easily follows.

qed

In [7] we also show;

Lemma 5.5 *If in theorem 5.4,* $N = \Omega$, *(i.e. each* f_i *is a convex embedding) then* f *is also a convex embedding , and* $f_i(\Omega) \to f(\Omega)$ *in local Hausdorff distance.*

The notion of *normalization* implicit in theorem 5.4 is very important; f is *normalized* at x if $f(x) = 0, df(x) = I$.

6 Application to Complex Analysis

Now what is the relevance of all this to several complex variables? The notion of a *blow-up by automorphisms* from [7] is our main application of the normal family theorem , in particular given a domain Ω with a compact quotient it shows how to construct a map from Ω to its rescale blow up , $\hat{\Omega}$; choose $x_i \to p$ radially in Ω and $\gamma_i \in \mathrm{Aut}(\ \Omega\ )$ such that $\gamma_i x_i \in K$, where K is a compact fundamental domain for Aut(Ω) . Now let $f_i = A_i \gamma_i^{-1}$, where A_i is affine, chosen such that f_i are normalized at some $x \in K$. By theorem 5.4, f_i form a normal family, also limiting maps f are convex holomorphic *embeddings*, and if $f_i \to f$, then $f(\Omega) = \lim A_i \Omega = \hat{\Omega}$, by lemma 5.5. We emphasize the transition here from holomorphic convergence to affine convergence.

The proof of theorem 2.2 is executed in three steps;

1. Ω is biholomorphic to its affine blow-up, $\hat{\Omega}$. This uses the blow-up by automorphisms defined above, and the normal family theorem of the previous section. Note that the normal family theorem gives an *existence theorem* for the *affine* blow-up.

   In [7] , the blow-up by automorphisms gives a *map* to $\hat{\Omega}$ and the Hurwitz theorem is used to show this is actually biholomorphic .

2. By analyzing the complex one-dimensional affine blow-up we see that $\hat{\Omega}$ can be chosen such that there is a complex affine line L such that $L \cap \hat{\Omega}$ is an affine upper-half-plane

3. Since $\hat{\Omega}$ is convex, and contains an affine upper-half-plane , it admits a translation group.

Now we recall the main ingredients in this proof and discuss some of the underlying intuitions, motivations and predecessors for each, the ingredients are;

1. rescaling and Hausdorff distance, *i.e.* the affine blow up,
2. the normal family theorem ,
3. constructing the map from Ω to $\hat{\Omega}$,
4. showing the map is biholomorphic ,
5. reduction of the problem to affine geometry of $\hat{\Omega}$,
6. showing $\mathrm{Aut}_0(\Omega\ )$ nontrivial.

Remark 6.1 We proceed with remarks on rescaling, and the affine blow up. The main point in using the affine blow-up, is to exploit the interplay of the geometry of $\partial\Omega$ near p, the geometry of $\hat{\Omega}$, and the algebraic properties of the sequence A_i.

The sequence A_i plays a role similar to the dilations of nilpotent lie groups that arise in the theory of singular integral operators due to Stein and his collaborators, however we don't need any boundary smoothness to produce it.

There is a basic relation of the affine blow-up to Poincare's theory of normal forms for diffeomorphisms and the Hartman-Grobman theory,[1]; *if* Ω *is invariant under a diffeomorphism* $\phi : \mathbf{C}^n \to \mathbf{C}^n$ *then the blow-up* $\hat{\Omega}$ *is invariant under the normal form of* ϕ. The estimates for affine blow-ups are quite trivial compared to those for normal forms for diffeomorphisms.

If the sequence

$$df(0)^k f^{-k} \to w$$

converges, and 0 is an attracting fixed-point for f, then

$$wfw^{-1} = \lim df(0)^k f df(0)^{-k} = df(0)$$

i.e. w conjugates f to its linear part, which is thus its Poincare normal form. In general the sequence does not converge, and the existence proof for normal forms is difficult. These ideas were already present in nineteenth century dynamical systems theory.

The idea of the affine blow-up elaborates on the classical idea of constructing tangent cones to varieties by rescaling. Pincuk used rescaling to produce a Siegel domain osculating a strictly pseudoconvex boundary point. In the works of Kuiper and Benzecri, as well as I. Graham some version of affine blow up is hidden in the estimates of intrinsic metrics. We would like to emphasize that affine blow-ups do not involve any automorphisms, this is why we isolate this concept here, rather than combining it with item three above, which does involve automorphisms.

One accomplishment of the blow-up technique is to dispense with all regularity hypotheses, replacing them with convexity hypotheses. Since every bounded symmetric domain has a convex embedding, but only the ball has a smooth embedding, this generalization has geometric significance. Affine geometry is a kind of 'common denominator' for the geometries of the various bounded symmetric domains, since the automorphism group of each has a minimal parabolic (maximal solvable) which acts affinely, in a canonical way, on the Siegel domain realization.

Regarding item 2, we have recently been informed that C. Fitzgerald may have some results along the lines of our distortion theorem for convex embeddings in several variables.

The theorem 3.5 is a generalization of Blaschke's Selection theorem, (completeness of the space of convex sets with respect to Hausdorff distance and compactness of bounded convex sets in $\mathbf{R}^n$ mod affine equivalence). There is an interesting connection between geometric compactness theorems such as theorem 3.5 and normal family theorems in complex analysis. In [7] we used normal family theorems in complex analysis to prove a geometric compactness theorem, lemma 5.5, related to Blaschke's Selection theorem. .

In [8] we use a geometric compactness theorem, theorem 3.5 to derive a normal family theorem. In both cases convexity is a useful hypothesis. We are not sure what the weakest hypothesis is, to guarantee the needed normal family theorems, it is weaker than convexity, but stronger than pseudoconvexity. In B. Wong's theorem strict pseudoconvexity is used.

In the complex one-dimensional case, a recent paper of Rodin and Sullivan, [32], provides an affine geometric proof of the Riemann mapping theorem. It should be possible to reprove distortion theorems as well, by applications of affine geometry.

Remark 6.2 We proceed with some remarks on the 'boundary localization technique', this is the construction of a map from Ω to $\hat{\Omega}$. In the special case of projective automorphisms this construction occurs in the work of Kuiper and Benzecri, and this is the earliest appearance we know of. In their version, one is given two domains, Ω_i such that

1. Ω_1 osculates Ω_2 at a mutual boundary point p and
2. $\exists \gamma_i^j \in Aut(\Omega_i)$ such that for $x \in \Omega_i$, $\gamma_i^j x \to p$, $j \to \infty$. *i.e.* p is an accumulation point for the orbit, in each domain.

Using estimates on an intrinsic metric, one extracts subsequences of automorphisms such that

$$f^k = \gamma_2^k (\gamma_1^k)^{-1} \to f$$

uniformly on compact sets.

Wong, using estimates on an intrinsic metric due to I. Graham, discovered how to use automorphisms and a hypothesis on the shape of the boundary to characterize biholomorphism types of bounded domains. This was refined and developed by many authors, especially Rosay and Greene-Krantz who dubbed it the 'boundary localization technique'. The *form* of this construction discovered by Greene-Krantz is almost identical to Benzecri's, the details are more involved because holomorphic geometry is 'less rigid' than projective geometry.

We would like to emphasize that these writers only compared pairs of domains where *both* have noncompact automorphism groups, whereas in [7] only one domain is assumed, a priori, to have a noncompact automorphism group. This makes a big difference in the type of results one can obtain. One should also note that we don't use intrinsic metrics to produce the map.

Greene and Krantz saw how to use this biholomorphic map systematically to study automorphism groups, however all of these authors were confined to studying very special cases, where $\hat{\Omega}$ is the ball, or an oval, because they had no general compactness theorem. Furthermore, their paradigm was always to compare a *well known model space* to a space osculating it, or exhausted by copies of it. They did not consider constructing a model space ($\hat{\Omega}$) out of the given space (Ω) by a blow-up, and investigating the structure of $\hat{\Omega}$. Benzecri does pose some questions in this spirit, he was specifically interested in the *space of pointed domains.*

The blow-up construction is best behaved for domains with very regular boundary. The following rough idea provides some motivation for the rescale blow-up with automorphisms;

1. Suppose $\gamma \in \mathrm{Aut}(\ \Omega\ )$ with no interior fixed point. If $\partial\Omega$ is real analytic then γ extends real analytically to $\partial\Omega$, and by the Brouwer fixed-point theorem γ has a fixed-point p on $\partial\Omega$. The extension theorem is due to Kohn, and Bell and Ligocka.

2. γ has a (formal) Poincare normal form at p. Note that Ω is convex and γ-invariant. If $A = d\gamma(p)$ is attracting then Ω is in the basin of attraction of γ.

3. Suppose A is attracting, then using a version of the normal family theorem $w_k = A^{-k}\gamma^{(k)} \to w$ real analytically as $k \to \infty$, and $w\gamma w^{-1} = A$, *i.e.* w conjugates γ to its linear part. In particular there are no resonances. $w(\Omega)$ is the blow-up of Ω by $\gamma^{(k)}$. Actually one has to be careful that w_k are properly normalized to apply the normal family theorem . It suffices to show $\gamma^k x \to p$ radially. This can be done with some further development of the affine geometry of Ω , closely related to intrinsic metrics for the holomorphic structure of Ω .

4. In the parabolic case where $d\gamma(p) = I$ it would be natural to try using projective transformations, A_k, to normalize w_k. We have not worked this out yet.

Using the affine blow up simplifies the technical details indicated above considerably, and facilitates broad generalizations. Groups of automorphisms are replaced by sequences of automorphisms , allowing one flexibility to choose sequences with convenient properties.

The work on affine geometry seems to have met complex analysis in the papers of J. Vey around 1972 and later in work of Kobayashi, [16]. These exploited analogies between the two fields, but didn't actually apply one to the other. We hope to show that there is a broad potential for applying affine geometry to solve many problems in the geometry of domains.

As an example we suggest the following problem; given a convex domain Ω and an automorphism γ such that for some $x \in \Omega$,

$$\gamma^k x \to \partial\Omega, \ k \to \infty$$

when can we embed $f : \Omega \to \mathbf{C}^n$ such that $f\gamma f^{-1}$ extends to an affine transformation on $\mathbf{C}^n$? Deeper applications should give results in complex analysis, with no affine geometry in the statement, but alot in the proof.

The details of the fourth item appear in the works of Fornaess and Sibony as well as Greene and Krantz. The arguements are elementary, but very tricky. This is not as important as the third item and we don't consider it in detail here.

As far as we know, the last two items are new. We will show in [8] that they have implications for the general theory of the geometry of domains, not only those domains with many automorphisms. In some sense this is because the domains we get by blowing up are dense in the space of convex domains mod affine equivalence. This contrasts with the old point of view where a dense open set of domains have no automorphisms.

To summarize, the affine blow up technique combines old ideas on rescaling and automorphisms with our normal family theorem in a new way. Together with items five and six, which to the best of our knowledge, are new observations, we can prove theorem 2.2, which came as a big surprise.

If theorem 2.2 was the only use of this technique then one may think we were at a dead end. We are confident that future developments will dissuade the reader from such a pessimistic diagnosis, and that elaborations of the technique, involving more affine geometry and theory of quasiisometries, will be useful in almost every aspect of the geometry of bounded domains.

7 More Applications

We proceed to give a brief account of how the works of B. Wong, Rosay, Greene and Krantz, Pinchuk, Bedford, Fornaess and Sibony and Kim are related to affine blow-ups. It seems that understanding this point leads to significant generalizations. The work of Kim provides an example, we discuss this later in the section. The rescale blow up has also been used in branches of geometry and group theory, not involving complex analysis, for example, in Gromov's work on nilpotent groups and in work of Mostow. We do not discuss this here.

In the mid-seventies B. Wong showed

Theorem 7.1 *A smoothly bounded domain Ω with a co-compact group of holomorphic automorphisms is biholomorphic to the ball.*

The technique was further developed by Rosay and others, especially Greene and Krantz. At present the proof has been simplified to the point that it boils down to an affine blow-up technique, along with some well known normal family theorems, and applications of the maximum principle. The boundary localization idea, as it is used here is already present in the works of Kuiper and Benzecri.

The proof of Wong's result is in three steps;

1. There is a $p \in \partial\Omega$ such that $\partial\Omega$ is strictly pseudoconvex at p.

2. We can assume for convenience that, at p, Ω osculates the ball to 3rd order. Thus the affine blow-up $\hat{\Omega}$ of Ω at p is projectively equivalent to the ball. In particular, the affine blow-up of Ω at a strictly pseudoconvex point is biholomorphic to the ball.

3. Since $\hat{\Omega}$ is Kobayashi hyperbolic and both Ω and $\hat{\Omega}$ have cocompact automorphism groups, Ω is biholomorphic to $\hat{\Omega}$. In fact, $\hat{\Omega}$ is a blow up of Ω at p by automorphisms.

For applications to complex analysis we only care if a blow-up is unique up to holomorphic equivalence. Thus we would classify Wong's theorem as a comparison theorem, the first two steps of the proof give existence of a blow-up, the third step is related to uniqueness theorems. Wong's original proof involved estimates for Kobayashi and Caratheodory volumes, due to I. Graham, [11]. The refined proof above doesnt require such estimates, the normal family theorem suffices, and the technical details are thus considerably simplified.

Rosay simplified Wong's proof considerably, and weakened the hypothesis to *noncompact automorphism group* while requiring the boundary be strictly pseudoconvex at an accumulation point for the group. The proof sketched above gives his result as well. Greene and Krantz extended this result to certain types of weakly pseudoconvex boundary points, that occur as accumulation points for the (noncompact) automorphism group of domains of the type $\{(z,w) : |z|^2 + |w|^{2k} = 1\}$. Their approach involved very long hard estimates. We will discuss this briefly below. It turns out that the affine blow-up, with the normal family theorem and a little more affine geometry allows one to circumvent these complications. A very recent paper, by Bedford and Pincuk, deals with a closely related problem.

Given a biholomorphic map $f : \Omega_1 \to \Omega_2$, there is a notion of *blowing up the map f* to a biholomorphic map $\hat{f} : \hat{\Omega}_1 \to \hat{\Omega}_2$, which we discuss in detail in [8] . This idea is implicitly contained in the paper of Pincuk. In the same paper Pincuk used *rescaling* to blow up a domain Ω in the special case of a strictly pseudoconvex point. This paper was brought to my attention recently by Greene and Kim.

Fornaess and Sibony considered exhaustions of a complex space Ω by a sequence of M_i each biholomorphic to a fixed M, such that $M_i \subset M_{i+1}$. In general Ω is not biholomorphic to M. They assume that M has a compact quotient, and prove that if Ω is hyperbolic then it is biholomorphic to M. (This is the preliminary part of their paper.) From our point of view the analysis proceeds in two steps; (We will assume that $M \subset \mathbf{C}^n$ is convex and affine-hyperbolic , it would suffice to assume that the closure of M is covered by charts in $\mathbf{C}^n$ such that ∂M corresponds to convex pieces of boundary in $\mathbf{C}^n$. For a more precise statement see the definition of *h-convex* in [7] .)

1. Ω is biholomorphic to a blow-up of M. Fix $x \in \Omega$, this determines $y_i \in M_i = M$, and Ω is biholomorphic to the blow-up of M by y_i. The interesting point here is that we can drop the hypothesis on Aut(Ω) in the first step, and still get an embedding of Ω to $\mathbf{C}^n$.

2. If M has a compact quotient, then any blow-up of M is biholomorphic to M. In particular, Ω is biholomorphic to M.

The technique in the first step, as usual, is to normalize the domain A_iM at A_iy_i, or the map f_i of A_iM into Ω , at x. To get convergence of the maps one needs the type of normal family theorem proved in [7] or [8] .

The harder part of Fornaess and Sibony's paper is the case where the Kobayashi metric of Ω has submaximal rank. They show that Ω fibers over a holomorphic retract of M. (The importance of 'holomorphic retracts' became clear afterwards, from the work of Lempert.) In the corank one case they get a complete structure theorem. When M is a convex domain, it would be interesting to drop the compact quotient hypothesis as above, and if Kobayashi metric of Ω has submaximal rank one should generalize blow-ups to 'degenerate blow-ups', where $A_i\Omega \to \hat{\Omega}$, but $\hat{\Omega}$ is hyperbolic of submaximal rank.

In the author's thesis and in a conversation with R. E. Greene (Berkeley, June/85) it was suggested that the blow-up technique was well suited to extending the results of Wong, and Greene and Krantz, to a classification of bounded convex domains with real analytic boundary and noncompact automorphism group. This has been achieved by K.T. Kim.

At least to the extent that every such domain has a noncompact realization of the form Im $w > F(z)$, with F weighted homogeneous. There is a further question as to which such domains have *smoothly bounded* realizations. We do not know whether Kim considers this, however [2] have results in this spirit.

The theorems of Greene and Krantz, and Kim, fit in the framework of affine blow-ups as follows;

1. Given a sequence of automorphisms γ_i and $x \in \Omega$ such that $\gamma_i x \to p \in \partial\Omega$ blow-up Ω by $x_i = \gamma_i x$. The normal family theorem above guarantees the existence of a blow-up $\hat{\Omega}$ biholomorphic to Ω .

2. If the blow-up is admissible at p then it is radial at p. This follows by the existence of a canonical 'affine normal direction' to $\partial\Omega$ at p. The 'affine normal direction' is constructed, for example, using cross-sections of Ω by hyperplanes parallel to $T_p\partial\Omega$. Some boundary regularity is required here, real analytic boundary certainly suffices.

3. The radial blow-ups actually converge in the real analytic sense, and kill off 'higher order terms'. This leaves a weighted homogeneous defining function for $\hat{\Omega}$.

4. Every Ω with weighted homogeneous convex defining function of the form Im $w > F(z)$, determines a convex domain with real analytic boundary and noncompact automorphism group. Classifying affine-hyperbolic convex domains with real analytic boundary and noncompact automorphism group, is much simpler. These are the Ω with convex defining function of the form Im $w > F(z)$, not necessarily weighted homogeneous. Here the noncompact group acts by translations. If Ω has a smoothly bounded embedding , one gets a more interesting analysis culminating in the fact that F is weighted homogeneous.

5. In general the blow-up is not admissible , for example in the case of parabolic approach. Nevertheless, we believe a closer analysis will show that any affine blow-up must converge real analytically, to a domain with weighted homogeneous polynomial defining function. See example (4.4). The technique there can be extended considerably; whereas in the proof we used affine automorphisms fixing p, it suffices to use an admissible blow-up at p taking q_i to $A_i q_i \to q \in \partial\hat{\Omega}$. These always exist.

This is analogous to reducing to $x_i \to p$ admissibly, following Rosay's simplification-generalization of Wong's theorem.

We do not know how Kim handles the non-admissible case (step 5 above). We have seen one preprint in which he handles the admissible case and states a 'Theorem A' which reduces the general case to the admissible case. His theorem A is proved in another preprint.

6. In the philosophy of Greene and Krantz, the second step should be to produce an invariant on Ω that forces $x_i \to \mathcal{S}$ 'admissibly', where $\mathcal{S}$ is a stratum of $\partial\Omega$ consisting of points with a condition on the germ of $\partial\Omega$. This is the hard step in the paper of Greene and Krantz, it is based on difficult d-bar estimates. We believe that the normal family theorem above or related compactness theorems enable one to use affine geometry to simplify (and we believe, avoid) the d-bar estimates. See [8] for some pertinent techniques. Though we have not seen the english translation, we understand that the paper of Bedford and Pincuk also succeeds in overcoming some of the problems involved in nonadmissible approach.

A basic problem related to the proof sketched above is; given two domains Ω_i defined by $\mathrm{Im}\, w > F_i(z)$, when are they biholomorphic ? The answer is iff there is a polynomial map $f : \Omega_1 \to \Omega_2$ which is 'weighted homogeneous, degree one'. Precise statements and proofs are in [8] .

The affine blow-up technique enables us to simplify and unify many results in the geometry of smoothly bounded domains. Because it only uses affine geometry and normal family theorems, we can extend many results to the context of convex, but nonsmooth, boundaries. Ultimately we expect the technique to extend to non-convex cases, such as the type of boundary points studied by Kohn and Nirenberg. Pseudoconvex domains, such as Teichmuller spaces, probably do not have nice blow-ups, but we do not yet understand them well enough to say for sure.

One of our goals here has been to bring to the readers attention some of the connnections between affine geometry, complex analysis, Poincare normal forms, and various papers many of which have come to our attention quite recently. There are bound to be some innaccuracies in my presentation of this material, for which I apologize. However I hope that the inclusion of these matters at least convinces the reader, (subject to appearance of the details), that the subject is not really so isolated, and that rather than being a dead end, it has barely gotten started.

References

[1] V.I. Arnold, 'Geometrical Methods in the Theory of O.D.Es', Springer, N.Y. (1983)

[2] E. Bedford and S. Pincuk, Domains in $\mathbf{C}^2$ with noncompact groups of holomorphic automorphisms, Mat. Sbornik(Russian) 135(177):2, 1988

[3] J. P. Benzecri, Sur les varietes localement affines et projectives, Bull Math Soc. France, 88(1960), 229-332

[4] W. Blaschke, Vorlesungen uber Differentialgeometrie und geometrische Grundlagen von Einsteins Relitivitatstheorie 3 vols. Julius Springer, Berlin, 1923.

[5] J.E. Fornaess and N. Sibony, Increasing sequences of complex manifolds. Math. Ann. 255(1981) no.3, pp.351-360

[6] S. Frankel, Bounded Convex Domains with Compact Quotient are Symmetric Spaces in Complex Dimension Two, 1986 Stanford thesis, supervisor Y.T. Siu.

[7] S. Frankel, Complex Geometry of Convex Domains that Cover Varieties, preprint (MSRI,nov.1987) to appear in Acta Math.

[8] S. Frankel, in preparation

[9] W. Goldman, to appear in Springer lecture notes

[10] W. Goldman, Convex real projective structures on compact surfaces, preprint

[11] I. Graham, Boundary behavior of the Caratheodory and Kobayashi metrics, in 'Several Complex Variables', (Proc. symp. Pure Math, vol XXX, part 1) p.149-152, AMS 1977

[12] R. Greene and S. Krantz, Characterization of certain weakly pseudoconvex domains with non-compact automorphism groups, preprint.

[13] M. Gromov, J. Lafontaine, P. Pansu, 'Structures Metriques sur les Varietes Riemanniennes' Cedic/ Fernand Nathan, Paris 1981

[14] V. Katz and E. B. Vinberg, Quasi-homogeneous cones, Math Zametki= Math Notes vol. 1 (1967) 231-235

[15] K.T. Kim, UCLA thesis, 1988

[16] S. Kobayashi, Intrinsic distances associated with flat affine or projective structures, J. Fac Sci Univ Tokyo 24 (1977) 129-135.

[17] S. Kobayashi, Invariant distances for projective structures, Ist. Naz. Alta Math Symp Math 26 (1982) 153-161

[18] J. L. Koszul, Varietes localement plat ct convexite, Osaka Math J., 2(1965), 285-290

[19] J. L. Koszul La forme hermitienne canaonique dcs cspaces homogene complexes Can J. Math 7 1955 p. 562.

[20] A. Koranyi and J. Wolf, Generalized Cayley transformations of bounded symmetric domains, Am. J. Math. (1965) p. 899.

[21] A. Koranyi and J. Wolf, Realization of hermitian symmetric spaces as generalized half-planes, Ann. Math. (2) 81 (1965) p. 265.

[22] N. Kuiper, Sur les surfaces localement affines, Colloque Int Geom Diff, Strasbourg, CNRS (1953) 79-87

[23] N. Kuiper, On convex locally projective spaces, Convegno Int Geom Diff, Italy (1954) 200-213

[24] Lay 'Convex Sets and their Applicatons' John Wiley, NY 1982.

[25] L. Lempert, La metrique de Kobayashi sur les domaines convexes de $\mathbf{C}^n$ Bull Soc Math de France 109 1981 p. 427.

[26] L. Lempert, Holomorphic retracts and intrinsic metrics in convex domains, Analysis Math. 8 (1982) p. 257.

[27] L. Lempert, Intrinsic metrics, in Proc. Symp. pure math AMS no. 41 ed. Y.T. Siu (1984).

[28] L. Lempert, International Congress Proceedings, 1986

[29] T. Ochiai, A lemma on open convex cones, J. Fac. Sci. U. Tokyo 12 1966 p. 231.

[30] I.I. Piiatetskii-Shapiro, 'Automorphic Functions and the Geometry of Clasical Domains' (english translation) Gordon Breach NY 1969

[31] S.I. Pinchuk, Holomorphic uniqueness of certain classes of domains in $\mathbf{C}^n$, (Russian) Mat Sbornik (N.S.) 111(153) 1980 no.1 p.67-94, 159.

[32] B. Rodin and D. Sullivan, Sphere packings and the Riemann mapping theorem, J. Diff. Geom. 26 no.2 (1987)p.349-360.

[33] J.P. Rosay, Characterisation de la boule parmi son groupe d'automorphismes, Annales de l'institut Fourier 29 1979 (4) p. 91.

[34] H. Rossi, The maximum modulus principle Ann. Math 72 1960 p. 1.

[35] B. Wong , Characterisation of the ball by its automorphism group, Inv math 41 1977 p.253

[36] Y. T. Siu, to appear in lect. notes of Inst. Mittag-Leffler

[37] B. Wong, The uniformization of compact Kähler surfaces of negative curvature, J. Diff. Geom. 16 (1981) no. 3 p. 407.

[38] S.T. Yau, Nonlinear Analysis in Geometry, L'enseignement Math. II, XXXIII, 1-2, 1987 p.109-158

[39] J. Vey, Sur une notion d'hyperbolicite sur les varietes localement plates, These, Univ de Grenoble (1968) C.R. Acad Sci Paris 266 (1968) 622-624

[40] J. Vey,Sur les automorphismes affines des ouverts convexes saillants, Ann Scuola Norm Sup Pisa 24 (1970) 641-665

[41] J. Vey Sur la division des domaines de Siegel, Ann Sci ENS 9 1976 p.203

MATHEMATICS DEPARTMENT
COLUMBIA UNIVERSITY
NEW YORK, NY 10027

Contemporary Mathematics
Volume **101**, 1989

COMPACTIFICATION OF COMPLETE KÄHLER–EINSTEIN MANIFOLDS OF FINITE VOLUME

Ngaiming Mok[1]

ABSTRACT Classically, the compactification of quotients X of bounded symmetric domains by torsion–free discrete arithmetic groups of automorphisms is achieved by algebraic methods. From the differential–geometric point of view they can be regarded as special examples of complete Kähler–Einstein manifolds of finite volume. Complex–analytically, with a few exceptions X are pseudoconcave manifolds. Quite recently there have been a number of studies on the compactification of complete Kähler manifolds using geometric and/or complex–analytic conditions. In this article we give a survey on research done in this direction. At the same time, we propose some possible applications of the methods and formulate a number of open problems and conjectures in related areas of research.

The algebraic theory of compactification of arithmetic quotients of bounded symmetric domains is a very classical one. Satake [39] (1956) was the first to give topological compactifications to certain arithmetic quotients of the Siegel space. This was used by Baily [6] (1958) to endow such compactifications with complex structures. In [40] (1960) Satake completed the topological compactification to all arithmetic quotients $X = \Omega/\Gamma$ of bounded symmetric domains Ω in general and the complex-analytic compactification was achieved by Baily–Borel [7] (1966). We call such compactifications Satake–Baily–Borel compactifications. They are obtained by adjoining to X arithmetic quotients of certain "rational boundary components", which are themselves bounded symmetric domains. The spaces obtained are singular in general. For example, when Ω is of rank one (i.e., the Euclidean unit ball) and of dimension ≥ 2 then the Satake–Baily–Borel compactifications $\overline{X}$ of $X = \Omega/\Gamma$ for Γ arithmetic are obtained by adjoining a finite number of points which are isolated singularities. It was desirable to endow $X = \Omega/\Gamma$ with natural non–singular compactifications (or compactifications with mild singularities). In [5] (1975)

1980 Mathematics Subject Classification Primary 52B35, 32C10

[1]Research partially supported by the National Science Foundation

Ash–Mumford–Rapoport–Tai completed the construction of toroidal compactifications. For any such $X = \Omega/\Gamma$, there exists a finite subgroup $\Gamma' \subset \Gamma$ such that the quotient $X' = \Omega/\Gamma'$ admits a non–singular toroidal compactification. Among other applications such compactifications enabled Mumford [32] to prove the Proportionality Principle for arithmetic quotients of bounded symmetric domains by using rather precise asymptotic descriptions of the canonical metrics. We consider from now on exclusively quotients $X = \Omega/\Gamma$ by arithmetic subgroups $\Gamma \subset \mathrm{Aut}(\Omega)$ acting without fixed points. We will call such an X an arithmetic variety. X, endowed with the quotient Kähler–Einstein metric, is necessarily of finite volume (cf. Raghanathan [37]). From the differential–geometric point of view they are special examples of complete Kähler–Einstein manifolds of finite volume. On the other hand Andreotti–Grauert [2] (1960) in a special case and Borel [9] (1966) showed that when X is irreducible and of dimension ≥ 2 it has the remarkable complex–analytic property of being pseudoconcave. In recent years there have been significant efforts in trying to generalize the compactification of arithmetic varieties to complete Kähler manifolds using differential–geometric conditions on curvature and volume and/or using the complex–analytic condition of pseudoconcavity. In this note we will give a survey on work done in this direction, incorporating results obtained after the author's talk on the subject in the AMS special session on Differential Geometry held in U.C.L.A. in November, 1987. We will also indicate some possible applications and proposed some open problems. For a systematic exposition of the basic methods cf. Mok [28, Chaps.III–IV].

§1 Complete Kähler manifolds of finite volume and pinched strictly negative curvature

In [44] (1982) Siu–Yau considered the question of compactifying complete Kähler manifolds of finite volume with strictly negative curvature pinched between two negative numbers. Their result completes the question of compactifying quotients of finite volume (in the Kähler–Einstein metric) of bounded symmetric domains by torsion–free discrete subgroups since it is known from Margulis' Arithmeticity Theorem (cf. Zimmer [48]) that irreducible non–arithmetic quotients exist only in the case of rank 1, where the sectional curvatures are strictly negative. They proved

Theorem 1 (Siu–Yau [44])

Let (X,ω) be a complete Kähler manifold of finite volume such that for some constant C

$$-C \leq \text{Sectional curvatures} \leq -1.$$

Then, X is biholomorphic to a Zariski–open subset of a non–singular projective–algebraic variety Z such that the complement $Z - X$ is an exceptional set of Z that can be blown down to a finite number of isolated singularities.

In the rank–1 situation the rational boundary components of Satake–Borel–Baily are single points. In Theorem 1 because of the curvature assumptions there is a geometric compactification of the universal covering in terms of equivalence classes of geodesics, called the Martin compactification, obtained by attaching a sphere at infinity to the cell $\tilde{X}$. Using Margulis' lemma they obtained rather precise description of of the "cusps" of X attached to certain points on the sphere at infinity. In case of non–arithmetic quotients $X = B^n/\Gamma$ one can obtain using Theorem 1 and the geometric description of the cusps and the associated groups of parabolic isometries exactly the same picture as in the arithmetic case: X can be compactified to a projective–algebraic manifold by adjoining a finite number of abelian varieties, which can be blown down to isolated singularities. Using this description one can extend the Proportionality Principle of Mumford for arithmetic varieties to the non–arithmetic case.

In [44] Siu–Yau first proved that (X,ω) is a strongly pseudoconcave manifold by using the geometric description of the cusps and Busemann functions. This allowed them to use an embedding theorem of Andreotti–Tomassini [3], based in part on the L^2–estimates of $\bar{\partial}$ of Andreotti–Vesentini [4] and Hörmander [17], to conclude that (X,ω) is biholomorphic to a domain on some projective–algebraic manifold Z. Because of strong psuedoconcavity there is a maximal compact subvariety of $Z - X$ which can be blown down to a finite number of isolated singularieties. Denoting by Z' the singular space thus obtained from Z, they proved that in fact X can be identified with the complement of the isolated singularities in Z' using the Ahlfors–Schwarz lemma of Yau [45].

§2 Embedding certain pseudoconcave Kähler manifolds of negative Ricci curvature

The notion of pseudoconcavity is important in complex analysis in part because pseudoconcave manifolds behave in many ways like compact complex manifolds. For instance, the embedding theorem of Andreotti–Tomassini [3] relies on Siegel's Theorem on pseudoconcave manifolds, proved by Andreotti [1] (1963). In other words, as in the case of compact complex manifolds, the field of meromorphic functions M(X) on an n–dimensional pseudoconcave manifold is of transcendence degree at most n over $\mathbb{C}$, and, when it is exactly n, M(X) is a finite extension field of a purely transcendental extension of $\mathbb{C}$ of transcendence degree n.

Very recently Nadel–Tsuji [33] (1988) considered the problem of compactifying pseudoconcave Kähler manifolds of negative Ricci curvature. They introduce the notion of very strongly q–pseudoconcave manifolds: a complex manifold X is very strongly q–pseudoconcave if there exists a smooth plurisubharmonic function ψ on X such that ψ is weakly plurisubharmonic outside a compact subset K, $-\psi$ is an exhaustion function (i.e., $\psi \longrightarrow -\infty$ at infinity), and such that outside K the Levi form of ψ has at least q positive eigenvalues. They proved

Theorem 2 (Nadel–Tsuji [33])
Let (X,ω) be an n–dimensional complete Kähler manifold of negative Ricci curvature. Assume that X is uniformized by a Stein manifold and that X is very strongly (n–2)–pseudoconcave. Then, X is biholomorphic to a quasi–projective variety.

Using Siegel's Theorem on pseudoconcave manifolds and the fact that on (X,ω) the canonical line bundle K_X is positive, one can embed X "birationally" into a projective–algebraic manifold Z. In other words, there exists a meromorphic mapping $F: X \to Z$ arising from $\Gamma(X,K_X^\nu)$ for some $\nu > 0$ such that F induces an isomorphism between M(X) and the field R(Z) of rational functions on Z. From the facts that X is very strongly (n–2)–pseudoconcave and that X contains no rational lines they deduce that X can in fact be biholomorphically embedded as an open subset of such a manifold Z. The crux of their argument was to show that X is Zariski–open in Z. By the assumption of pseudoconcavity there exists on $Z - X$ a maximal compact divisor D, which we may assume to be a union of smooth divisors intersecting at worst at normal crossings. By removing D and a finite union V of hyperplane sections one obtains by simultaneous uniformization of Ahlfors–Bers (Griffiths [16]) a quasi–projective manifold $W' \subset Z - D - V$ with $K_Z+[D]+[V]$ ample such that W' is uniformized by a bounded domain of holomorphy Ω and an open subset $W = X - V$ on W'. They proved that $W = W'$ by showing that the complete Kähler–Einstein metrics μ and μ' on W and W' resp. of constant Ricci curvature -2π have the same volume. (The existence of μ and μ' is guaranteed by Cheng–Yau [12] and Mok–Yau [29].) The key point is to obtain a precise lower bound for Volume(W,μ) by adapting an asymptotic Weyl formula of Demailly to complete Kähler manifolds. Let $\Gamma^2(W,K_W^\nu)$ denote the space of square–integrable holomorphic sections of the ν–th power of the canonical line bundle K_W of (W,μ). They obtained a Riemann–Roch inequality of the form

Proposition 1
In the notations above we have

$$\liminf_{\nu\to\infty} \frac{1}{\nu!} \dim_{\mathbb{C}}\Gamma^2(W,K_W^\nu) \geq \text{Volume}(W,\mu).$$

To compare Volume(W,μ) to Volume(W',μ') they used pseudoconcavity to show that $\Gamma^2(W,K_W^\nu) \subset \Gamma(Z,(K_Z+[D]+[V])^\nu)$ for the compact manifold Z. By the construction of the complete Kähler–Einstein metric (W',μ') (Kobayashi [21] and Cheng–Yau [13]) $(\mu')^n$ extends to a good metric on the compact manifold Z for the line bundle $K_Z+[D]+[V]$ in the sense of Mumford [32], so that Volume(W',μ') equals $\lim_{\nu\to\infty} \frac{1}{\nu!} \dim_{\mathbb{C}}\Gamma(Z,(K_Z+[D]+[V])^\nu)$, as a consequence of the Riemann–Roch Theorem and the Kodaira Vanishing Theorem on compact manifolds. The resulting inequality Volume(W',μ') $\geq$ Volume(W,μ) contradicts the Ahlfors–Schwarz lemma for volume forms (Yau [45]) unless strict equality holds.

Let $N = \Omega/\Gamma$ be an irreducible arithmetic variety of dimension ≥ 2, the proof of the pseudoconcavity of N by Andreotti–Grauert [2] and Borel [9] actually shows that N is very strongly (n–2)–pseudoconcave. Thus, Theorem 2 yields a generalization of the fact that arithmetic varieties can be complex–analytically compactified.

We remark that the original Kähler metric ω of negative Ricci curvature on X was only needed to obtain a holomorphic embedding of X onto an open subset of Z using L^2–estimates of $\bar{\partial}$. It does not enter into the volume estimates later. What we need in Theorem 2 is actually the fact that X carries some positive line bundle. At the same time, it was not assumed that (X,ω) is of finite volume. It is not difficult to prove using the existence of ψ and extension theorems for closed positive currents that given Theorem 2, any Kähler metric on X is of finite volume.

§3 A local compactification theorem on bounded domains

In this section we discuss a theorem of "local Zariski–openness" on a bounded domain of holomorphy, given by

Theorem 3 (Mok [25])

Let (Ω,ω) be a bounded domain of holomorphy in $\mathbb{C}^n$ equipped with the canonical complete Kähler–Einstein metric on Ω of Ricci curvature -2π. Suppose b is a point on $\partial\Omega$ such that for some open neighborhood U of b in $\mathbb{C}^n$, $\Omega \cap U$ is of finite volume with respect to ω. Then, $U - \Omega$ is a complex–analytic subvariety of U of pure codimension 1.

Theorem 3 is almost an immediate consequence of the volume estimates for complete Kähler–Einstein metrics ω on bounded domains of holomorphy Ω. Roughly speaking, the volume form must blow up near the boundary as fast as in the case of the Poincaré metric on the punctured disc near the puncture. On the other hand using the assumption of finite volume and the Ahlfors–Schwarz lemma we showed that the Kähler–Einstein form ω extends trivially as a closed positive (1,1) current T to U. Writing $\omega^n = V.d\lambda$, $d\lambda$ denoting the Euclidean volume form, we obtain two potentials for ω on $\Omega \cap U$, given by $\log(V)$ and a solution φ of the equation $\sqrt{-1}\partial\bar{\partial}\varphi = T$ on U (shrinking U if necessary). The difference $\psi = \varphi - \log(V)$ then gives a plurisubharmonic function on U whose $(-\infty)$–set is precisely $U - \Omega$. We then apply a criterion of Bombieri [8] on characterizing analytic subvarieties in terms of the $(-\infty)$–set of plurisubharmonic functions with Lelong number $\geq c$ for some positive number c. The volume estimate of Mok–Yau [29] can now be used to show that the latter criterion is satisfied by ψ.

The point of including the simple Theorem 3 here is to illustrate one possible approach to proving Zariski–openness in a geometric situation, i.e., to find plurisubharmonic potential functions with positive Lelong numbers at points exterior to the domain. Such potentials should arise from the Kähler metric. We note here that the argument of Theorem 3 does not work for an arbitrary domain Ω on a compact manifold Z, unless we assume conditions of negative curvature on Z. In fact, the volume estimate fails as one can see by blowing up points on $\partial\Omega$ and lifting Ω to the blown–up manifold Z'. One can easily construct counter–examples to the analogue of Theorem 3 for unbounded domains using Bieberbach maps (cf. [25]).

In view of Theorem 3 one sees that the key point in Theorem 2 of Nadel–Tsuji was to show that the complete Kähler–Einstein manifold (W,μ) is of finite volume. A precise *a–priori* knowledge of its volume is not really necessary.

§4 Compactifying complete Kähler–Einstein manifolds of bounded curvature and of finite volume

As a differential–geometric generalization to the compactification of arithmetic varieties it is proved

Theorem 4 (Mok–Zhong [30])

Let (X,ω) be a complete Kähler manifold of finite volume and of negative Ricci curvature. Suppose furthermore that the sectional curvatures are bounded and that the even Betti numbers are finite. Then, X is biholomorphic to a quasi–projective manifold.

Conjecturally the topologoical condition is redundant. Theorem 4 in its present form is nonetheless strong enough to give a generalization of the fact that arithmetic varieties can be compactified. In fact, by a theorem of Gromov et. al. (cf. Ballmann–Gromov–Schroeder [48]) a real–analytic complete Riemannian manifold (M,g) of finite volume and bounded seminegative Riemannian sectional curvature is necessarily diffeomorphic to the interior of a compact manifold with boundary.

The underlying complex manifold in Theorem 4 is no longer pseudoconcave and the field M(X) may in general be of infinite transcendence degree over $\mathbb{C}$. The starting point of our approach is to prove a Siegel's Theorem with growth. Denote by K the canonical line bundle of X. For $\alpha > 0$ and ν a positive integer denote by $\Gamma^{\alpha}(X,K^{\nu}) \subset \Gamma(X,K^{\nu})$ the vector space of holomorphic sections s of class L^{α}. Define

$\Gamma^0(X,K^\nu) = \cup_{\alpha>0}\Gamma^\alpha(X,K^\nu)$. The system $\{\Gamma^0(X,K^\nu)\colon \nu > 0\}$ of pluricanonical sections is closed under multiplication. We proved dimension estimates on $\Gamma^0(X,K^\nu)$, yielding Siegel's Theorem for the field $R(X)$ of "rational" functions arising as quotients of such sections. A strengthening of the argument yielded in addition Bézout estimates for the intersection of zero–divisors $[Zs]$ of $s \in \Gamma^0(X,K^\nu)$. As in the pseudoconcave case Siegel's Theorem for $R(X)$ allowed us to embed X "birationally" into a projective–algebraic manifold Z. Here by a "birational" embedding we mean a meromorphic mapping $F\colon X \to Z$ arising from $\Gamma^0(X,K^\nu)$ for some $\nu > 0$ such that F induces an isomorphism between the field of "rational" functions $R(X)$ on the Kähler manifold X and the field of rational functions $R(Z)$ on the projective–algebraic manifold Z. For the sake of simplified presentation we assume that X is already embedded holomorphically as an open subset of Z. Our proof of the Zariski–openness of X in Z breaks down essentially into two steps. We established this first under the additional assumption that X is locally Stein in Z. We then gave an argument which worked precisely at those boundary points $b \in \partial\Omega$ at which locally b is an interior point of its hull of holomorphy.

Assume first of all that X is locally Stein in Z. Our approach is to slice Z by hyperplane sections to get Riemann surfaces S and to show that $C = S \cap X$ is a finite Riemann surface such that the complement in S consists of a finite number of points whose cardinality is uniformly bounded independent of S. Using the Cohn–Vossen inequality [14] the problem is reduced to a uniform estimate of the Gauss–Bonnet integral of the complete Hermitian Riemann surfaces $(C,\omega|_C)$. This was achieved by a conversion to a problem of Bézout estimates in the projectivized tangent bundle $\mathbb{P}T_X$ in the same way that one converts the Gauss–Bonnet integral of a minimal surface in $\mathbb{R}^3$ to the area of the image under the Gauss map. To obtain the Bézout estimate on $\mathbb{P}T_X$ we needed the assumption that (X,ω) is of bounded sectional curvature. Using this result we applied Oka's theorem [34] [35] on characterizing domains of holomorphy and Skoda's L^2–estimate on the ideal problem [43] to prove that $Z - X$ is a hypersurface in Z. On the other hand, if $b \in \partial X$ is a point lying in the interior of the hull of holomorphy $H(U \cap X)$ for a contractible Stein coordinate neighborhood U of b in Z, we prove that $U - X$ is complex-analytic at b by producing a plurisubharmonic potential function ψ on U with Lelong number ≥ 1 at every point of $U - X$ and using Bombieri's criterion as described in §3. The two arguments can be put together to show that X is Zariski-open in Z. In general we only have a "birational" embeding $F\colon X \to Z$. The arguments presented above can be improved to show that there exists an "algebraic" subvariety V of X and a Zariski–open subset W of Z such that F maps $X - V$ biholomorphically onto W. To desingularize F in a finite number of steps we borrowed a topological argument of Demailly [49].

§5 Some applications and open problems

The proof of Theorem 4 yields a finiteness theorem on the group of bimeromorphic transformations Bimer(X). More generally, we have

Theorem 5 (Mok [28, Chap.III])
Let (X,ω) be a complete Kähler manifold of finite volume and bounded negative Ricci curvature. Then, the group Bimer(X) of bimeromorphic transformations of X is finite.

In particular, Theorem 5 answers in the affirmative a question raised in Yau ([46], Problem 36). The proof of Theorem 5 only relied on the Siegel's Theorem with growth in the proof of Theorem 4 and an argument of Sakai [38]. It applies therefore to the rather general case as given in the theorem.

A second possible application of the proof of Theorem 4 is to a question of strong rigidity of quotients of bounded symmetric domains of finite volume $N = \Omega/\Gamma$. Suppose now N is globally irreducible and of complex dimension ≥ 2. In the compact case given a compact Kähler manifold X homotopic to N, X is biholomorphic to N up to conjugation of irreducible local factors, as a consequence of the Strong Rigidity Theorem for Kähler manifolds of Siu [41] [42], Jost–Yau [18] and Mok [24], which improves upon the Hermitian case of Mostow's Strong Rigidity Theorem [31]. When N is non–compact the same theorem remains true if X is Hermitian locally symmetric and of finite volume by Prasad [36] in the rank–1 case and by Margulis' Superrigidity Theorem in case of rank ≥ 2 (cf. Zimmer [47]). In the general Kähler case and for N of arbitrary rank the arguments of Jost–Yau [19] [20] and Mok [26] yield the Strong Rigidity Theorem for X biholomorphic to a quasi–projective variety provided that there exists a projective–algebraic compactification $\overline{X}$ with $\overline{X} - X$ of codimension ≥ 3 in $\overline{X}$. The condition on the codimension of $\overline{X} - X$ is a very strong condition complex–analytically since it implies in particular that X is pseudoconcave. It is quite plausible that this additional assumption can be removed from Jost–Yau [20]. Given that and a strengthening of Theorem 4 by removing the topological condition, one would yield a satisfactory generalization of the Hermitian case of the Strong Rigidity Theorem of Margulis–Prasad, which we formulate as a conjecture

Conjecture 1
Let $N = \Omega/\Gamma$ be a globally irreducible quotient of finite volume of a bounded symmetric domain of complex dimension ≥ 2. Suppose (X,ω) is a complete Kähler manifold of finite volume, bounded sectional curvature and negative Ricci curvature homo-

topic to N. Then, up to conjugation on irreducible local factors of N, X is biholomorphic to N.

Given $N = \Omega/\Gamma$ as in Conjecture 1, the compactification theorem of Satake–Borel–Baily gives a compactification $\overline{N}_{min}$ such that $\overline{N}_{min} - N$ is of complex codimension ≥ 2 in $\overline{N}_{min}$. Here we use the notation $\overline{N}_{min}$ and call it a minimal compactification because by a theorom of Borel [10] any compactification of N dominates $\overline{N}_{min}$. It is a very interesting problem to know if the existence of such a compactification can be deduced from the complex–analytic condition of very strong q–pseudoconcavity as in Theorem 2 of Nadel–Tsuji or from curvature conditions in the locally irreducible case. (One cannot explain the case of irreducible quotients of polydiscs of finite volume using curvature alone since the curvature tensor is the same as in the case of products S of Riemann surfaces of finite volume, where S is certainly not pseudoconcave.)

In these directions it is tempting to conjecture

Conjecture 2
Let (X,ω) be a complete Kähler manifold of finite volume such that for some constant C

$$-C \leq \text{holomorphic bisectional curvatures} \leq -1$$

Then, X can be compactified to a Moišezon space Z obtained by adjoining to X a finite number of isolated singularities of Z.

In case holomorphic bisectional curvatures are replaced by Riemannian sectional curvatures, Conjecture 2 is Theorem 1 of Siu–Yau. For complex surfaces a case–by–case study based on Theorem 4 gives some evidence that Conjecture 2 might be true in 2 dimensions. Nonetheless no conceptual approach has been developed to address this problem even in this case.

In another direction, one might ask

Problem 1
Suppose X is an n–dimensional very strongly q–pseudoconcave manifold, $0 \leq q \leq n-2$, carrying a positive line bundle and uniformized by a Stein manifold, can X be compactified to a Moišezon space Z such that $Z - X$ is a complex subvariety of Z of codimension $\geq n - q$ in Z?

A positive answer to Problem 1 is also related to the analogous question of finding Moišezon compactifications Z with $Z - X$ of large codimension in Z under geometric conditions on sectional curvatures. The two problems are related via the Busemann functions of (X,ω) (cf. e.g., Nadel– Tsuji [32]).

Regarding Theorem 4 prior to Mok–Zhong [30] there was a method in the special case of two dimensions and nonpositive curvature to desingularize the birational embedding $F: X \longrightarrow Z$ by showing directly that the union V of the base point locus and the ramification locus of F contains at most of a finite number of irreducible curves V_i. This was done by proving that the Gauss–Bonnet integral on each V_i is necessarily an integer. This approach led to the independent question of asking when analytic Chern numbers X (defined as integrals of Chern–Weil forms over X using the Kähler metric ω) and of its "algebraic" subvarieties are integers. Such problems on integrality and topological invariance of certain analytic characteristic numbers in the Riemannian context were studied in Cheeger–Gromov [11] under assumptions of finite volume, bounded sectional curvature and existence of certain profinite normal coverings (as is satisfied under the additional assumption of nonpositive sectional curvature). In the present Kähler case in particular the answer to the following problem is unknown

Problem 2
Let (X,ω) be a complete Kähler manifold of finite volume, bounded sectional curvature and negative Ricci curvature. Are all analytic Chern numbers integers?

In the special case of complex surfaces and under the additional assumption that (*) (X,ω) is of Riemannian sectional curvature ≤ 0 and of holomorphic sectional curvature ≤ -1, Yeung [47] gave an affirmative answer to Problem 2. He also obtained the same conclusions for certain "algebraic" submanifolds S of a 3–dimensional complete Kähler manifold (X,ω) satisfying the hypotheses of Problem 2 and (*). In the latter case the additional difficulty is an estimate on the second fundamental form as $(S,\omega|_S)$ is no longer of bounded sectional curvature. An affirmative answer to Problem 2 would be related to the problem of parametrizing complex manifolds X admitting Kähler metrics satisfying the hypotheses as stated in Problem 2. Let $\mathcal{F}_n$ be the family of all underlying complex manifolds X of n–dimensional complete Kähler manifolds (X,ω) satisfying the geometric hypotheses of Problem 2. Let $\{c_I\}$ be an enumeration of the set of Chern numbers on n–dimensional manifolds and denote by $c_I(X,\omega)$ the analytic Chern number computed using the Kähler metric ω. Conjecturally, we have

Conjecture 3
For any two real numbers α, β with $\alpha < \beta$ denote by $\mathcal{F}_n(\alpha,\beta) \subset \mathcal{F}_n$ the subset consisting of all n–dimensional complex manifolds X admitting a Kähler–Einstein

metric ω of bounded curvature such that $\alpha < c_I(X,\omega) < \beta$ for all analytic Chern numbers $c_I(X,\omega)$. Then, X can be parametrized effectively by a quasi–projective variety. In particular, there are at most a finite number of diffeomorphism types for the underlying smooth manifold $X \in \mathscr{F}_n(\alpha,\beta)$.

In the case of compact manifolds Conjecture 3 is a consequence of Matsusaka's big theorem (cf. Lieberman–Mumford [22]), which shows that there exists a positive integer p depending only on α, β and the complex dimension such that X can be embedded into a projective space $\mathbb{P}^N$, the Riemann–Roch Theorem and the Kodaira Vanishing Theorem for compact manifolds, which together give a formula for N in terms of Chern numbers of X.

Related to the statement of Theorem 4 the following more general problem is open:

Problem 3
Let (X,ω) be a complete Kähler manifold of finite volume and of bounded negative Ricci curvature. Is X necessarily biholomorphic to a quasi–projective manifold?

The proof of Siegel's Theorem on $R(X)$ works without the assumption of bounded curvature. It also shows that slices of (X,ω) by "hyperplane sections" are of finite volume. By slicing by Riemann surfaces this comes close to showing that the birational embedding $F: X \longrightarrow Z$ misses at most a pluripolar set. An answer to Problem 4 in the special case when (X,ω) is Kähler–Einstein and uniformized by a bounded domain of holomorphy would already be very interesting. In the positive direction one could try to imitate the proof of Theorem 3 of Nadel–Tsuji and substitute the extension argument $\Gamma^2(W,K_W^\nu) \subset \Gamma(Z,(K_Z+[D]+[V])^\nu)$ using pseudoconcavity by some argument relying on the special choice of embedding using Siegel's Theorem for the "rational" functions $R(X)$. It is fair to say that up to this point that there is no overwhelming evidence either in the affirmative or in the negative direction. It would be equally interesting to construct a counter–example (which seems an even more formidable task should it exist).

Finally, one should be able to improve upon the method of proof of Theorem 4 to prove a non–compact analogue of the Kodaira Embedding Theorem in the following form

Conjecture 4
Let (X,ω) be a complete Kähler manifold of finite volume and bounded sectional curvature. Suppose there exists on (X,ω) a Hermitian holomorphic line bundle

(L,θ) of positive and bounded curvature. Then X is biholomorphic to a quasi-projective variety.

In Mok–Zhong [30] we established Conjecture 4 under the additional assumptions that $(L\otimes K_X^{-1},\theta.\omega^n)$ is of positive curvature for the canonical line bundle K_X, $\dim_{\mathbb{C}}(X) = n$, and that the even Betti numbers of X are finite. There is first of all the difficulty of removing the topological assumption. Furthermore, removing the assumption that $(L\otimes K_X^{-1},\theta.\omega^n)$ is of positive curvature presents a siginificant additional difficulty because of the appearance of the Ricci term in the L^2–estimate of $\bar{\partial}$.

REFERENCES

[1] Andreotti, A. Théorèmes de dépendance algébrique sur les espaces pseudoconcaves, Bull. Soc. math. France 91(1963), 1–38.

[2] Andreotti, A. & Grauert, H. Algebraische Körper von automorphen Funktionen, Nachr. Akad. Wiss. Göttingen, math.–phy. Klasse (1961), 39–48.

[3] Andreotti, A. & Tomassini, G. Some remarks on pseudoconcave manifolds, in Essays on Topology and Related Topics dedicated to G. de Rham, ed. by A. Haefliger and R. Narasimhan, Springer Verlag, Berlin–Heidelberg–New York 1970, pp. 85–104.

[4] Andreotti, A. & Vesentini, E. Carleman estimates for the Laplace–Beltrami operator on complex manifolds, Inst. Hautes Études Sci. Publ. Math. 25(1965), 81–130.

[5] Ash, A., Mumford, D., Rapoport, M. & Tai, Y. Smooth compactifications of locally symmetric varieties, Math. Sci. Press, Brookline 1975.

[6] Baily, W.L.Jr. On Satake's compactification of V_n, Amer. J. Math. 80(1958), 348–364.

[7] Baily, W.L.Jr. & Borel, A. Compactification of arithmetic quotients of bounded symmetric domains, Ann. math. 84(1966), 442–528.

[8] Bombieri, E. Algebraic values of meromorphic maps, Invent. Math. 10(1970) 267–287.

[9] Borel, A. Pseudo–concavité et groupes arithmétiques, in Essays on Topology and Related Topics dedicated to G. de Rham, ed. by A. Haefliger and R. Narasimhan, Springer Verlag, Berlin–Heidelberg–New York 1970, pp. 70–84.

[10] Borel, A. Some metric properties of arithmetic quotients of symmetric spaces and an extension theorem, J. Diff. Geom. 6(1972), 542–560.

[11] Cheeger, J. & Gromov, M. On the characterisitc numbers of complete manifolds of bounded curvature and finite volume, in Differential Geometry and Complex Analysis, II. E. Rauch Memorial Volume, Springer Verlag, Berlin–Heidelberg–New York 1985, pp. 115–154.

[12] Cheng, S.–Y. & Yau, S.–T. On the existence of a complete Kähler metric on non–compact manifolds and the regularity of Fefferman's equation, Comm. Pure Appl. Math. 33(1980), 507–544.

[13] Cheng, S.–Y. & Yau, S.–T. Inequality between Chern numbers of singular Kähler surfaces and characterization of orbit spaces of discrete groups of SU(2,1), Contemporary Mathematics 49, AMS 1986, pp. 31–43.

[14] Cohn–Vossen, S. Kürzeste Wege und Totalkrümmung auf Flächen, Comp. Math. 2(1935), 69–133.

[15] Demailly, J.P. Champs magnétiques et inégalités de Morse pour la $\bar{\partial}$–cohomologie, Ann. Inst. Fourier 35–4(1985), 189–225.

[16] Griffiths, P.A. Complex–analytic properties of certain Zariski open sets on algebraic varieties, Ann. Math. 94(1971), 69–100.

[17] Hörmander, L. L^2–estimates and the existence theorems for the $\bar{\partial}$–operator, Acta Math. 114(1965), 89–152.

[18] Jost, J. & Yau, S.–T. A strong rigidity theorem for a certain class of compact analytic surfaces, Math. Ann. 271(1985), 143–152.

[19] Jost, J. & Yau, S.–T. The strong rigidity of locally symmetric complex manifolds of rank one and finite volume, Math. Ann. 275(1986), 291–304.

[20] Jost, J. & Yau, S.–T. On the rigidity of certain discrete groups and algebraic varieties, Math. Ann. 278(1987), 481–496.

[21] Kobayashi, R. Existence of Kähler–Einstein metrics on an open algebraic manifold, Osaka J. Math. 21(1984), 399–418.

[22] Lieberman, D. & Mumford, D. Matsusaka's big theorem, AMS Proc. Symp. Pure Math., Vol.29(1975), 513–530.

[23] Mok, N. An embedding theorem of complete Kähler manifolds of positive bisectional curvature onto affine algebraic varieties, Bull. Soc. math. France 112(1984), 197–258.

[24] Mok, N. The holomorphic or anti–holomorphic character of harmonic maps into irreducible compact quotients of polydiscs, Math. Ann. 272(1985), 197–216.

[25] Mok, N. Complete Kähler–Einstein metrics on bounded domains locally of finite volume at some boundary points, to appear in Math. Ann.

[26] Mok, N. Strong rigidity of irreducible quotients of polydiscs of finite volume, to appear in Math. Ann.

[27] Mok, N. Compactification of complete Kähler surfaces of finite volume satisfying certain curvature conditions. Preprint.

[28] Mok, N. Topics in Complex Differential Geometry. Preprint.

[29] Mok, N. & Yau, S.–T. Completeness of Kähler–Einstein metrics on bounded domains and the characterization of domains of holomorphy by curvature conditions, Proc. Sym. Pure Math., Vol. 39(1983), Part I, pp. 41–59.

[30] Mok, N. & Zhong, J.–Q. Compactifying complete Kähler–Einstein manifolds of finite typological type and bounded curvature. Preprint.

[31] Mostow, G.D. Strong rigidity of locally symmetric spaces, Ann. Math. Studies 77, Princeton U. Press, Princeton 1973.

[32] Mumford, D. Hirzebruch's proportionality theorem in the non–compact case, Invent. Math. 42(1977), 239–272.

[33] Nadel, A. & Tsuji, H. Compactification of complete Kähler manifolds of negative curvature, to appear in J. Diff. Geom.

[34] Oka, K. Domaines pseudoconvexes, Tohoku Math. J. 49(1942), 15–52.

[35] Oka, K. Domaines finis sans point critique intérieur, Japan J. Math. 23(1953), 97–115.

[36] Prasad, G. Strong rigidity of Q–rank 1 lattices, Invent. math. 21(1973), 255–286.

[37] Raghanathan, M.S. Discrete Subgroups of Lie Groups, Springer Verlag, Berlin–Heidelberg–New York 1972.

[38] Sakai, F. Kodaira dimension of complement of divisors, Complex Analysis and Algebraic Geometry, a collection of papers dedicated to K. Kodaira, Iwanami, Tokyo 1977.

[39] Satake, I. On the compactification of the Siegel space. J. Indian Math. Soc. 20(1956), 259–281.

[40] Satake, I. On compactifications of the quotient spaces for arithmetically defined discontinuous groups, Ann. Math. 72(1960), 555–580.

[41] Siu, Y.–T. The complex–analyticity of harmonic maps and the strong rigidity of compact Kähler manifolds, Ann. Math. 112(1980), 73–111.

[42] Siu, Y.–T. Strong rigidity of compact quotients of exceptional bounded symmetric domains, Duke Math. J. 48(1981), 857–871.

[43] Skoda, H. Applications des techniques L^2 à la théorie des idéaux d'un algèbre de fonctions holomorphes avec poids, Ann. Sci. Éc. Norm. Sup. (4) 5(1972), 548–580.

[44] Siu, Y.–T. & Yau, S.–T. Compactification of negatively curved complete Kähler manifolds of finite volume, Ann. Math. Studies 102(1982), 363–380.

[45] Yau, S.–T. A general Schwarz lemma for Kähler manifolds, Amer. J. Math. 100(1978), 197–203.

[46] Yau, S.–T. Problem section, Ann. Math. Studies 102(1982), 669–706.

[47] Zimmer, R.J. Ergodic Theory and Semisimple Groups, Monographs in Mathematics, Birkhäuser, Boston–Basel–Stuttgart 1984.

[48] Ballmann, W., Gromov, M. & Schroeder, V. Manifolds of Nonpositive Curvature, Progress in Mathematics, Vol. 61, Birkhäuser, Boston–Basel–Stuttgart 1985.

[49] Deamilly, J.P. Mesures de Monge–Ampère et caractérisation géometrique des variétés algébriques affines, Bull. Soc. Math. France, Mémoires, No. 19(1985), 123 pages.

COLUMBIA UNIVERSITY, NEW YORK CITY, NY 10027

Contemporary Mathematics
Volume **101**, 1989

Topological types of isolated hypersurface singularities.

STEPHEN S.-T. YAU[1]

§1. INTRODUCTION

No matter whether you are a topologist, algebraist or geometer, one of the fundamental goals is to find a necessary and sufficient condition for two given objects to be isomorphic in the given category. In the theory of isolated hypersurface singularities, the two fundamental problems are as follows: Let $(V,0)$ and $(W,0)$ be two isolated hypersurface singularities in $\mathbf{C}^{n+1}$.

Problem 1. Give a simple algebraic criterion for $(\mathbf{C}^{n+1},V,0)$ to be homeomorphic to $(\mathbf{C}^{n+1},W,0)$.

Problem 2. Give a simple algebraic criterion for $(\mathbf{C}^{n+1},V,0)$ to be biholomorphic to $(\mathbf{C}^{n+1},W,0)$.

One supposed that the first problem would be easier than the second one, but it turned out to be contrary. In 1982 [15], Mather and the author solved the second problem completely. We showed that two isolated hypersurface singularities in $\mathbf{C}^{n+1}$ are biholomorphically equivalent if their corresponding moduli algebra (a finite dimensional commutative local $\mathbf{C}$-algebra) are isomorphic. On the other hand, the progress on the first problem was not as fast as one wants although many well known mathematicians including Milnor and Zariski worked on it. Actually even the Zariski multiplicity problem whether multiplicity of isolated hypersurface singularity is an invariant of topological type, was solved completely only for $n = 1$ case. Recently there are some progress in this problem for $n = 2$ case

1980 *Mathematics Subject Classification* (1985 Revision). Primary 32B99; Secondary 32C40.

[1]This work is partiall supported by N.S.F. grant.

(cf. [27], [29], [30]). In fact we formulated a conjecture for problem 1 in case $n = 2$ (for the statement, see §5). Xu and the author [27] proved this conjecture for quasi-homogeneous singularities. In this article we shall only discuss the recent progress on problem 1 above and its related Zariski multiplicity problem. We shall point out that the results in [27] are in fact true for a more general class of semiquasi-homogeneous singularities in the sense of Arnold [1].

In §2, we discuss problem 1 for plane curve singularities. We also describe Milnor's theory on singular points of complex hypersurface [16]. This is the most important theory which provides necessary tools to study problem one. In §3, we prove that the charasteristic polynomial of the singularity as well as the homotopy groups of the link of singularity are invariants of topological type. The former statement is a consequence of an important theorem of Lê Dung Tráng [10], although the proof here is slightly different from his, where we use the notion of Whitehead product. In §4, we shall discuss a deep result of Lê-Ramanujan [12] and its corollary. We point out that under the hypothesis of this corollary, we can actually determine the analytic type of singularities [15], [28]. In §5, we describe the recent work [27] of Xu and the author, where we solved problem 1 in case of quasi-homogeneous surface singularities. The proof depends on the fundamental results of Neumann [18], Orlik-Wagreich [19] and Varchenko [26]. We point out here that the same proof applies even for semi-quasi-homogeneous surface singularities. So the problem 1 is solved in this case also (cf. Theorem 5.9). In §6, we discuss the Zariski multiplicity conjecture with emphasis on two-dimensional case. As a consequence of the proof of Theorem 5.2 and Theorem 5.3, Xu and the author [27] proved the Zariski multiplicity conjecture in the quasi-homogeneous case. We point out here that the same proof also works for semi-quasi-homogeneous singularities (cf. Theorem 6.1). Finally we discuss our recent result [30] on Zariski's multiplicity conjecture. We proved that Zariski's multiplicity conjecture is true if one of the arithmetic genus of the singularity is small (i.e. not more than two).

I would like to thank A. Libgober for some discussion on preparing this article.

§2. Topological types of isolated hypersurface singularities

Definition 2.1. Let $(V_1, 0)$ and $(V_2, 0)$ be two isolated hypersurface singularities in $\mathbf{C}^{n+1}$. We say that $(V_1, 0)$ and $(V_2, 0)$ have the same topological type if $(\mathbf{C}^{n+1}, V_1, 0)$ is homeomorphic to $(\mathbf{C}^{n+1}, V_2, 0)$.

Even for $n = 1$, it took more than forty years for people to completely understand the topological type of plane curve singularities. Let f be the defining function of the plane curve singularity $(V, 0)$. Then f is reduced, i.e. in its decomposition in irreducible analytic functions in $\mathbf{C}\{X, Y\}$, it is square free. Suppose now that f is irreducible in $\mathbf{C}\{X, Y\}$, i.e. the analytic local ring $\mathcal{O} = \mathbf{C}\{X, Y\}/(f)$ is an integral domain. Then we have

THEOREM 2.1. *(Puiseux) The normalization $\overline{\mathcal{O}}$ of $\mathcal{O}$ is a regular analytic local ring and $\overline{\mathcal{O}}$ is a finite $\mathcal{O}$-module.*

Let x and y be the images of X and Y in $\mathcal{O}$. The maximal ideal $(x, y) = \mathcal{M}$ of $\mathcal{O}$ generates a principal ideal $\mathcal{M}\overline{\mathcal{O}}$, because $\overline{\mathcal{O}} \cong \mathbf{C}\{t\}$. Suppose that $\mathcal{M}\overline{\mathcal{O}} = x\overline{\mathcal{O}}$, i.e. by definition x is a transversal parameter. Then we may choose the uniformizing paramter t of $\overline{\mathcal{O}}$ so that

$$
\begin{aligned}
x &= t^n \\
y &= \sum_{\nu \geq n} a_\nu t^\nu .
\end{aligned}
\tag{2.1}
$$

We call (2.1) a Puiseux expansion of f at 0.

Consider now the completion $\mathcal{O}_1 = \mathbf{C}[[X, Y]]/(f)$ of $\mathcal{O}$, where $\mathbf{C}[[X, Y]]$ is the ring of complex formal power series in two variables. Let us call K the field of quotients of $\mathcal{O}_1$. Actually if t is the uniformizing parameter of $\overline{\mathcal{O}}_1$, one has $K = \mathbf{C}((t))$ and K is a cyclic extension of $\mathbf{C}((x))$.

Let G be the Galvis of K over $\mathbf{C}((x))$. Define $G_i = \{\sigma \in G : v(\sigma y - y) \geq i\}$ where v is the valuation of K. Then

$$G_0 = G \supset G_1 \supset G_2 \supset \dots .$$

Necessarily for some $k > 0$, $G_k = \{1\}$. Let $\beta_1, \dots, \beta_g$ be defined by

$$
\begin{aligned}
\beta_1 &= \sup\{i : G_i = G\} \\
\beta_{j+1} &= \sup\{i : G_i = G_{\beta_j + 1}\} \\
\beta_g &= \sup\{i : G_i \neq \{1\}\}.
\end{aligned}
$$

We shall call $\beta_1, \dots, \beta_g$ the Puiseux exponents of f at 0. We can notice that there is a uniquely determined sequence of pairs (m_i, n_i) $(i = 1, \dots, g)$ of relatively prime numbers such that:

$$\beta_1 = \frac{m_1}{n_1} \cdot n$$

$$\beta_2 = \frac{m_2}{n_1 n_2} \cdot n$$

$$\vdots$$

$$\beta_i = \frac{m_i}{n_1 \ldots n_i} \cdot n$$

and necessarily $n_1 \cdots n_g = n$ otherwise the multiplicity of f at 0 should be strictly less than n. The pairs (m_i, n_i) $(i = 1, \ldots, g)$ are called the Puiseux pairs of f at 0. Notice that $\beta_1 < \ldots < \beta_g$ implies $m_i n_{i+1} < m_{i+1}$ for $i = 1, \ldots, g-1$.

In 1929, K. Brauner proved in [4] the following theorem.

THEOREM 2.2. *Let $f(X, Y)$ be analytically irreducible at 0 and $f(0) = 0$. Let n be its multiplicity at 0 and $\beta_1, \ldots, \beta_g$ be the Puiseux exponents of f at 0. Then the plane curve singularity defined by f has the same topology type as the curve singularity defined by*

$$\begin{cases} x = t^n \\ y = t^{\beta_1} + \ldots + t^{\beta_g}. \end{cases}$$

In 1932, W. Burau in [5] and O. Zariski in [32] proved that the converse of Theorem 2.2 is also true.

THEOREM 2.3. *Let $f(X, Y)$ be analytically irreducible at 0 and $f(0) = 0$. Then the Puiseux exponents of f at 0 are invariants of topological type of $(V, 0)$ where $V = \{f(X, Y) = 0\}$.*

Finally, M. Lejeune in [14] and O. Zariski in [35] proved the following theorem.

THEOREM 2.4. *Let $f(X, Y)$ be reduced at 0 and $f(0) = 0$. Then the topology type of the plane curve singularity defined by f is determined by the topology type of every irreducible component of f at 0 and all the pairs of intersection multiplicity of these components.*

These together with the theorem of J. Reeve [22], which asserts that the intersection multiplicity of two plane curves is the same as the linking number of the corresponding knots, give a complete understanding of the topological type of plane curve singularities.

In 1968, Milnor [16] made fundamental contribution in understanding the topology of isolated complex hypersurface singularities. Let us recall his beautiful theory briefly as below.

THEOREM 2.5. *Let V be a complex algebraic subvariety in $\mathbf{C}^{n+1}$ and $S(V)$ be the singular set of V. Let $f : V \to \mathbf{C}$ be an algebraic function on V. Then the restriction of f to $V - S(V)$ has only a finite number of critical values.*

Corollary 2.6. Let $f : \mathbf{C}^{n+1} \to \mathbf{C}$ be a polynomial function. Then there exist $t_1, \ldots, t_r \in \mathbf{C}$ such that for all $t \in \mathbf{C} - \{t_1, \ldots, t_r\}$ the hypersurface defined by $f = t$ is nonsingular.

Corollary 2.7. Let V be complex algebraic subvariety of $\mathbf{C}^{n+1}$. Let x_0 be either a simple point of V or an isolated point of the singular set $S(V)$. Then every sufficiently small sphere S_ϵ centered at x intersects V traversely in a smooth manifold.

Let B_ϵ (resp. B_ϵ^0) denote the closed (resp. open) ball consisting of all x with $\|x - x_0\| < \epsilon$ (resp. $< \epsilon$). Again let x_0 be either a simple point or an isolated singular point of V.

Proposition 2.8. For all sufficiently small strictly positive real numbers ϵ_1, ϵ_2, $(S_{\epsilon_1}, S_{\epsilon_1} \cap V)$ is diffeomorphic to $(S_{\epsilon_2}, S_{\epsilon_2} \cap V)$ as pair. Moreover, $(B_{\epsilon_1}, B_{\epsilon_1} \cap V)$ is homeomorphic to $(B_{\epsilon_1}, C(K_{\epsilon_1}))$ as pair, where $K_{\epsilon_1} = S_{\epsilon_1} \cap V$ and $C(K_{\epsilon_1})$ is the real cone over K_{ϵ_1} which is the union of all line segments jointing points $k \in K_{\epsilon_1}$ to the base point x_0.

We are now ready to state the Milnor's fibration theorem.

THEOREM 2.9. *Let $f : (\mathbf{C}^{n+1}, 0) \to (\mathbf{C}, 0)$ be a complex polynomial. Let V be the hypersurface defined by $f = 0$. Then there exists $\epsilon_0 > 0$ such that for all ϵ with $0 < \epsilon \le \epsilon_0$, the differentiable mapping $\varphi_\epsilon : S_\epsilon - V \cap S_\epsilon \to S^1$ defined by $\varphi_\epsilon(z) = f(z)/|f(z)|$ for all $z \in S_\epsilon - V$, is a locally trivial differentiable fibration.*

In [16], Milnor gave another presentation of this fibration.

THEOREM 2.10. *For $\epsilon > 0$ sufficiently small and $\epsilon \gg \eta > 0$, the mapping $\psi_{\epsilon,\eta} : B_\epsilon^0 \cap f^{-1}(\partial D_\eta) \to \partial D_\eta$ induced by f, where B_ϵ^0 is the interior of B_ϵ and $\partial D_\eta = \{z \in \mathbf{C} : |z| = \eta\}$, is a smooth fibration isomorphic to φ_ϵ in Theorem 2.9 by an isomorphism which preserves the arguments.*

Corollary 2.11. Let $\epsilon_0 > 0$ as in Theorem 2.9. Fix ϵ with $0 < \epsilon \le \epsilon_0$. Then the Milnor fiber $F_\theta = \varphi_\epsilon^{-1}(e^{i\theta})$ is parallelizable and has the homotopy type of a n-dimensional finite CW-complex.

Milnor [16] also proved the following.

THEOREM 2.12. *The topological space $K_\epsilon = S_\epsilon \cap V$ is $n-2$ connected.*

Given any locally trivial fibration $\phi : E \to S^1$ over the circle, the natural action of a generator of $\pi_1(S^1)$ on the homology of the fiber is described by automorphism $h_* : H_*F_0 \to H_*F_0$. Here h denotes the characteristic homeomorphism of the fibre $F_0 = \phi^{-1}(1)$. It is obtained, using the covering homotopy theorem, by choosing a continuous one-parameter family of homeomorphisms. $h_t : F_0 \to F_t$ for $0 \le t \le 2\pi$, where h_0 is the identity and $h = h_{2\pi}$ is the required characteristic homemorphism. h induces on the homology group of F_0 the isomorphisms which is by definition the local monodromy at V at 0.

By a theorem of Milnor and a theorem of Palamodov [21] we have the following theorem.

THEOREM 2.13. *If 0 is an isolated critical point of f, for $\epsilon > 0$ small enough, the fibers of φ_ϵ have the homotopy type of a bouquet of μ spheres of dimension n with*

$$\mu = \dim_{\mathbf{C}} \mathbf{C}\{z_0, \dots, z_n\}/(\frac{\partial f}{\partial z_0}, \dots, \frac{\partial f}{\partial z_n}).$$

Remark. A bouquet of spheres is the topological space union of spheres having a single point in common. The μ above is called Milnor number.

§3. INVARIANT OF THE TOPOLOGICAL TYPES

In this section, we shall give necessary conditions for two isolated hypersurface singularities which have the same topological type. Let is first recall the important notion of Whitehead product in algebraic toplogy [7]. Consider a given space X and a given basic point x_0 in X. Let $m \ge 1$ and $n \ge 1$ be given integers. For any two given elements $\alpha \in \pi_m(X, x_0)$, $\beta \in \pi_n(X, x_0)$, the Whitehead product of α and β is an element $[\alpha, \beta]$ of $\pi_{m+n-1}(X, x_o)$, which is defined as follows.

Let us choose representative maps $f : (I^m, \partial I^m) \to (X, x_0)$, $g : (I^n, \partial I^n) \to (X, x_0)$ for α, β resepctively. Since $I^{m+n} = I^m \times I^n$, we have $\partial I^{m+n} = (I^m \times \partial I^n) \cup (\partial I^m \times I^n)$. Here I^n is the n-cube. Hence we define a map $h : \partial I^{m+n} \to X$ by taking for each point (s, t) in ∂I^{m+n}

$$h(s,t) = \begin{cases} f(s) & \text{if } t \in \partial I^n \\ g(t) & \text{if } s \in \partial I^m. \end{cases}$$

Since the point $r_0 = (0, \dots, 0)$ of ∂I^{m+n} is in $\partial I^m \times \partial I^n$, we have $h(r_0) = x_0$. Since ∂I^{m+n} is homeomoephic to S^{m+n-1}, h represents an element γ of $\pi_{m+n-1}(X, x_0)$. It can be shown that γ depends only on the elements α and β. So we may define $[\alpha, \beta] = \gamma$. We shall list some properties of Whithead products:

[W1] If $\alpha \in \pi_1(X, x_0)$ and $\beta \in \pi_1(X, x_0)$, then $[\alpha, \beta]$ is the commutator $\alpha\beta\alpha^{-1}\beta^{-1}$ of $\pi_1(X, x_0)$.

[W2] If $\alpha \in \pi_m(X, x_0)$ and $\beta \in \pi_i(X, x_0)$ with $m > 1$, then $[\alpha, \beta]$ is the element $\beta\alpha - \alpha$ of $\pi_m(X, x_o)$ where $\beta : \pi_m(X, x_0) \to \pi_m(X, x_0)$ is a group automorphism.

[W3] If $m > 1$, then the assignment $\alpha \to [\alpha, \beta]$ for a given $\beta \in \pi_n(X, x_0)$ defines a homomorphism

$$\beta_* : \pi_m(X, x_0) \to \pi_{m+n-1}(X, x_0).$$

[W4] If $m + n > 2$, then, for every $\alpha \in \pi_m(X, x_0)$ and $\beta \in \pi_n(X, x_0)$ we have $[\beta, \alpha] = (-1)^{mn}[\alpha, \beta]$.

[W5] If $\sigma : I \to X$ is a path joining x_0 to x_1, then, for every $\alpha \in \pi_m(X, x_1)$ and $\beta \in \pi_n(X, x_1)$, we have $\sigma_{m+n-1}[\alpha, \beta] = [\sigma_m(\alpha), \sigma_n(\beta)]$.

[W6] If $\phi : (X, x_0) \to (Y, y_0)$ is a map, then, for every $\alpha \in \pi_m(X, x_0)$ and $\beta \in \pi_n(X, x_0)$, we have $\phi_*[\alpha, \beta] = [\phi_*(\alpha), \phi_*(\beta)]$.

[W7] For any $\alpha \in \pi_m(X, x_0)$, $\beta \in \pi_n(X, x_0)$, $\gamma \in \pi_q(X, x_0)$, the following Jacobi identity holds:

$$(-1)^{mq}[[\alpha, \beta], \gamma] + (-1)^{nm}[[\beta, \gamma], \alpha] + (-1)^{qn}[[\gamma, \alpha], \beta] = 0.$$

Milnor's theory indeed allows us to understand the topological types of isolated hypersurface singularities a lot better than before. In fact the following important theorem was first proved by Lê Dung Tráng [10], although the proof given here is slightly different from his.

THEOREM 3.1. *Suppose that the two isolated hypersurface singularities $(V, 0)$ and $(\widetilde{V}, 0)$ have the same topological types. Then they have the same Milnor number μ and their local monodromy are conjugated to each other.*

Proof. By Proposition 2.8, there exists $\epsilon_0 > 0$ with the following properties: If $\epsilon_0 > \epsilon > 0$, then $B_\epsilon - V \cap B_\epsilon$ is homotopy equivalent to $S_\epsilon - V \cap S_\epsilon$ and $B_\epsilon - \widetilde{V} \cap B_\epsilon$ is homotopy equivalent to $S_\epsilon - \widetilde{V} \cap S_\epsilon$. Moreover for any $\epsilon_0 > \epsilon_1 > \epsilon_2 > 0$, $S_{\epsilon_1} - S_{\epsilon_1} \cap V$ and $S_{\epsilon_1} - S_{\epsilon_1} \cap \widetilde{V}$

are diffeomorphic to $S_{\epsilon_2} - S_{\epsilon_2} \cap V$ and $S_{\epsilon_2} - S_{\epsilon_2} \cap \widetilde{V}$ respectively. We shall also assume that ϵ_0 is so chosen such that Theorem 2.9 is applicable for both $(V, 0)$ and $(\widetilde{V}, 0)$. Since $(V, 0)$ and $(\widetilde{V}, 0)$ have the same topological type, there exist neighborhoods $U, \widetilde{U}$ of 0 and homeomorphism $\psi : U \to \widetilde{U}$ such that $\psi(U \cap V) = \widetilde{U} \cap \widetilde{V}$ and $\psi(0) = 0$. We shall assume that $B_{\epsilon_0} \subseteq U \cap \widetilde{U}$.

Let $\epsilon_0 > \epsilon_4 > 0$. Since ψ is continuous, there is $\epsilon_0 > \epsilon_3 > 0$ such that $\psi(B_{\epsilon_3}) \subseteq B_{\epsilon_4}$. $\psi(B_{\epsilon_3})$ is an open neighborhood of 0. So we can find $\epsilon_0 > \epsilon_2 > 0$ such that $B_{\epsilon_2} \subset \psi(B_{\epsilon_3})$. Since ψ is a homoeomorphism, we can find $\epsilon_0 > \epsilon_1 > 0$ such that $\psi(B_{\epsilon_1}) \subset B_{\epsilon_2} \subset \psi(B_{\epsilon_3})$. Let x be a point in $B_{\epsilon_1} - V$ and $y = \psi(x) \in B_{\epsilon_4} - \widetilde{V}$. We have

$$\begin{array}{ccccccc}
\pi_i(\psi(B_{\epsilon_1}) - \widetilde{V}, y) & \to & \pi_i(B_{\epsilon_2} - \widetilde{V}, y) & \to & \pi_i(\psi(B_{\epsilon_3}) - \widetilde{V}, y) & \to & \pi_i(B_{\epsilon_4} - \widetilde{V}, y) \\
\uparrow \cong & & & & \uparrow \cong & & \\
\pi_i(B_{\epsilon_1} - V, x) & & \xrightarrow{\quad\cong\quad} & & \pi_i(B_{\epsilon_3} - V, x). & &
\end{array}$$

Since $\pi_i(B_{\epsilon_1} - V, x) \to \pi_i(B_{\epsilon_3} - V, x)$ and $\pi_i(B_{\epsilon_2} - \widetilde{V}, y) \to \pi_i(B_{\epsilon_4} - \widetilde{V}, y)$ are isomorphisms, we see easily that $\pi_i(B_{\epsilon_1} - V, x) \to \pi_i(B_{\epsilon_2} - \widetilde{V}, y)$ is an isomorphism for all i.

Since $S_{\epsilon_1} - S_{\epsilon_1} \cap V \to S^1$ is a locally trivial fibration with fiber F_θ, we have the following exact sequence

$$\pi_{n+1}(S^1) \to \pi_n(F_\theta) \to \pi_n(S_{\epsilon_1} - S_{\epsilon_1} \cap V) \to \pi_n(S^1) \to$$

$$\cdots \to \pi_1(F_\theta) \to \pi_1(S_{\epsilon_1} - S_{\epsilon_1} \cap V) \to \pi_1(S^1) \to 0.$$

It follows that $\pi_n(S_{\epsilon_1} - S_{\epsilon_1} \cap V) \cong \pi_n(F_\theta)$ and $\pi_1(S_{\epsilon_1} - S_{\epsilon_1} \cap V) \cong \pi_1(S^1)$. Since $B_{\epsilon_1} - B_{\epsilon_1} \cap V$ is homotopy equivalent to $S_{\epsilon_1} - S_{\epsilon_1} \cap V$, $\pi_n(B_{\epsilon_1} - B_{\epsilon_1} \cap V) \cong \pi_n(S_{\epsilon_1} - S_{\epsilon_1} \cap V) \cong \pi_n(F_\theta)$ and $\pi_1(B_{\epsilon_1} - B_{\epsilon_1} \cap V) \cong \pi(S_{\epsilon_1} - S_{\epsilon_1} \cap V) \cong \pi_1(S^1)$. By Hurewicz theorem $\pi_n(F_\theta)$ is naturally isomorphic to $H_n(F_\theta)$ because F_θ is $(n-1)$-connected. Therefore by [W3] the generator h_* of $\pi_1(S^1)$ acts on $H_n(F_\theta)$ as homomorphism. Since Whitehead product is functorial by [W6], we have the following commutative diagram

$$\begin{array}{ccc}
H_n(F_\theta) & \xrightarrow{\quad h_* \quad} & H_n(F_\theta) \\
\downarrow & & \downarrow \\
H_n(\widetilde{F}_\theta) & \xrightarrow{\quad \tilde{h}_* \quad} & H_n(\widetilde{F}_\theta).
\end{array}$$

However by [W2], h_* is precisely the monodromy automorphism minus the identity map on $H_n(F_\theta)$. The theorem follows immediately.

Q.E.D.

Remark 3.2. The fact that Milnor number is an invariant of topological type was first observed by Teissier [25].

Definition 3.3. Let $(V,0) \subseteq (\mathbf{C}^{n+1},0)$ be an isolated hypersurface singularities. The generator $\pi_1(S^1)$ induces the monodromy automorphism $h^* : H^n(F_\theta, \mathbf{C}) \to H^n(F_\theta, \mathbf{C})$. Then the characteristic polynomial $\Delta_V(z)$ of the singularity $(V,0)$ is $\det(zI - h^*)$.

Corollary 3.4. Let $(V,0) \subseteq (\mathbf{C}^{n+1},0)$ be an isolated hypersurface singularites. Then the characteristic polynomial $\Delta_V(z)$ is an invariant of topological type of $(V,0)$.

Definition 3.5. Let $(V,0)$ be an isolated singularity in $(\mathbf{C}^{n+1},0)$. Denote $K_\epsilon = V \cap S_\epsilon$. By proposition 2.8, K_ϵ is independent of ϵ as a differentiable manifold. We shall denote it by K_V from now on. K_V is called the link of the singularity $(V,0)$.

THEOREM 3.6. *Let $(V,0)$ and $(\widetilde{V},0)$ be two isolated hypersurface singularities. If $(V,0)$ and $(\widetilde{V},0)$ have the same topological type, then $\pi_i(K_V) \cong \pi_i(K_{\widetilde{V}})$ for all i.*

Proof. Since $(V,0)$ and $(\widetilde{V},0)$ have the same topological type, there exist neighborhoods $U, \widetilde{U}$ of 0 and homeomorphism $\psi : U \to \widetilde{U}$ such that $\psi(U \cap V) = \widetilde{U} \cap \widetilde{V}$ and $\psi(0) = 0$. Let $\epsilon_0 > 0$ be sufficiently small so that $B_{\epsilon_0} \subset U \cap \widetilde{U}$ and Proposition 2.8 and Theorem 2.9 are applicable for both $(V,0)$ and $(\widetilde{V},0)$.

Let $\epsilon_0 > \epsilon_4 > 0$ be given. Since ψ is continuous, there is $\epsilon_0 > \epsilon_3 > 0$ such that $\psi(B_{\epsilon_3}) \subseteq B_{\epsilon_4}$. $\psi(B_{\epsilon_3})$ is an open neighborhood of 0. We can find $\epsilon_0 > \epsilon_2 > 0$ such that $B_{\epsilon_2} \subset \psi(B_{\epsilon_3})$. Since ψ is a homeomorphism, we can find $\epsilon_0 > \epsilon_1 > 0$ such that $\psi(B_{\epsilon_1}) \subset B_{\epsilon_2} \subset \psi(B_{\epsilon_3})$. Let x be a point in $B_{\epsilon_1} \cap V - \{0\}$ and $y = \psi(x)$ in $\psi(B_{\epsilon_1} \cap V - \{0\})$. We have the following commutative diagram

$$\begin{array}{ccc}
\pi_i(\psi(B_\epsilon)\cap\widetilde{V}-\{0\},y)\to\pi_i(B_{\epsilon_2}\cap\widetilde{V}-\{0\},y)\to & \pi_i(\psi(B_{\epsilon_3})\cap\widetilde{V}-\{0\},y)\to & \pi_i(B_{\epsilon_4}\cap\widetilde{V}-\{0\},y) \\
\uparrow \cong & & \uparrow \cong \\
\pi_i(B_{\epsilon_1}\cap V-\{0\},x) & \xrightarrow{\cong} & \pi(B_{\epsilon_3}\cap V-\{0\},x).
\end{array}$$

Since $\pi_i(B_{\epsilon_1} \cap V - \{0\}, x) \to \pi_i(B_{\epsilon_3} \cap V - \{0\}, x)$ and $\pi_i(B_{\epsilon_2} \cap \widetilde{V} - \{0\}, y) \to \pi_i(B_{\epsilon_4} \cap \widetilde{V} - \{0\}, y)$ are isomorphisms, we see easily that $\pi_i(B_{\epsilon_1} \cap V - \{0\}, x) \to \pi_i(B_{\epsilon_2} \cap \widetilde{V} - \{0\}, y)$ is an isomorphism. Observe that $B_{\epsilon_1} \cap V - \{0\}$ and $B_{\epsilon_2} \cap \widetilde{V} - \{0\}$ are homotopy equivalent to K_V and $K_{\widetilde{V}}$ respectively. Thus we have shown $\pi_i(K_V)$ is isomorphic to $\pi_i(K_{\widetilde{V}})$ for all i.

Q.E.D.

§4. Families of n-dimensional isolated hypersurface singularities.

Consider an anlaytic family of n-dimensional hypersurfaces having an isolated singularity at the origin where the Milnor number of the singularity at the origin does not change in this family. Under this hypothesis with $n = 1$, H. Hironaka conjectured that the topological type of the singularity does not change. In 1973, Lê and Ramanujam [12] gave a proof of this conjecture in the more general case of C^∞ family of n-dimensional hypersurface of dimension $n \neq 2$. The hypothesis $n \neq 2$ comes from the fact that they are using h-coberdism theorem. In fact they proved the following theorem.

THEOREM 4.1. *Let $F(t,z)$ be a polynomial in $z = (z_0, \ldots, z_n)$ with coefficients which are smooth complex valued functions of $t \in I = [0,1]$ such that $F(t,0) = 0$ and such that for each $t \in I$, the polynomials $\frac{\partial F}{\partial z_i}(t,z)$ in z have an isolated zero at 0. Assume moreover that the integer*

$$\mu_t = \dim_{\mathbf{C}} \mathbf{C}\{z\}/(\frac{\partial f}{\partial z_0}(t,z), \ldots, \frac{\partial f}{\partial z_n}(t,z))$$

is independent of t. Then the monodromy fibrations of the singularities of $F(0,z) = 0$ and $F(1,z) = 0$ at 0 are of the same fiber homotopy. If further $n \neq 2$, these fibrations are even differentiably isomorphic and the topological types of the singularities are the same.

Unfortunately it is still an open problem (actually a difficult one) whether two isolated hypersurface singularities having the same topological type can be connected by a μ-constant family. Nevertheless Lê-Ramanujan's theorem gives an important tool to study the topological types of isolated hypersurface singularities.

In order to get some conditions which will connect two given isolated hypersurface singularities by a μ-constant family, we need a few definitions and facts from commutative algebra.

Definition 4.2. Let A be a noetherian normal local domain of dimension $n+1$ with maximal ideal $\mathcal{M}$. Let I be an $\mathcal{M}$-primary ideal. An element $f \in A$ is said to be integral over I if it satisfies an equation

$$f^k + a_1 f^{k-1} + \cdots + a_k = 0, \quad a_i \in I^i.$$

The set $\overline{I}$ of elements of A integral over I form an ideal containing , called the integral closure of I.

We are now ready for the next corollary which again was due to Lê-Ramanujan [12]. For any $f \in \mathbf{C}[z_0, \ldots, z_n]$ with an isolated singular point at origin, define

$$I_1(f) = \sum_{i=0}^{n} \frac{\partial f}{\partial z_i} \mathbf{C}\{z_0, \ldots, z_n\}$$

$$I_2(f) = (f)\mathbf{C}\{z_0, \ldots, z_n\} + I_1(f)$$

$$I_3(f) = \overline{I_1(f)} = \text{integral closure of } I_1(f).$$

Since f is integral over $I_1(f)$, it follows that $I_1(f) \subseteq I_2(f) \subseteq I_3(f)$. Define the Artinian local algebras $A_i(f)$ by

$$A_i(f) = \mathbf{C}\{z_0, \ldots, z_n\}/I_i(f).$$

Corollary 4.3. Suppose $f, g \in \mathbf{C}[z_0, \ldots, z_n]$ with isolated singular point at origin. Assume that for some i $(1 \leq i \leq 3)$ we are given an isomorphism over $\mathbf{C}$, $\lambda : A_i(f) \cong A_i(g)$. Then the singularities of $f = 0$ and $g = 0$ at 0 are of the same topological type.

Remark 4.4. In fact, under the assumption that $A_2(f)$ is isomorphic to $A_2(g)$ as $\mathbf{C}$-algebra, then Mather and the author [15] have shown that the singularities of $f = 0$ and $g = 0$ at 0 are actually of the same analytic type.

Remark 4.5. We can given $A_1(f)$ a $\mathbf{C}\{t\}/(t^{n+1})$ algebra structure by defining $\overline{a(t)} \cdot \overline{u} = \overline{a(f)u}$ for any $a(t) \in \mathbf{C}\{t\}$ and any $u \in \mathbf{C}\{z_0, \ldots, z_n\}$. Here $\overline{a(t)}$ and $\overline{u}$ are the image of $a(t)$ and u in $\mathbf{C}\{t\}/(t^{n+1})$ and $A_1(f)$ respectively. By a theorem of Briancon and Skoda, the above definition is well defined. In [28], (cf. also [24]), we proved that if $A_1(f)$ is isomorphic to $A_1(g)$ as $\mathbf{C}\{t\}/(t^{n+1})$ algebra, then there is a $\mathbf{C}$-algebra automorphism ϕ of $\mathbf{C}\{z_0, \ldots, z_n\}$ such that $f = g_0\phi$.

§5. Classification of topological type s for surface singularities.

As we saw in §2, the topological types of plane curve singularities were completely understood by the end of sixties. However, after almost twenty years, there was no progress in understanding the topological types of surface singularities. In fact, there was not even a conjecture what the result should be. Recently, by working on Zariski multiplicity problem, we come up with the following conjecture.

Conjecture. Let $(V, 0)$ be an isolated hypersurface singularity in $(\mathbf{C}^3, 0)$. Then the topological type of $(V, 0)$ determines and is determined by the characteristic polynomial $\Delta_V(z)$ of $(V, 0)$ and the fundamental group the $\pi_1(K)$ of the link of $(V, 0)$.

Remark 5.1. We know that the topological type of $(V, 0)$ determines $\Delta_V(z)$ and $\pi_1(K)$ by Theorem 3.1 and Theorem 3.6.

Recall that a hypersurface singularity $(V, 0) = \{(z_0, \ldots, z_n) : f(z_0, \ldots, z_n) = 0\} \subseteq \mathbf{C}^{n+1}$ is quasi-homogeneous if f is in the Jacobian ideal of f, i.e. $f \in (\frac{\partial f}{\partial z_0}, \ldots, \frac{\partial f}{\partial z_n})$. Recently Xu and the author [27] proved that the above conjecture is true for quasi-homogeneous surface singularities. Namely we proved the following theorem.

THEOREM 5.2. *Let $(V, 0)$ and $(W, 0)$ be two isolated quasi-homogeneous surface singularities in $\mathbf{C}^3$. Then $(\mathbf{C}^3, V, 0)$ is homeomorphic to $(\mathbf{C}^3, W, 0)$ if and only if $\pi_1(K_V) \cong \pi_1(K_W)$ and $\Delta_V(z) = \Delta_W(z)$.*

In fact Xu and the author have also proved the following theorem which is of independent interest.

THEOREM 5.3. *Let $(V, 0)$ and $(W, 0)$ be two isolated quasi-homogeneous surface singularities having the same topological type. Then $(V, 0)$ is connected to $(W, 0)$ by family of constant topological type. In fact $(V, 0)$ is connected to one of the following seven class by a family of constant topological type:*

class I	$V(I) = \{z_0^{a_0} + z_1^{a_1} + z_2^{a_2} = 0\}$.
class II	$(II) = \{z_0^{a_0} + z_1^{a_1} + z_1 z_2^{a_2} = 0\} \quad a_1 > 1$
class III	$V(III) = \{z_0^{a_0} + z_1^{a_1} z_2 + z_2^{a_2} z_1 = 0\} \quad a_1 > 1,\ a_2 > 1$
class IV	$V(IV) = \{z_0^{a_0} + z_0 z_1^{a_1} + z_1 z_2^{a_2} = 0\}$
class V	$V(V) = \{z_0^{a_0} z_1 + z_1^{a_1} z_2 + z_0 z_2^{a_2} = 0\}$
class VI	$V(VI) = \{z_0^{a_0} + z_0 z_1^{a_1} + z_0 z_2^{a_2} + z_1^{b_1} z_2^{b_2} = 0\}$
	where $(a_0 - 1)(a_1 b_2 + a_2 b_1) = a_0 a_1 a_2$
class VII	$V(VII) = \{z_0^{a_0} z_1 + z_0 z_1^{a_1} + z_0 z_2^{a_2} + z_1^{b_1} z_2^{b_2} = 0\}$
	where $(a_0 - 1)(a_1 b_2 + a_2 b_1) = a_2(a_0 a_1 - 1)$.

Definition 5.4. A polynmomial $h(z_0, \ldots, z_n)$ is weighted homogeneous of type $(w_0, \ldots, w_n)$, where $(w_0, \ldots, w_n)$ are fixed positive rational numbers, if it can be expressed as a linear combination of monomials $z_0^{i_0} z_1^{i_1} \ldots z_n^{i_n}$ for which $i_0/w_0 + \ldots + i_n/w_n = 1$. $(w_0, w_1, \ldots, w_n)$ is called the weights of polynomials h. Let $w_i = u_i/v_i$ be the reduced fraction of w_i i.e. u_i and v_i are integers with $(u_i, v_i) = 1$.

By the theorem of Saito [23], we may assume from now on that $w_i \geq 2$ for $i = 0, \ldots, n$. Saito also proved that quasi-homogeneous function with isolated singularity at 0 can be put into weighted homogeneous polynomial by a biholomorphic change of coordinates. The following proposition which is a consequence of Milnor and Orlik [17] is due to Yoshinaga [31].

Proposition 5.5. Let $f(x_0, \ldots, x_n)$ (respectively $g(x_0, \ldots, x_n)$) be a weighted homogeneous polynomial with weights $(\frac{u_0}{v_0}, \ldots, \frac{u_n}{v_n})$ (respectively $(u'_0/v'_0, \ldots, u'_n/v'_n)$) where u_i/v_i (respectively u_i/v_i) is the reduced fraction of w_i (respectively w'_i). Assume that f (respectively g) has an isolated singularity at origin. Then $\Delta_f(z) = \Delta_g(z)$ if and only if the following two conditions are satisfied:

(1) $\{2, u_0, \ldots, u_n\} = \{2, u'_0, \ldots, u'_n\}$

(2) For any $u \in \{2, u_0, \ldots, u_n\}$ $\prod_{u_i = u} (1 - \frac{u_i}{v_i}) = \prod_{u'_j = u} (1 - \frac{u'_j}{v'_j})$ where the product over an empty set is assumed to be one.

The proof of Theorem 5.2 and Theorem 5.3 made use of the fundamental results of Neumann [18] and Orlik-Wagreich [19], the above proposition 5.5 and the deep theory below due to Varchenko [26].

Let $\mathbf{N} \subset \mathbf{R}_+$ be the set of all nonnegative integers and of all nonnegative real numbers. Let $f = \Sigma a_k x^k$, $a_k \in \mathbf{C}$, $k \in \mathbf{N}^{n+1}$, be an element in $\mathbf{C}\{x_0, \ldots, x_n\}$ and supp f be the set $\{k \in \mathbf{N}^{n+1} : a_k \neq 0\}$. We denote by $\Gamma_+(f)$. The convex hull of the set $\bigcup_{k \in \text{supp } f} (k + \mathbf{R}_+^{n+1})$, in $\mathbf{R}_+^{n+1}$. The polyhedron $\Gamma(f)$ which is the union of all compact facets of $\Gamma_+(f)$ will be called Newton's diagram of the power series f. The polynomial $\sum_{k \in \Gamma(f)} a_k x^k$ will be called the main part of the power series f. Let γ be a closed facet of $\Gamma(f)$. Let us denote the polynomial $\sum_{k \in \gamma} a_k x^k$ by f_γ. The main part of the power series f will be called nondegenerate if for any closed facet $\gamma \in \Gamma(f)$ the polynomials $(x_0 \frac{\partial f_\gamma}{\partial x_0}), \ldots, (x_n \frac{\partial f_\gamma}{\partial x_n})$ have no common zero in $\{\in \mathbf{C}^{n+1} : x_0 \ldots x_n \neq 0\}$.

We shall define the notion of characteristic polynomial $\Delta_\Gamma(z)$ associated with the Newton's diagram $\Gamma(f)$. Let

$$\Delta_\Gamma(z) = \left[\prod_{\ell=1}^{n+1} \Delta^\ell(z)^{(-1)^{n+\ell+1}} \right] (z-1)^{(-1)^{n+1}}$$

where Δ^ℓ is a polynomial defined as below. Δ^ℓ is defined by the $(\ell - 1)$ dimensional facets of the intersections of $\Gamma(f)$ with all possible ℓ-dimensional coordinate planes.

Let L be ℓ-dimensional affine subspace of $\mathbf{R}^{n+1}$ such that $L \cap \mathbf{Z}^{n+1}$ is ℓ-dimensional lattice. By definition let the ℓ-dimensional volume of the cube (spanned by any basis of $L \cap \mathbf{Z}^{n+1}$) be equal to one.

Now we shall define Δ^ℓ. Let $I \subseteq \{0,.1,\dots,n\}$ and $|I| = \ell$ where $|I|$ is the number of the elements of I. Let us consider the pair L_I, $L_I \cap \Gamma(f)$, where $L_I = \{k \in \mathbf{R}^{n+1} : k_i = 0 \quad \forall i \notin I\}$. Let $\Gamma_1(I),\dots,\Gamma_{j(I)}(I)$ be all $(\ell-1)$-dimensional facets of $L_I \cap \Gamma(f)$ and $L_1,\dots,L_{j(I)}$ be the $(\ell-1)$-dimensional affine subspaces, containing them respectively.

Let $\sum_{i\in I} a_i^j k_i = m_j(I)$ be the equation of L_j in L_I where a_i^j, $m_j(I) \in \mathbf{N}$ and the greatest common divisor of the numbers a_i^j, $i \in I$, is equal to one. The numbers $a_i^j, m_j(I)$ are defined by these conditions uniquely. The numbers $m_j(I)$ will take part in the definition of Δ^ℓ. Another definition of $m_j(I)$ is the following. Consider the quotient of the lattice $\mathbf{Z}^{n+1} \cap L_I$ by the subgroup generated by vectors of $\mathbf{Z}^{n+1} \cap L_j$. This is a cyclic group of order $m_j(I)$. Let $V(\Gamma_j(I))$ be the $(\ell-1)$-dimensional volume of $\Gamma_j(I)$ in L_j. Let

$$\Delta^\ell(z) = \prod_{I,|I|=\ell} \prod_{j=1}^{j(I)} (z^{m_j(I)} - 1)^{(\ell-1)!V(\Gamma_J(I))}.$$

It was observed by Varchenko [26] that $m_j(I)(\ell-1)!V(\Gamma_j(I))$ is equal to $\ell!$ multiplied by the ℓ-dimensional volume of the cone over $\Gamma_j(I)$ with vertex at origin. Accordinmg to this remark degΔ^ℓ contains the following geometric meaning. Let $\Gamma_-(f)$ be the cone over $\Gamma(f)$ with vertex at the origin. Then degΔ^ℓ is the sum of ℓ-dimensional volumes of the intersections of $\Gamma_-(f)$ with all possible ℓ-dimensional coordinate planes, multiplied by $\ell!$. The following theorem due to Varchenko is of fundamental importance.

THEOREM 5.6. *Let f belong to the square of the maximal ideal of $\mathbf{C}\{x_0, x_1, \dots, x_n\}$ and let the main part of the power series f be nondegenerate. Then the characteristic polynomial of the monodromy of f at the origin is equal to the characteristic polynomial of the Newton diagram of f.*

In fact in our original proof of Theorem 5.2 and Theorem 5.3, we did not make use of proposition 5.5, we only need Theorem 5.6.

Let $f(z_0,\dots,z_n)$ be a weighted homogeneous function with weights $(w_0,\dots,w_n)$. Then there exist non-zero integers $q_0,\dots,q_n$ and a positive integer d so that

$$f(t^{q_0}z_0,\dots,t^{q_n}z_n) = t^d f(z_0,\dots,z_n).$$

In fact let $\langle w_0,\dots,w_n\rangle$ denote the smallest positive integers d such that there exists, for each i, an integer q_i, so that $q_i w_i = d$. These are the q_i and d above.

Definition 5.7. A function f is semi-quasi-homogeneous if $f = f_0 + f'$, where f_0 is quasi-homogeneous of degree d and has an isolated singularity at 0 and all the monomials of f' are of degree greater than d.

In [1] Arnold gave normal forms for semi-quasi-homogeneous function in the following manner.

THEOREM 5.8. *A semi-quasi-homogeneous function f with weighted homogeneous part f_0 is biholomorphically equivalent to the normal form $f_0 + c_1e_1 + \ldots + c_re_r$ where the c_i are numbers and the e_i are basis monomials of the Milnor algebra*

$$\mathbf{C}\{z_0, \ldots, z_n\}/(\frac{\partial f_0}{\partial z_0}, \ldots, \frac{\partial f_0}{\partial z_n})$$

of the function f_0 of degree greater than $d =$ degree of f_0.

By the remark 2.5 of Lê-Ramanjan [12], the singularities $f = 0$ and $f_0 = 0$ at 0 have the same topological type.

Consequently we have the following theorem.

THEOREM 5.9. *Theorem 5.2 and Theorem 5.3 are true for semi-quasi-homogeneous singularities.*

§6. ZARISKI MULTIPLICITY PROBLEM

In his retiring presidential address to the American Mathematical Society in 1971, Zariski [34] asked whether $(V, 0)$ and $(W, 0)$ have the same multiplicity if they have the same topological type. He expected that topologists would be able to answer his question in relatively short order. However the question appears much harder than what Zariski thought. Even special cases of Zariski's problem have proved to be extremely difficult. Only recently Greuel [6] and O'shea [20] proved independnetly that topological type constant family of isolated quasi-homogeneous singularities are equi-multiple. For quasi-homogeneous surface singularites, Laufer [8] explained the constant multiplicity for topological type constant family of singularities from a different viewpoint. However, it was not known that whether two quasi-homogeneous singularities having the same topological type can be put into a topological type constant family. In [27], Xu and the author proved Theorem 5.3. Thus Zariski problem is solved affirmatively in this case. Actually we solved the problem directly without using the result of Greuel and O'shea. In fact, in view of Theorem 5.9 above, we deduce the following theroem.

THEOREM 6.1. *Let $(V,0)$ and $(W,0)$ be two isolated semi-quasi-homogeneous singularities in $\mathbf{C}^3$. If $(\mathbf{C}^3, V, 0)$ is homeomorphic to $(\mathbf{C}^3, W, 0)$ as germs, then V and W have the same multiplicity at the origin.*

Let $(V,0)$ be a dimension two isolated hypersurface singularity. Lê and Teissier [13] observed that A´Campo's work [2] can often be used to give positive results towards Zariski's question. Let $C(V,0)$ be the reduced tangent cone. Let $P \subset (V,0)$ denote the hypersurface in $\mathbf{CP}^2$ over which $C(V,0)$ is a cone. Then, the work of A´Campo shows that the multiplicity of $(V,0)$ is determined by the topological type of $(V,0)$ in case the topological Euler number $\chi(\mathbf{P}C(V,0))$ is non-zero. The same arguments also show that, for isolated hypersurface two-dimensional singularities, the embedded topology and the multiplicity determine $\chi(\mathbf{P}C(V,0))$. However, so far, by using A´Campo's result, one can only prove that a surface in $\mathbf{C}^3$ having at 0 a singularity of multiplicity 2 cannot have the same topological type at 0 as another surface singularity of multiplicty different from 2.

For plane curve singularites, the Zariski question was known to be true. The reason that the Zariski question could be answered was that the topological types of plane curve singularities were well understood. If $(m_1, n_1), \ldots, (m_g, n_g)$ are the Puiseux pairs for plane irreducible curve singularity, then one knows that $n_1 \cdots n_g$ is the multiplicity of the singularity (cf. [9]).

Definition 6.2. Let $(V,0)$ be a normal two dimensional singularity. Let $\pi : (M, A) \to (V,0)$ be a resolution with exceptional set A. The geometric genus of a normal two dimensional singularity $(V,0)$ is the integer

$$P_g(V,0) = \dim_{\mathbf{C}} R^1\pi_*(\mathcal{O}_M)_0.$$

The arithmetic genus of a normal two dimensional singularity $(V,0)$ is the integer

$$p_a(V,0) = \underset{D}{sup} p_a(D),$$

where D is a positive cycle and $p_a(D)$ is the virtual genus of D on M.

In [30], we first observed the following theorem.

THEOREM 6.3. *Let $(V,0)$ be an isolated hypersurface two dimensional singularities. Then $p_g(V,0)$ and $p_a(V,0)$ are invariants of topological type of $(V,0)$.*

As a result of the above observation, we have proved the following special case of Zariski's multiplicity conjecture.

THEOREM 6.4. *Let $(V,0)$ and $(W,0)$ be two isolated two-dimensional hypersurface singularities in $\mathbf{C}^3$ having the same topological type. If $p_a(V,0) \leq 2$, then $\nu(V,0) = \nu(W,0)$ where $\nu(V,0)$ and $\nu(W,0)$ are the multiplicities of $(V,0)$ and $(W,0)$ respectively.*

REFERENCES

1. Arnold, V.I., Normal forms of functions in neighborhood of degenerate critical points, Russian Math. Surveys, 29 (1974), 10–50.

2. A'Campo, N., La fonction zeta d'une monodromie, Comment. Math. Helv., 50 (1975), 233–248.

3. Benson, M., and Yau, S.S.-T, Equivalence between isolated hypersurface singularities. (preprint).

4. Brauner, K., Zur Geometrie der Funktionen Zweier komplexen Veränderlicken, Abh. Math. Sem. Hamburg, 6 (1928), 1–54.

5. Burau, W., Kennzeichnung der Schlauchknoten, Abh. Math. Sem. Hamburg, 9 (1932), 125–133.

6. Greuel, G.M., Constant Milnor number implies constant multiplicity for quasi-homogeneous singularities, Manuscripta Math. 56 (1986), 159–166.

7. Hu, S.T., *Homotopy Theory*, Academic Press, New York (1959).

8. Laufer, H., Tangent cones for deformations of two-dimensional quasi-homogeneous singularities (preprint).

9. Lê Dũng Tráng, Sur les noeuds algebriques, Compositio Mathematica, 25 (1972), 282–322.

10. Lê Dũng Tráng, Topologie des singularités des hypersurfaces complexes, Astérisque 7 et 8 (1973), 171–182.

11. Lê Dũng Tráng, Three lectures on local monodromy, Lecture Notes Series No. 43, Aarhus Universitet, Sept. 1974.

12. Lê Dũng Tráng and Ramanujam, C.P., The invariance of Milnor's number implies the invariance of the topological type, Amer. Journ. Math., vol. 98 (1976), 67–78.

13. Lê Dũng Tráng and Teissier B., Report on the problem sessions, Proc. Symp. Pure Math., 40 part 2 (1983), 105–116.

14. Lejeune, M., Sur ℓ' equivalence des singularité des courbes algebroides planes, Coefficients de Newton, Centre de Math. del ℓ' Ecole Polytechnique, 1969.

15. Mather, J. and Yau, S.S.-T., Classification of isolated hypersurface singularities by their moduli algebras, Invent. Math 69 (1982), 243–251.

16. Milnor, J., Singular points of complex hypersurfaces, Ann. of Math. Stud., 61, Princeton University Press, 1968.

17. Milnor, J., and Orlik, P., Isolated singularities defined by weighted homogeneous polynomials, Topology, vol. 9 (1970), 385–393.

18. Neumann W., A calculus for plumbing applied to the topology of complex surface singularities and degenerating complex curves, Trans. Amer. Math. Soc., 268 (1981), 299–344.

19. Orlik, P. and Wagreich, P., Isolated insgulariteis of algebraic surfaces with $\mathbf{C}^*$-action, Ann. Math., 93 (1971), 205–228.

20. O'Shea, D., Topological trivial deformations of isolated quasi-homogeneous hypersurface are equi-multiple, Proceedings A.M.S. vol. 101 (1987), 260–262.

21. Palamodov, V.I., Multiplicity of a holomorphic transformation, Func. Analysis and Appl. I (1967), 218–226.

22. Reeve, J., A summary of results in the topological classification of plane algebroid singularities, Rend. Sem. Math. Toringo, 14 (1954/1955), 159–187.

23. Saito, K., Quasi-homogene isolierte Singularitäten von Hyperflächen, Invent. Math., 14 (1971), 123–142.

24. Scherk, J., A propos d'un théoréme de Mather et Yau, Note présentee par B. Madgrange, C.R. Acad. Sc. Paris, *t.* 296, (28 Mars 1983).

25. Teissier, B., Deformation á type topologique constant I, II: in Séminaire Douady-Verdier 1971–72, Astérisque 16 (Societé Mathématique de France), (1974), 215–249.

26. Varchenko, A.N., Zeta function of monodromy and Newton's diagram, Invent. Math., 37 (1976), 253–262.

27. Xu, Yijing and Yau, S.S.-T., Classification of topological types of isolated quasi-homogeneous two dimensional hypersurface singularites. (preprint)

28. Yau, S.S.-T, Criteria for right-left equivalence and right equivalence of holomorphic functions with isolated critical points, Proceedings of Symposia in Pure Mathematis, A.M.S. vol. 41, (1984), 291–297

29. Yau., S.S.-T. Topological types and multiplicities of isolated quasi-homogeneous surface singularities, (to appear), Bulletin A.M.S.

30. Yau, S.S.-T, The multiplicity of isoalted two dimensional hypersurface singularities: Zariski problem, (preprint).

31. Yoshinaga, E., Topological types of isolated singularities defined by weighted homogeneous polynomials, J. Math. Soc. Japan, 35 (1983), 431–436.

32. Zariski, O. On the topology of algebroid singularities, Amer. J. Math. 54 (1932), 433–465.

33. Zariski, O., Contribution to the problem of equi-singularity in question on algebraic varieties, C.I.M.E. Varenna 1969, Edizioni Cremonese, Roma (1970), 265–343.

34. Zariski, O., Some open questions in the theory of singularities, Bull. A.M.S. 77 (1971), 481–491.

35. Zariski, O., General theory of saturation and saturated local rings, II, Amer. J. Math., 93 (1971), 872–964.

DEPARTMENT OF MATHEMATICS, STATISTICS, AND COMPUTER SCIENCE
UNVERSITY OF ILLINOIS AT CHICAGO
CHICAGO, ILLINOIS, 60680

Department of Mathematics, Statistics, and Computer Science
Box 4348, m/c 249
University of Illinois at Chicago
Chicago, Illinois, 60680

Contemporary Mathematics
Volume **101**, 1989

Complex Foliations

T. Duchamp M. Kalka *

June 30, 1988

Abstract

Complex foliations have surfaced as a useful tool for studying certain problems in complex variables. Previous applications have generally involved Monge-Ampère foliations, which are foliations associated to solutions of the Monge-Ampère equation. However, many properties of Monge-Ampère foliations extend to a larger class of complex foliations, which we call symmetric foliations and which are characterized by a symmetry property of their anti-holomorphic twist tensor. We review here some of these applications and disuss our work on the geometry of complex foliations, with particular emphasis on properties of symmetric foliations.

A *complex foliation* is a foliation of a complex manifold by complex submanifolds. If, in addition, the tangent bundle of the foliation is a holomorphic subbundle of the tangent bundle of the manifold then the foliation is said to *holomorphic*.

Complex foliations arise naturally in several complex variables, particularly in the study of the complex homogeneous Monge-Ampère equation; and this paper begins with a short review of some of these applications. The *complex Bott connection* is introduced in Section 2 and used to give generalizations of some of the results of [BK], [BB] and [B]. The local geometry of complex foliations of complex surfaces is particularly rigid and is discussed in the final section of the paper.

1 Complex Monge-Ampère Foliations

To give the reader a feeling for the role of complex foliations in the study of several complex variables, we will review here a few applications from the literature.

1980 Mathematics Subject Classification (1985 Revision). 53C12
This paper is in final form and no version of it will be submitted for publication elsewhere.
*The second author's research is partially supported by Louisiana Board of Regents grant #TUU1-091-06.

A Maximum Principle

It has been recognized for some time that the complex Monge-Ampère equation is an important tool in the study of problems in several complex variables and in complex geometry. For a smooth function, u, on a complex n-dimensional manifold the equation in question is given by the formula

$$(\partial\overline{\partial}u)^n \equiv \underbrace{\partial\overline{\partial}u \wedge \ldots \wedge \partial\overline{\partial}u}_{n \text{ terms}} = 0.$$

In local coordinates, $z = (z^i)$, the equation assumes the form

$$\det\left(\frac{\partial^2 u}{\partial z^i \partial \overline{z}^j}\right) = 0.$$

Work of Bedford and Taylor [BT] has shown that the operator $(\partial\overline{\partial})^n$ can profitably be thought of as a non-linear analogue in n complex variables of the Laplacian $\Delta = \frac{1}{4}\frac{\partial^2}{\partial z \partial \overline{z}}$.

Among the difficulties encountered in studying the Monge-Ampère equation is the fact that it is neither linear nor elliptic. One approach, introduced in [BK], is to associate a complex foliation to a solution of the homogeneous Monge-Ampère equation; properties of the foliation can then be used to deduce properties of the solution.

Specifically, let M be a complex manifold of complex dimension n, let q be a positive integer less than n and consider a smooth real valued, plurisubharmonic function u satisfying the conditions

$$(\partial\overline{\partial}u)^{q+1} \equiv 0, \text{ and } (\partial\overline{\partial}u)^q \neq 0.$$

To such a function u one can associate a foliation $\mathcal{F}$ of M by $(n-q)$ dimensional complex submanifolds, the *Monge-Ampère foliation associated to u*. The leaves of $\mathcal{F}$ are the maximal dimensional complex manifolds on which the function u is pluriharmonic, i.e. if N is a leaf of $\mathcal{F}$ then the equation $i_N^*(\partial\overline{\partial}u) = 0$ holds.

To prove this just define $T\mathcal{F}$, the tangent bundle to the leaves of the foliation, to be the annihilator bundle of the complex Hessian of u,

$$\mathrm{Ann}(\partial\overline{\partial}u) = \{X \in T^{1,0}(M) | \partial\overline{\partial}u(X, \overline{Y}) = 0 \quad \forall Y \in T^{1,0}(M)\}.$$

It is not difficult to see that $T\mathcal{F}$ is a complex subbundle of rank $n-q$ and is an integrable real vector subbundle of TM. It follows that $\mathcal{F}$ is a complex foliation. A complex foliation which is locally the Monge-Ampère foliation associated to a plurisubharmonic function u is said to be a *Monge-Ampère foliation.*

In [BK] Monge-Ampère foliations are used to establish a weak maximum principle for the gradient of a solution of the equation $(\partial\overline{\partial}u)^n = 0$ on a bounded domain $D \subset \mathbb{C}^n$:

Theorem (Bedford-Kalka). *If $u \in C^3(\overline{D})$ is a real, not necessarily plurisubharmonic, solution of the equation, $(\partial\overline{\partial})^n u = 0$, then the equality*

$$\max_{z\in\overline{D}} \left|\frac{\partial u(z)}{\partial z^j}\right| = \max_{z\in\partial D} \left|\frac{\partial u(z)}{\partial z^j}\right|$$

holds.

Because each solution of the Monge-Ampère equation equation yields a complex foliation, it is natural to ask if all complex foliations arise (locally) in this way. It is shown in [BK] that all codimension 1 complex foliations are (locally) Monge-Ampère. However, necessary conditions are given in [BK] which show that in higher codimensions the generic complex foliation is *not* Monge-Ampère. As presented in [BK], the conditions are somewhat *ad hoc*; however, as will be seen below they have a natural interpretation within the context of the Bott connection.

Holomorphic Rigidity of "Annuli"

In [BK] the Hermitian metric on the normal bundle of a Monge-Ampère foliation is not used; as far as we know, the first place it is used in the the paper of Bedford and Burns [BB] where it is used to prove a rigidity theorem.

Specifically they consider domains of the form $\Omega = D_1 \backslash \overline{D}_0$ where $D_0 \subset\subset D_1 \subset\subset \mathbb{C}^n$ are strongly pseudoconvex domains with smooth boundary in $\mathbb{C}^n$ and $\overline{D}_0$ is connected and holomorphically convex in D_1. Suppose further that u is a solution of the following Dirichlet problem:

$$(1.1)\qquad \begin{cases} u & \text{plurisubharmonic} \\ u = 1 & \text{on } \partial D_0 \\ u = 0 & \text{on } \partial D_1 \\ (\partial\overline{\partial}u)^n = 0 & \text{on } \Omega. \end{cases}$$

The (unique) solution of (1.1) achieves the supremum of the Chern, Levine, Nirenberg norm for the homology class $\Gamma_\Omega = [\partial D_0] = [\partial D_1]$. The Chern, Levine, Nirenberg norm is defined as follows: Let $\mathcal{P}$ denote the class of $C^2(\Omega)$ plurisubharmonic functions u, with $0 < u < 1$ and $(\partial\overline{\partial}u)^n = 0$. If $\Gamma \in H_{2n-1}(\Omega, \mathbb{R})$ is a homology class, then the norm of Γ is defined as

$$(1.2)\qquad N(\Gamma) = \sup_{\mathcal{P}} \int_\Gamma d^c v \wedge (dd^c v)^{n-1},$$

where $d^c = i(\partial - \overline{\partial})$. The rigidity theorem of Bedford-Burns is the following:

Theorem (Bedford-Burns). *Let Ω_j, $j = 1, 2$, be two simply connected domains of the type described above and suppose the solutions u_j of (1.1)] on Ω_j are in $C^4(\overline{\Omega}_j)$ and satisfy $(\partial\overline{\partial}u_j)^{n-1} \neq 0$. If $N([\Gamma_{\Omega_1}]) = N([\Gamma_{\Omega_2}])$ and $f : \Omega_1 \to \Omega_2$ is a holomorphic mapping with $H^{2n-1}(f) \neq 0$, then f is a biholomorphism.*

The assumption that $(\partial\bar{\partial}u_j)^{n-1} \neq 0$ is necessary to guarantee the existence of a non-singular foliation. The strong pseudoconvexity and the Dirichlet boundary condition then guarantee that the leaves of the foliation are transverse to both the inner and the outer boundaries. Transversality is used to prove a uniqueness result for suprema of solutions of (1.2)]. Finally, that f is a biholomorphism follows from uniqueness.

It is natural to try to generalize the theorem by relaxing the assumption that $(\partial\bar{\partial}u)^{n-1} \neq 0$—in fact if D_0 has more than one component the condition $(\partial\bar{\partial}u)^{n-1} \neq 0$ is obstructed topologically, for it implies that the Monge-Ampère foliation associated to u is transverse to $\partial\Omega$, yielding a field of 1-planes on $\partial\Omega$, thus forcing the vanishing of its Euler characteristic.

Such a generalization is also proved in [BB] under the assumption that each of the functions u_j is a real analytic:

> **Theorem (Bedford-Burns).** *Let Ω_1, Ω_2 be as in the statement of (1.2)] and assume that the solution of (1.1)] on Ω_j is real analytic on Ω_j and C^3 on the inner boundary of Ω_j. Let Γ_j denote the homology class defining the outer boundary of Ω_j, and suppose $N(\Gamma_1) = N(\Gamma_2)$. If $f_*(\Gamma_1) = \alpha\Gamma_2 + \sum_{i=1}^k c_i\gamma_i$, where $\alpha \neq 0, c_i \geq 0$ and γ_i denotes the homology classes defined by the inner boundary components of Ω_2, then f is a covering map.*

The proof is similar to the proof with the non-degeneracy assumption: leaves of the foliation must be shown to intersect both the inner and outer boundary components of Ω. The details of the proof are somewhat involved, for relaxing the non-degeneracy condition, $(\partial\bar{\partial}u)^{n-1} \neq 0$, raises the possibility that the foliation defined by u is singular; and we will not attempt to give it here.

However, it is worth noting that the proof requires the introduction of the *anti-holomorphic twist tensor*. This is a measure of the degree to which a complex foliation fails to be holomorphic, and is the fundamental local invariant of a complex foliation. Bedford and Burns derive a formula for the Ricci tensor of the form $\omega = dd^c u$ in terms of the anti-holomorphic twist tensor which they then use to prove the following result:

> **Theorem (Bedford-Burns).** *The Ricci form of ω is non-negative and vanishes if and only if the foliation $\mathcal{F}$ is holomorphic.*

We will see in Section 2 that this theorem holds for a much larger class of foliations.

Curvature of Monge-Ampère Equations

The proof given by Burns [B] of the uniformization theorem of Stoll [S] is another example of an application of complex foliations to complex variables. Stoll's theorem is the following:

Theorem (Stoll). *Let M be a connected n-dimensional complex manifold, $\tau : M \to [0, \infty)$ a smooth exhaustion which is strictly plurisubharmonic and such that $dd^c \log \tau \geq 0$, $(dd^c \log \tau)^n = 0$ on $M_* = M \backslash \{\tau = 0\}$. Then there is a biholomorphic map*

$$\Phi : \mathbb{C}^n \to M,$$

such that $\Phi^* \tau = ||z||^2$.

Before discussing Burns' proof, a remark on the common structure of all the proofs of this theorem (see [B], [S], [W]) is in order: They use the fact that $\log \tau$ is a solution of the Monge-Ampère equation on M_*. The resulting foliation can then be used to study the geometry of M. It is easy to see that the the form $\partial\bar{\partial}\tau$ is non-degenerate and gives M the structure of a Kähler manifold. The map Φ in the theorem is the exponential map in this metric. It is not difficult to see that Φ is holomorphic in directions tangent to the foliation. *To show that Φ is holomorphic in the normal directions is equivalent to showing that the foliation is holomorphic.*

Burns gives two proofs that the foliation is holomorphic. We sketch here the second. Choose coordinates, (w, z), $w \in \mathbb{C}$, $z \in \mathbb{C}^{n-1}$, such that the set $z = 0$ is a leaf of the foliation. The Ricci form of $dd^c \log \tau$ on the restriction the normal bundle of the leaf is given by the formula

$$\varphi = \frac{i}{2} \log \det(\log \tau)_{i\bar{j}}\, dw \wedge d\bar{w}, \qquad 1 \leq i, j \leq n-1.$$

To prove that the foliation is holomorphic, first Burns proves the *Shottky-Landau Theorem for Foliations*:

Theorem (Burns). *Let $\mathcal{F}$ be a foliation of the complex manifold M defined by a plurisubharmonic solution u of the Monge-Ampère equation with $(dd^c u)^{n-1} \neq 0$. Let L be a leaf of $\mathcal{F}$ with Ricci form φ, then on the subset $\{\varphi > 0\} \subset L$ the inequality,*

$$\mathrm{Ric}(\varphi) \geq \frac{2}{n-1} \varphi$$

holds.

An application of the Ahlfors Lemma is then used to conclude that if L is uniformized by $\mathbb{C}$ then the foliation is holomorphic at all points on L.

The proof of Stoll's theorem, now follows from the observation that the vector field, $X = \tau^{i\bar{j}} \tau_{\bar{j}} \frac{\partial}{\partial z^i}$ is tangent to the foliation and complete. The complex flow of X yields a uniformization of each leaf by $\mathbb{C}$.

2 The Bott Connection

In all of the above only foliations defined by solutions of the Monge-Ampère equation were considered. However, much of the analysis does not require this much; moreover future applications may involve complex foliations that to not arise from solutions of the Monge-Ampère equation. For this reason we initiated in [DK1], [DK2] a study of the geometry of complex foliations without the Monge-Ampère assumption.

Before going into details it is perhaps worthwhile to outline the philosophy of our approach. For any smooth foliation there is a naturally defined process of covariant differentiation of normal vectors with respect to vectors tangent to the leaves and given by Lie differentiation (see below for more details). Because the covariant derivative is only defined in directions tangent to leaves, the connection is called a *partial connection* (see [KT]). The Cartan structure equations and Chern-Weil theory of connections and characteristic classes can be developed for partial connections, with one modification: Instead of using the ordinary de Rham complex of differential forms, one must use instead the so-called relative de Rham complex (see [H], [KT], [V] for details):

$$\Omega^{\bullet}_{\mathcal{F}} \equiv \Omega^{\bullet}(M)/\Gamma(M,Q^*) \wedge \Omega^{\bullet-1}(M)$$

with differential, $d_{\mathcal{F}}$ induced by the ordinary differential (the integrability condition, $d\Gamma(M,Q^*) \subset \Gamma(M,Q^*) \wedge \Gamma(M,T^*M)$, insures that $d_{\mathcal{F}}$ is well-defined. The relative de Rham complex makes precise the notation of computing modulo normal forms and computing exterior derivatives in directions tangent to leaves.

It is not hard to show that relative de Rham complex yields a resolution of the sheaf, $C^{\infty}_{\mathcal{F}}$, of germs of smooth functions which are locally constant along the leaves of $\mathcal{F}$. Hence, the cohomology of the relative de Rham complex computes, $H^{\bullet}(M,C^{\infty}_{\mathcal{F}})$; and a (complex) partial connection on a complex vector bundle, E, defines *relative Chern Classes*, $c_k(E,\mathcal{F}) \in H^{2k}(M,C^{\infty}_{\mathcal{F}})$, which, of course, are independent of the connection chosen.

The remarks just made apply to any smooth foliation. What makes them particularly relevant to the study of complex foliations is that in this case there is a natural choice of complex connection, which we call the *complex Bott connection*, and which is flat precisely when the foliation is holomorphic. This raises the possibility of relating the local structure of complex foliations to cohomological invariants and is the basis of much our work.

To continue the development it is necessary to be more precise. If $\mathcal{F}$ is a foliation on a manifold there is an exact sequence of bundles:

$$0 \longrightarrow T\mathcal{F} \overset{i}{\hookrightarrow} TM \overset{\pi}{\longrightarrow} Q \longrightarrow 0 \tag{2.3}$$

where Q is the normal bundle of the foliation.

A *partial connection* [KT] on a (real) vector bundle $E \overset{\pi}{\longrightarrow} (M,\mathcal{F})$ on a foliated manifold $(M, \mathcal{F})$ is an $\mathbb{R}$-linear map

$$\nabla : \Gamma(M, E) \to \Gamma(M, E \otimes T\mathcal{F})$$

which satisfies the condition

$$\nabla(fs) = f\nabla s + e \otimes d_{\mathcal{F}} f$$

for $s \in \Gamma(M, E)$. Its *curvature 2-form* is the relative 2-form defined by the usual formula:

$$R_\nabla(X, Y)s = (\nabla_X \nabla_Y - \nabla_Y \nabla_X - \nabla_{[X,Y]})s$$

for $X, Y \in \Gamma(M, T\mathcal{F}), s \in \Gamma(M, E)$.

In the case where M is a complex manifold and $\mathcal{F}$ a complex foliation, then the complex structure tensor on M gives both Q and $T\mathcal{F}$ the structure of complex vector bundles and (2.3) respects complex structures. The standard splitting of complex tangent vectors according to type induces splittings,

$$Q \otimes \mathbb{C} = Q_{(1,0)} \oplus Q_{(0,1)} \text{ and } T\mathcal{F} \otimes \mathbb{C} = T\mathcal{F}_{(1,0)} \oplus T\mathcal{F}_{(0,1)}$$

and there are complex vector bundle isomorphisms $Q \cong Q_{(1,0)}$ and $T\mathcal{F} \cong T\mathcal{F}_{(1,0)}$. There are two partial connections on the normal bundle of a complex foliation:

Definition 2.4 (i) For $X \in \Gamma(M, T\mathcal{F} \otimes \mathbb{C}), Y \in \Gamma(M, Q \otimes \mathbb{C})$ $Y = \pi(\widetilde{Y})$ the partial connection defined by the formula

$$\widetilde{\nabla}_X Y = \pi[X, \widetilde{Y}]$$

is called the *Bott connection*.
(ii) the *complex Bott connection* ∇ on $Q_{(1,0)}$ is defined by the formula,

$$\nabla_X Y = \pi_{(1,0)} \left(\widetilde{\nabla}_X Y\right)$$

where $X \in \Gamma(M, T\mathcal{F} \otimes \mathbb{C})$, $Y \in \Gamma(M, Q_{(1,0)}$ and $\pi_{(1,0)} : Q \otimes \mathbb{C} \to Q_{(1,0)}$ is the projection map.

All of the important geometric properties of complex foliations can be derived from the structure equations of these connections. To describe these let ω and Λ be the q-by-q matrices of relative connection 1-forms,

(i) $\nabla e = e \otimes \omega$

(ii) $\widetilde{\nabla} e = e \otimes \omega + \bar{e} \otimes \overline{\Lambda}$

for $e = (e_1, \ldots, e_q)$ a local framing for $Q_{(1,0)}$.

It is helpful to use local coordinates (w, z) centered at point $x_0 \in M$ and chosen so that a coordinate patch is of the form $\Delta^p \times \Delta^q$ where Δ^k is the unit polydisc in $\mathbb{C}^k$ and so that $\Delta^p \times \{0\}$ is contained in the leaf of $\mathcal{F}$ containing x_0. Such coordinates are said to be *adapted (to the leaf through x_0)*. In such coordinates a local framing for the bundle $T\mathcal{F}_{(1,0)}$ is given by vector fields of the form

$$X_{(j)} = \frac{\partial}{\partial w^j} + \lambda_j^\alpha \frac{\partial}{\partial z^\alpha} \qquad j = 1, \ldots, p = n - q.$$

where the functions λ_j^α vanish on the set $z = 0$. A local framing for the conormal bundle, $Q^{(1,0)}$ is given by the forms,

$$\theta^\alpha \equiv dz^\alpha + \lambda_j^\alpha dw^j,$$

and a local framing for the bundle $T_{\mathcal{F}}^{(1,0)}$ is given by the relative forms,

$$[dw^j] \equiv i^* dw^j.$$

Using the integrability conditions $[X_{(j)}, X_{(k)}] = 0$, the following formulas are easily derived:

$$\omega = \left(\omega_\alpha^\beta\right) = \left(-\frac{\partial \lambda_j^\beta}{\partial z^\alpha}[dw^j]\right) \text{ and } \Lambda = \left(\Lambda_{\overline{\alpha}}^\beta\right) = \left(\frac{\partial \lambda_j^\beta}{\partial \overline{z}^\alpha}[dw^j]\right).$$

From these and the integrability conditions one can easily verify the following structure thoerem for the complex Bott connection"

Proposition 2.5 *Let $\theta = (\theta^1, \ldots, \theta^q)$ be the dual coframe to a local framing e of $Q_{(1,0)}$. Let ω and Λ be as in (2.6) and let Ω denote the curvature matrix of ∇ on $Q_{(1,0)}$. Then the following structure equations are satisfied*

(i) $d_{\mathcal{F}}\theta = -\omega \wedge \theta - \Lambda \wedge \overline{\theta}$

(ii) $\Omega \equiv d_{\mathcal{F}}\omega + \omega \wedge \omega = -\Lambda \wedge \overline{\Lambda}$

(iii) $d_{\mathcal{F}}\Lambda = -\omega \wedge \Lambda - \Lambda \wedge \overline{\omega}.$

Remark 2.6 The (real) Bott connection is easily shown to be flat. However, it generally will not preserve the complex structure on the normal bundle. The complex Bott connection is defined via projection from the real Bott connection in such a way as to automatically preserve the complex structure on the normal bundle. The price paid is that the complex connection developes curvature.

The situation is analogous to the case of a Riemannian manifold embedded in Euclidean space. The tangent bundle of Euclidean space has a flat connection which restricts to the submanifold to give a flat connection; this connection does not respect the splitting into tangential and normal directions along the submanifold. However, by projecting the covariant derivative of a tangential vector to the tangential direction and that of a normal vector to a normal vector one obtains a connection that does preserve the splitting. The resulting connection generally has curvature and it can be written as a quadratic expression in the second fundamental form.

The quantity, $\tau = \Lambda^{\alpha}_{\overline{\beta}} \otimes \overline{\theta^{\beta}} \otimes e_{\alpha}$ is a tensor, called the *antiholomorphic torsion tensor of $\mathcal{F}$.* (That is is in fact a tensor follows from the change of frame formula

$$\Lambda' = g^{-1} \Lambda \overline{g}$$

for $(e', \overline{e}') = (eg, \overline{eg})$.) The tensor τ was first introduced in [BB] in the case of Monge-Ampère foliations under the name *anti-holomorphic twist.*

Observe that the foliation, $\mathcal{F}$, is holomorphic if and only if the vector fields, $X_{(j)}$ above are holomorphic. The importance of antiholomorphic twist is the easily demonstrated fact that *a complex foliation is holomorphic if and only if its holomorphic twist tensor vanishes.* For this reason a complex foliation is said to be *holomorphic at a point* if the tensor τ vanishes at that point.

We now want to examine the condition that the foliation $\mathcal{F}$ arise (locally) from a solution of the Monge-Ampère equation. Our goal is to find weak necessary conditions which a sufficient to force the foliation to be holomorphic.

Suppose, therefore, that $\mathcal{F}$ is locally at Monge-Ampère foliation and let u be a locally defined solution of the Monge-Ampère equation giving rise to the $\mathcal{F}$. The form $\sigma = (\sqrt{-1}/2)\partial\overline{\partial}u$ is a closed, non-negative real form of type (1,1) and the identity

$$T\mathcal{F} = \{X \mid X \lrcorner\, \sigma = 0\}$$

holds. In particular, it follows that if X is a vector field tangent to $\mathcal{F}$ then the Lie derivative, $\mathcal{L}_X \sigma$, vanishes. Let h denote the Hermitian form on the normal bundle Q associated to σ. The the following proposition holds.

Proposition 2.7 *The Lie derivative $\mathcal{L}_X \sigma$ vanishs for all vector fields, X tangent to $\mathcal{F}$ if and only if each of the following conditions holds:*

(i) $$\nabla h = 0$$

(ii) $$\frac{\sqrt{-1}}{2} \Lambda^{\gamma}_{\overline{\alpha}} U_{\gamma\overline{\beta}} \theta^{\alpha} \wedge \overline{\theta^{\beta}} = 0$$

It is natural to extract each of the two conditions of the above proposition. A *Hermitian* foliation is then a complex foliation for which condition (i) holds. This differential condition on a foliation is shown in [DK1] to be equivalent to a finite set of *algebraic* conditions on h. They coincide with conditions derived in [BK] and shown to be necessary conditions for $\mathcal{F}$ to be Monge-Ampère.

A *symmetric foliation* is one for which condition (ii) is satisfied. It is not hard to show that if $\mathcal{F}$ is a symmetric foliation then in the neighborhood of each point there is a framing θ^{α} with respect to which the matrix $(\Lambda^{\alpha}_{\overline{\beta}})$ is symmetric.

Except in the codimension one case, where all foliations are locally Monge-Ampère, a complete set of necessary and sufficent conditions for a foliation to be Monge-Ampère is yet to be found. But the condition that $\mathcal{F}$ be symmetric *is* a easily checked. It is reasonable, therefore, to determine which properties of Monge-Ampère foliations hold for symmetric (or Hermitian) foliations. We close this section with two theorems of this nature. Both make use of the *relative*

Chern forms of a complex foliation, which are defined in the usual way by the formula

$$\sum C_j(\mathcal{F})t^j = \det(I - \frac{t}{2\pi\sqrt{-1}}\Omega) = \det(I + \frac{t}{2\pi\sqrt{-1}}\Lambda \wedge \overline{\Lambda}).$$

In particular, the first Chern form of $\mathcal{F}$ can be written in the form,

$$C_1(\mathcal{F}) = \frac{1}{2\pi\sqrt{-1}} \sum_{\alpha,\beta} \Lambda^{\alpha}_{\overline{\beta}} \wedge \overline{\Lambda^{\beta}_{\overline{\alpha}}}$$

By the remarks made at the beginning of this section, the Chern forms define cohomology classes,

$$c_k(\mathcal{F}) \equiv [C_k(\mathcal{F})] \in H^{2k}(M, C^{\infty}_{\mathcal{F}})$$

Because the foliation $\mathcal{F}$ is holomorphic if and only if the tensor Λ vanishes, it follows that the Chern forms (and, consequently, the relative Chern classes) of a holomorphic foliation vanish. But, in the case of symmetric foliations various converses hold. They all rely of the fact that for symmetric foliations the first relative Chern form of $\mathcal{F}$ is non-positive and vanishes precisely when and where the foliation is holomorphic.

The following theorem (see [DK1] for the proof and several related results) uses the fact that if N is a leaf of $\mathcal{F}$ then there is a natural map $H^2(M, C^{\infty}_{\mathcal{F}}) \to H^2(N, \mathbb{C})$ which sends $c_1(\mathcal{F})$ to the first Chern class of the normal bundle of N in M.

Theorem 2.8 *Let $\mathcal{F}$ be a symmetric foliation of a complex manifold and N is a compact leaf of $\mathcal{F}$ which supports a Kähler metric then $\mathcal{F}$ is holomorphic at all points of N if and only if the first Chern class of the normal bundle of N vanishes. In particular, suppose M is a Kähler manifold and $\mathcal{F}$ is a compact symmetric foliation of M. Then $\mathcal{F}$ is holomorphic if and only if the identity, $c_1(\mathcal{F}) = 0$, holds.*

Remark 2.9 In practice, the condition, $c_1(\mathcal{F}) = 0$, is sometimes difficult to check. However, in [DK1] we show that the vanishing of the first Chern class of the normal bundle $Q \to M$ implies the vanishing of $c_1(\mathcal{F})$.

A final application of the structure equations of a complex foliations is the following generalization of Burns' Schottky-Landau theorem [B]. Because the proof will not appear elsewhere, we will give a fairly complete presentation. Let $\mathcal{F}$ be a symmetric, codimension q foliation of a complex manifold M by complex curves. Following Burns, define the relative $(1,1)$-form $\varphi = -\pi C_1(\mathcal{F})$ by

$$\varphi = \frac{i}{2}\text{tr}(\Omega).$$

Recall that because $\mathcal{F}$ is symmetric this form is non-negative and vanishes precisely at those points where $\mathcal{F}$ is holomorphic.

On the set $\varphi > 0$, the form defines a Riemannian metric on the leaves of $\mathcal{F}$ and the Ricci form, $Ric(\varphi)$, is the relative 2-form whose restriction to each leaf satisfies the equation

$$Ric(\varphi) = -K\frac{\varphi}{2},$$

where K is the Gauss curvature. In coordinates,

$$Ric(\varphi) = \frac{i}{2}\partial_{\mathcal{F}}\overline{\partial}_{\mathcal{F}}\log(S)$$

where

$$\varphi = \frac{i}{2}S\, dw \wedge d\overline{w} \bmod Q^*$$

and (w, z^α) are coordinates adapted to $\mathcal{F}$. (Here we have used the splitting according to type to decompose the relative de Rham differential in the form $d_{\mathcal{F}} = \partial_{\mathcal{F}} + \overline{\partial}_{\mathcal{F}}$.)

Theorem 2.10 *Let $\mathcal{F}$ be a symmetric codimension $n-1$ foliation on an n dimensional manifold, M. Then the inequality,*

$$Ric(\varphi) \geq \frac{2}{(n-1)}\varphi$$

holds on the set $\varphi > 0$, i.e. on the set where $\mathcal{F}$ fails to be holomorphic.

Proof. The proof closely follows that of Burns [B]. Let N be a leaf of $\mathcal{F}$ and let w be a holomorphic coordinate on N. Observe that the restriction of the relative de Rham differential to N is just the ordinary differential on N and that it suffices to prove the theorem at the point $w = 0$ of N.

Choose a co-frame $\{\theta^\alpha\}$ for $Q^{(1,0)}$ such that the following conditions are satisfied:

(i) The torsion matrix Λ is symmetric.

(ii) The covariant derivatives with respect to the complex Bott connection, $\nabla\theta^\alpha$, all vanish at the point $w = 0$.

Then on N the connection and torsion matrices can be written in the forms:

$$\omega = G\, dw - \overline{H}\, d\overline{w} \text{ and } \Lambda = A\, dw$$

where $G(w)$, $H(w)$ and $A(w)$ are complex, q-by-q, matrix-valued functions. Note that the function $S(w)$, defined above, is given by the formula

$$S(w) = \mathrm{tr}(A\overline{A}).$$

We have to prove the inequality,

$$\log(S)_{w\overline{w}} \equiv \frac{SS_{w\overline{w}} - |S_w|^2}{S^2} \geq \frac{2}{q}S.$$

To do this compute S_w and $S_{w\overline{w}}$ at the point $w = 0$.

Before starting the computation some preliminary observations are required. First note that conditions (i) and (ii) above are equivalent to the conditions,

$$A^t = A \text{ and } G(0) = H(0) = 0,$$

where $()^t$ denotes transpose. Next observe that the restriction to N of the structure equations,

$$\Omega \equiv d_{\mathcal{F}}\omega + \omega \wedge \omega = -\Lambda \wedge \overline{\Lambda} \text{ and } d_{\mathcal{F}}\Lambda = -\omega \wedge \Lambda - \Lambda \wedge \overline{\omega}$$

assume the form:

$$G_{\overline{w}} + \overline{H_{\overline{w}}} + G\overline{H} - \overline{H}G = A\overline{A} \text{ and } \frac{\partial A}{\partial \overline{w}} = \overline{H}A + A\overline{G}.$$

The quantity $S_w(0)$ can now be easily computed. Just use equation second structure equation to eliminate derivatives with respect to $\overline{w}$:

$$S_w(w) = \operatorname{tr}(A_w\overline{A}) + \operatorname{tr}(A\overline{A_{\overline{w}}}) = \operatorname{tr}(A_w\overline{A}) + \operatorname{tr}(A(H\overline{A} + \overline{A}G))$$

At $w = 0$, the connection coefficients vanish, yielding the following identity:

$$S_w(0) = \operatorname{tr}(A_w\overline{A}).$$

To compute $S_{w\overline{w}}(0)$, differentiate the formula for S_w using the vanishing condition, $H(0) = G(0) = 0$, and the matrix identity, $\operatorname{tr}(ABC) = \operatorname{tr}(BCA) = \operatorname{tr}(CAB)$ which holds for any set of q-by-q matrices A, B, C. The computation at the point $w = 0$ proceeds as follows:

$$\begin{aligned}
S_{w\overline{w}}(0) &= \operatorname{tr}(A_w\overline{A})_{\overline{w}} + \operatorname{tr}(A(H\overline{A} + \overline{A}G))_{\overline{w}} \\
&= \operatorname{tr}(A_{w\overline{w}}\overline{A} + A_w\overline{A_w}) + \operatorname{tr}(AH_{\overline{w}}\overline{A} + A\overline{A}G_{\overline{w}}) \\
&= \operatorname{tr}((\overline{H}A + A\overline{G})_w\overline{A} + A_w\overline{A_w}) + \operatorname{tr}(AH_{\overline{w}}\overline{A} + A\overline{A}G_{\overline{w}}) \\
&= \operatorname{tr}((\overline{H_{\overline{w}}}A + A\overline{G_{\overline{w}}})\overline{A} + A_w\overline{A_w}) + \operatorname{tr}(AH_{\overline{w}}\overline{A} + A\overline{A}G_{\overline{w}}) \\
&= \operatorname{tr}((\overline{H_{\overline{w}}} + G_{\overline{w}})A\overline{A}) + \operatorname{tr}((\overline{G_{\overline{w}}} + H_{\overline{w}})\overline{A}A) + \operatorname{tr}(A_w\overline{A_w}) \\
&= \operatorname{tr}(A\overline{A}A\overline{A} + \overline{A}A\overline{A}A + \operatorname{tr}(A_w\overline{A_w}) \\
&= 2\operatorname{tr}(A\overline{A}A\overline{A}) + \operatorname{tr}(A_w\overline{A_w}).
\end{aligned}$$

Because $A(w)$ is symmetric, the quantity $\operatorname{tr}(A_wA_w)$ is non-negative and the Schwartz inequality gives the inequality,

$$|\operatorname{tr}(A_w\overline{A})|^2 \leq \operatorname{tr}(A_w\overline{A_w})\operatorname{tr}(A\overline{A}) = \operatorname{tr}((A_w\overline{A_w})S.$$

Combine the previous formulas for $S_w(0)$ and $S_{w\overline{w}}$ to compute as follows:

$$\begin{aligned} \log(S)_{w\overline{w}} &= \frac{1}{S^2}(S(2\,\mathrm{tr}(A\overline{A}A\overline{A}) + \mathrm{tr}(A_w\overline{A_w})) - |\mathrm{tr}(A_w\overline{A})|^2 \\ &\geq \frac{1}{S^2}(S(2\,\mathrm{tr}(A\overline{A}A\overline{A}) + \mathrm{tr}(A_w\overline{A_w})) - S\mathrm{tr}(A_w\overline{A_w})) \\ &\geq \frac{1}{S}((2\,\mathrm{tr}(A\overline{A}A\overline{A}) + \mathrm{tr}(A_w\overline{A_w})) - \mathrm{tr}(A_w\overline{A_w})) = \frac{1}{S}(2\,\mathrm{tr}A\overline{A}A\overline{A}). \end{aligned}$$

The result now follows by applying the inequality,

$$(\mathrm{tr}A\overline{A}A\overline{A}) \geq \frac{1}{q}(\mathrm{tr}A\overline{A}))^2 = \frac{1}{q}S^2,$$

to the last expression. ▌

Corollary 2.11 *Let $\mathcal{F}$ be a symmetric complex foliation with one dimensional leaves. Then $\mathcal{F}$ is holomorphic along every leaf which is uniformized by $\mathbb{C}$.*

Proof. The proof given in Burns [B] applies without change. ▌

Remark 2.12 In the case where M is a complex surface we can do better. In this case all inequalities in the above proof are in fact equalities and we arrive at the result

$$Ric(\varphi) = 2\varphi.$$

wherever $\mathcal{F}$ is not holomorphic its leaves are endowed with a metric of constant curvature equal to -4.

3 Foliations of Surfaces

The remark at the end of Section II suggests that it might be possible to solve the Cartan equivalence problem for complex foliations of surfaces by complex curves. This is carried out in [DK2], whose main results we now describe.

Because all holomorphic foliations are locally equivalent, we only consider non-holomorphic foliations. We, therefore, assume that the torsion tensor τ is non-vanishing. It can then be used to select a distinguished class of framings of the holomorphic tangent bundle of M:

Definition 3.13 An *adapted frame* at a point $p \in M$ consists of a pair of independent vectors $X, Y \in T_{1,0}M_p$ such that

(i) $Y \in T\mathcal{F}$

(ii) $\tau(Y \otimes \overline{\pi(X)}) = \pi(X)$.

The bundle of all adapted frames is denoted by the symbol $P = P(M, \mathcal{F})$.

The main result of [DK2] can now be stated.

Theorem 3.14 *Let $(M, \mathcal{F})$ be a pair consisting of a complex surface together with a foliation by complex curves which is not holomorphic at any point. Then the manifold $P(M, \mathcal{F})$ has a complex structure with respect to which $pr : P \to M$ is a holomorphic fibration. There is a global framing $(\theta, \eta, \varphi, \psi)$ of the cotangent bundle of P by forms of type $(1,0)$. If $M', \mathcal{F}')$ is another such foliation and $f : (M, \mathcal{F}) \to (M', \mathcal{F}')$ is an isomorphism of complex foliations, the induced map $\tilde{f}^* : T^*P' \to T^*P$ satisfies*

$$\tilde{f}^*(\theta', \eta', \varphi', \psi') = (\theta, \eta, \varphi, \psi).$$

This theorem is proved in a more or less standard way by applying the Cartan method of equivalence to the space of adapted frames and reducing its structure group to the identity. This results in the following structure equations:

$$\begin{cases} d\theta &= -\omega \wedge \theta + \eta \wedge \overline{\theta} \\ d\eta &= -\psi \wedge \theta - (\varphi - \overline{\varphi}) \wedge \eta \\ d\varphi &= \psi \wedge \overline{\theta} - 2\theta \wedge \overline{\psi} - \eta \wedge \overline{\eta} - 3A\theta \wedge \overline{\eta} \\ d\psi &= -\eta \wedge \overline{\psi} - \psi \wedge \overline{\varphi} - 3A\eta \wedge \overline{\eta} + (B\overline{\theta} + 2\overline{C}\eta + C\overline{\eta} + 2A\psi + A\overline{\psi}) \wedge \theta, \end{cases}$$

where the functions A, B, C are the fundamental invariants, defined by the structure equations.

In fact more is true. The bundle, P, has the structure of a right principal G-bundle, where G is the subgroup of $GL(2, \mathbb{C})$ consisting of matrices of the form,

$$\begin{bmatrix} a & 0 \\ b & a/\overline{a} \end{bmatrix},$$

let H denote the subgroup of G consisting of matrices of the form

$$\begin{bmatrix} a & 0 \\ 0 & a/\overline{a} \end{bmatrix}$$

and set $E = P/H$. Then $P \to E$ is a principal H-bundle and a section of the bundle $E \to M$ can be identified with a splitting of the exact sequence of complex vector bundles

$$0 \to T\mathcal{F} \to TM \to Q \to 0.$$

Moreover, the group G embeds as a subgroup of $SL(3, \mathbb{R})$ and the framing of the theorem used to define an $s\ell(3, \mathbb{R})$-valued 1-form, ω, on P. Using this embedding Theorem 3.13 can be recast in the following form:

Theorem 3.15 *The $s\ell(3, \mathbb{R})$-valued 1-form ω is a Cartan connection on the right principal H-bundle $P \to E$.*

To proof this result it is necessary to first identify the flat foliation, i.e. the unique folitation for which the identities $A = B = C = 0$ hold. Consider the manifold $M = \mathbb{C}\mathbb{P}^2 \backslash \mathbb{R}\mathbb{P}^2$ and the foliation can be described as follows. Every real line in $\mathbb{R}\mathbb{P}^2$ is the restriction of exactly one complex line in $\mathbb{C}\mathbb{P}^2$ and there is exactly one such line through each point of M. This defines $\mathcal{F}$. That $(M, \mathcal{F})$ is flat follows from the fact that $SL(3, \mathbb{R})$ acts transitively and effectively on $(M, \mathcal{F})$ and has dimension 8, the dimension of $P(M, \mathcal{F})$. One then shows that $P(M, \mathcal{F})$ is diffeomorphic to $SL(3, \mathbb{R})$ and that, under this identification, equations (3.5) reduce to the structure equations for the Lie algebra $s\ell(3, \mathbb{R})$. The proof of the theorem is then a relatively straightforeward computation.

Remark 3.16 *Preliminary calculations indicate that a smilar structure theorem holds for foliations of higher codimension. However, a complete solution of the equivalence problem will require the determination of the possible normal forms for the antiholomorphic twist tensor, and we have not yet succeeded in doing this.*

References

[BB] E. Bedford and D. Burns, *Holomorphic mappings of annuli and the associated extremal function,* Ann. Scuola Norm. Pisa, VI (1979), 381–414.

[BK] E. Bedford and M. Kalka, *Foliations and the complex Monge-Ampère equation,* Comm. Pure Appl. Math. XXX (1977), 543–572.

[BT] E. Bedford and B. A. Taylor, *A new capacity for plurisubharmonic functions,* Acta. Math. **149** (1982), 1–40.

[B] D. Burns, *Curvature of Monge-Ampère foliations and parabolic manifolds,* Ann. of Math. **115** (1982), 261–274.

[DK1] T. Duchamp and M. Kalka, *Invariants of complex foliations and Monge-Ampère equation,* Mich. Math. J. **35** (1988), 91–115.

[DK2] T. Duchamp and M. Kalka, *The equivalence problem for complex foliations of complex surfaces,* Ill. J. of Math. (to appear).

[H] Heitsch, J., *A cohomology for foliated manifolds,* Comment. Math. Helv. **50** (1975), 197–218.

[KT] Kamber, F. and Tondeur, Ph., *Foliated Bundles and Characteristic Classes*, Lecture Notes in Math., **493**, Springer, Berlin, 1975.

[S] W. Stoll, *The characterization of strictly parabolic manifolds,* Ann. Scuola Norm. Pisa VII, (1980), 87–154.

[V] I. Vaisman, *Cohomology and Differential Forms*, Dekker, New York, 1983.

[W] P. M. Wong, *The geometry of the homogeneous Monge-Ampère equation*, Inv. Math. **67** (1982), 261–274.

Department of Mathematics
University of Washington GN-50
Seattle, WA 98195

Department of Mathematics
Tulane University
New Orleans, LA 70118